MORE THAN
MARADONA

MORE THAN MARADONA

Kirsten Schlewitz

HIGH
PRESS
PUBLISHING

Published by High Press Publishing Limited in 2024.

www.highpress.co

Kirsten Schlewitz has asserted her right to be identified as author of this work.

Editor: Anushree Nande
Illustrator: Marcus Marritt
Proof reader: Daniel Ford
Designer: Darren Exell

Marcus Marritt has asserted his right under the Copyright, Designs and Patents Act 1988, to be identified as the author of the illustrations.

A CIP catalogue record for this book is available from the British Library upon request.

Printed and bound by CPI Group (UK) Ltd, Croydon, CR0 4YY.

ISBN 978-1-7384496-0-6 (Hardback)
ISBN 978-1-7384496-1-3 (Paperback)
ISBN 978-1-7384496-2-0 (eBook)

Contents

Acknowledgements

Whew! I hardly know where to begin! Writing a book has been a dream of mine since I was a child; to be able to write it about Napoli, my beloved club, and integrate the history of Naples, a city that captivates me, is almost beyond comprehension.

First thanks, then, to Chris Nee, the man who, all those years ago, as we traveled to a distant village to watch non-league football, turned to me and asked whether I'd heard about a brilliant young midfielder named Marek Hamšík.

Hugs and kisses to Aaron Campeau—who knew a little Aston Villa site would result in a tome on Napoli history? You're the best best friend a girl could ask for, even from 7,500 miles (wait, kilometers) away.

Graham MacAree, this book exists in large part so I can say, "I told you so." In seriousness, thanks for the belief in my calcio career, and for being willing to dive down the long weird Napoli hole with me.

Conor Dowley, who helped me launch The Siren's Song, the—unfortunately now defunct—blog that refused to take Napoli fandom too seriously.

All the Effortistas out there: I only hope I can provide 1/1,000th of the inspiration to you that you've given to me.

I'd be nowhere without Dr. Henry Bell, Michele Borelli, Ken Cioffredi, Marco D'Onofrio, and Raffa Rispo. Beyond interviews, beyond a sounding board, beyond the meltdowns over Napoli's performance (or lack thereof), y'all supported me, unhesitantly, in this crusade to spread The Word of Napoli. We never imagined it would become this easy, did we? There are not enough words, written, spoken, or performed, to express my appreciation for your friendship and the work that you put into the Far From Vesuvius network. Forza Napoli Sempre!

Of course, a special thanks to Tim Gentles for believing in me. For someone with significant health issues, knowing that a publisher was willing to be patient while I produced the best possible product made all the difference. The Drillboard mission is unbeatable and I'm thrilled to be part of the family.

I am particularly grateful to Daniel Ford for final edits, ensuring consistency in phrasing, style, and punctuation, and designing a cohesive layout. As my voice differs from that of a "typical football journo," I sincerely appreciate his willingness to maintain that voice while also verifying facts and confirming readability.

A huge grazie to Marcus Marritt for the gorgeously perfect chapter illustrations and cover design.

Anushree Nande, Jessie Losch, Sonja Missio, and Megan Smith, my squad. You were there when only conversations and Lorenzo Insigne gifs could pick me up off the floor. #JellyfishForLife

For Anu, my favorite editor in the world, without whom I'd be nowhere, a special thanks. We did it! It was a slog, and I'm sure you never, ever want to think about Serie A rules and relegations again, but without you I'd have drowned in the Bay of Naples.

My uncle, Richard Buchheim, who died during the writing of this book, and his mother, Betty Buchheim, my Mam, who died long ago. Both remain in the unshakeable voice in my head that tells me to keep going, even through the pain.

My parents, who never failed to encourage my writing, even when they began to worry about job prospects, and who supported my wanderlust heart, even when it led me far away. Barb Schlewitz, anyone who's ever been within earshot of you during a Seattle Seahawks game totally gets why I can't remain silent during a match. I know you fear me ever writing a book again, but I'm sorry, I make no promises. Dan Schlewitz, you put a ball in my hand, a bigger ball at my feet, and made sure I was aware of that even bigger ball I stood on and the heavens that surrounded it. Because of you I never doubted that a woman's place is in the stadium.

Above all, credit for this book goes to Uros Popovic, my researcher, my friend, my strength, my inspiration, my partner. Thanks for making sure I didn't get beaten up in the away section of the San Paolo when Napoli beat your Fiorentina 3–0. Thanks double for watching my mouth when my jet-lagged self saw them play out that goalless draw against Red Star at our Marakana home in Belgrade. Thanks triple for being willing to stay by my side, even when it only affords you front-row seats to a cascade of hysterical tears caused by watching Marek score the goal that puts Napoli back in the UEFA Champions League. Grazie mille for traveling with me to celebrate the third scudetto. You've never looked hotter than when wearing a Napoli shirt, singing Napoli chants.

While I have done my best to confirm facts, accounts do differ; in an attempt to support the choices I've made, extensive citations are available at the end of the book. Any errors that remain are mine and mine alone.

Preface

He came to them like an "angel descending from heaven," but soon he would become their god. In July 1984, over 70,000 SSC Napoli fans packed into the San Paolo stadium to await their savior, each paying 1,000 lire for the privilege. The drought had gone on too long; in the 58 years since Napoli's official founding, the team had never won the title. The scudetto had almost always been captured by a team from the much richer North. The fans knew of the machinations behind the scenes that had made this moment possible. Many had even given money themselves to make it happen. And they were convinced their patron saint, Saint Gennaro, had blessed this union. As Diego Armando Maradona, already one of the best players in the world at 24 years old, climbed out of the helicopter and waved to the crowd, the resulting cries could likely be heard on the island of Capri, over 20 miles away.

Seven years later, he would slink away in disgrace, his cocaine habit having caught up with him and a positive test after a match leading to a suspension of 15 months. But in those seven years, the stumpy 5'7" man with the mess of black curls and captivating smile had led Napoli to the victories the fans had expected. Even his outrageous antics, from the cocaine to the booze to the suspected orgies, would not turn the fans against him. With Maradona, Napoli won two scudetti, finished second twice, won the Coppa Italia, the UEFA Cup, and the Italian Supercoppa. The fans lucky enough to experience it would never forget the celebrations after the first scudetto win, in which impromptu street parties and celebrations broke out all over the city and the coffins of Juventus and Milan were burned to cheers from the crowd, while those too young to have watched the magic that is Maradona would listen rapturously to tales told about Napoli's first title win.

The rest of the world may have felt Maradona left Naples in disgrace, but nothing could be further from the truth. He was the man that brought them what are still the only two titles the club has won. Until 2017, he also held the club record for most goals scored. Even now, his presence is everywhere in Naples, in the murals on the walls, in the small shrines tucked into niches in the crooked streets, in the "Church of Maradona" where an altar to him remains. It seems that, until another player arrives to bring Napoli their third scudetto, Maradona will remain the city's primary deity.

Even those who only watch football when they tune in to the World Cup every four years will almost certainly have heard of Maradona. When Argentina play, there is no way to avoid talking about him, whether it is in comparison to their current shining star, Lionel Messi, or a discussion about the "Hand of God" goal against England in the 1986

World Cup, or the amazing dribble that extended almost 70 yards and put five England players in a twist, or even his inane performance as coach to the Argentina men's national team in 2010. Should casual watchers want more, they'd likely check out his Wikipedia page and learn what a hero he was to Napoli. But that's where most knowledge of the Naples club begins and ends: with the city's god, Diego Armando Maradona.

But for anyone who dares peek behind the curtain, they will be rewarded with a club history of such richness and depth, such corruption and scandal, such passion and loyalty, that they'll unlikely be able to tear themselves away from the story. That's the hope, anyway, as this book will provide a much deeper dive into Napoli's nearly unbelievable history.

For the most part, the book will follow a standard chronology: from the early 1900s when the first iteration of what is now SSC Napoli was formed; the influence fascism had on the Italian game; their stumbles through the years of World War II; the stars of the 1950s and 1960s who helped Napoli clinch their first silverware; the desire, almost dangerous, of the city and club to win the title in the 1970s; the coming of Maradona and the players who supported him; the tailspin of the 1990s; the bankruptcy and subsequent relegation in the 2000s; and finally to the success of the 2010s, the modern heroes that fueled the way, and the path Napoli intend to follow over the next decade. The majority was written before the 2022–2023 season began, but the epilogue covers Napoli's almost magical road to their third scudetto.

The intention of this book is not to provide a dry, game-by-game analysis in which each player is scrutinized and each refereeing decision questioned (though there must be some of this; it is a calcio book after all). Instead, I hope to weave together Italian history and Neapolitan culture, as the ups and downs that Napoli have experienced over the years go hand-in-hand with historical events and the way Neapolitans live out their daily lives.

The past 120 years span a number of exciting, astonishing, and downright unbelievable episodes of Napoli history, those that lead to Maradona, those that flow from his presence at the club, and those that are only tangentially related to the Neapolitans' idol. All are worthy of being told. Let's dive in…

For Marek, who started it all

CHAPTER 1

A Short History of Naples

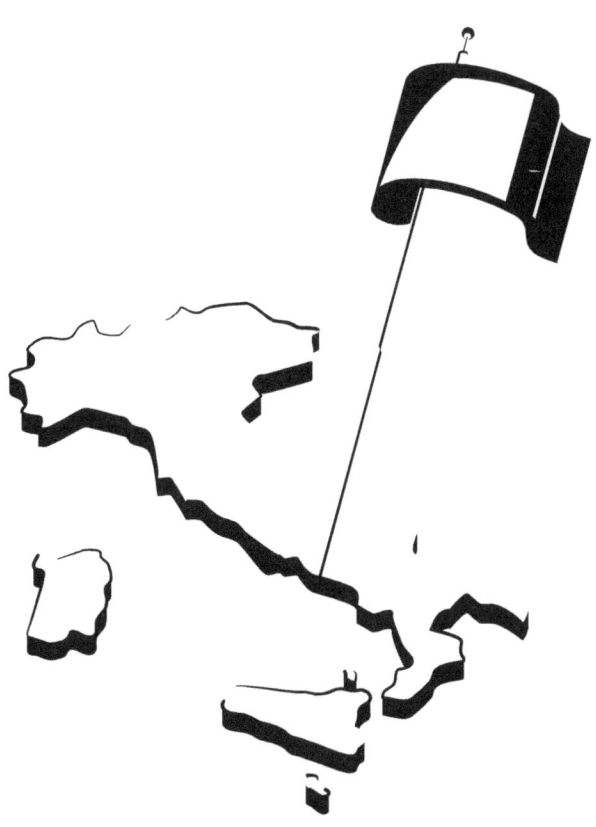

1

There is no way to tell the story of the club now known as SSC Napoli without first outlining the history of its city. While further details will come to light throughout this book, it's best to provide a short overview before we shift to the details of the club. Without this history, it would be difficult to understand how the culture and customs of Naples have shaped the many iterations of SSC Napoli and the fans that stand behind their club through it all.

Mount Vesuvius, an active volcano on the shores of the Bay of Naples, has been a constant throughout the history of the city, and remains emblematic of its people today. For a city with such a tumultuous history, it helps to have a symbol to focus on. Yet much else has remained the same for Neapolitans throughout thousands of years: the importance of family, a reliance on the Church and faith, a desire to share a cup of coffee or a good meal with friends, and an eagerness to embrace visitors to the city. And while *calcio* entered Naples' history relatively recently, almost as soon as it emerged, it became a strong tradition throughout the area.

Naples, created to endure

Mythical Naples

Knowing the myths surrounding the founding of Naples helps explain one element of SSC Napoli—and also echoes the irreconcilable differences in stories told about their own history. One of their nicknames, the *Partenopei*, means "the sirens." While Napoli's siren song has captured many a fan over the years, the word dates back to the mythical sirens of the Bay of Naples, a story perhaps best known by English students studying Homer's *The Iliad* and *The Odyssey*.

In one version, Demeter, goddess of the harvest and divine protector of the agricultural lands of Campania, engaged in a love affair with Zeus, who sits at the top of the hierarchy of the Greek gods. Beautiful Persephone, the product of this love affair, spent her days playing in the fields of the region. Hades, acting on his attraction to Persephone, seized her from one such field and carried her away to his kingdom, the underworld.

A panicked Demeter searched everywhere for her daughter. Fortunately, Persephone's friends and handmaidens were nearby. In one version of the myth, her mother turned Persephone's friends into birds, able to search more quickly when they took flight. To provide light to these sisters as they searched, Demeter created shining beacons, now known as the volcanoes of Southern Italy: Etna, of course, along with Ischia, Stromboli, and the Phlegrean Fields. The largest, Vesuvius, was meant to illuminate all of Campania.[1]

Demeter called off the search when she learned that Hades had made her daughter the queen of the underworld. However, she managed to negotiate an arrangement with him that Persephone would spend half the year with her, and half with him. Her arrival at her mother's house heralded the arrival of spring. In most Persephonian myths, the importance lies here, with the goddess returning to bring the harvest once again.

The myth's importance in Neapolitan culture lies with her friends, the nymphs Demeter sent out to find her daughter. When they could not do so, Demeter cursed them to live on an island in the Bay of Naples, forever singing songs to entice sailors to land on their shores—where they would seduce, then devour them. Parthenope, the daughter of the god Achelous and the muse Terpsichore,[2] is the best known of the cursed sirens. It is Parthenope who sings to Odysseus in *The Odyssey*, but ultimately fails to lure him to shore after he has his men lash him to the ship's mast. Odysseus' escape sealed Parthenope's fate, as she threw herself in the ocean to drown. Her body washed ashore on the island of Megaride in the Bay of Naples, and was taken to the mainland to be enshrined.[3] The former island, now a peninsula, is now the center of yet another myth, which would emerge in the Middle Ages.

Another founding story centers, unsurprisingly, on Mount Vesuvius, the adored yet feared symbol of the city. In this Roman myth, a centaur named Vesuvius falls in love with Parthenope, a beautiful mermaid. Eros, the god of love, throws his magic dart at Parthenope, ensuring Vesuvius' love is reciprocated.[4] Jupiter, king of the gods, becomes jealous of the couple, and turns the centaur into Mount Vesuvius and Parthenope into the city of Naples. Hence, the anger of the volcano that manifests in its frequent eruptions and lava flows: Vesuvius can see his love, but not be with her.[5]

In both myths Napoli fans can find something of themselves. Often the team play beautifully, seductively, luring in supporters before breaking their hearts by falling apart when it matters most. It's also easy—for those not born supporting Napoli—for new fans to identify with being shot by Cupid's arrow, only to encounter sadness and heartbreak when the team seems stuck, unable to reach their potential.

Greek Naples

The myth of the siren of Parthenope wove itself into the lives of the Greek settlers near the Bay of Naples. They first arrived in about 700 BCE, colonizing the island of Ischia and then moving inland.[6] They established the town of Cumae near the Phlegrean Fields, an area of volcanic activity filled with geysers and hot springs.[7] More specifically, Cumae was near the Solfatara volcano, believed to be the entrance to the underworld. So close to the alleged kingdom of Hades and Persephone—naturally the locals adhered to the cult of the siren.

The Cumaens struck out further, building a walled settlement on the hill of Pizzofalcone where the siren's tomb was located. The settlement was first called Paleopolis or "old city," but soon absorbed the name of the venerated siren, Parthenope.[8] Despite the strong walls the Cumaens built, they were overtaken by the local Etruscans in 524 BCE, and many residents fled Parthenope.[9] The Greeks then returned and overthrew the Etruscans in 474 BCE. So confident were they that the threat had been eliminated that they established a settlement along the shore of the Bay of Naples, the "new city" of Neapolis, now known to English speakers as Naples.

The Greek influence over Naples—language, religion, customs—held fast throughout its first millennium, even surviving the dominion of Rome.[10] Just before this time, approximately 400 BCE, Naples became the principal Greek port in the Mediterranean, trading in wine, grain, olives, and lemons.[11] The city's inhabitants worshiped Aphrodite, goddess of love; Demeter, goddess of the harvest; and Dionysius, the god of wine.[12] This trio's blessings imbued the city with a seductive spirit. The Romans allowed Naples to remain a center of Greek culture. In the '30s, the city was a beloved haunt of the poet Virgil, who built an elegant villa in the city. Others followed, seeking to escape the noise of Rome for the beauty and tranquility of Naples, building villas stretching from the Sorrentine peninsula to the ancient port of Misenum on the Gulf of Pozzuoli.

Those who know Naples today will not be surprised to learn that the ancient city was the center of Epicureanism, a philosophy that embraces pleasure as the highest good.[13] The citizens desired freedom from physical and moral pain as well as the stress of daily life and politics. Often this is visible in modern-day Naples, in the form of men sitting outside a cafe chatting and smoking, in the Sunday lunches that remain family gatherings, in the slow strolls of people down the *lungomare* (promenade). But one of the few places one won't find an absence of pain and stress is with football fans before the whistle blows at the end of a Napoli game.

Roman Naples

Naples, content under Greek rule, fought hard after the Romans declared war in 328 BCE. So hard, in fact, that the result was the *foedus neapolitanum* treaty that established that Naples would become a federated city, in which the citizens of Naples would retain a certain independence, including being permitted to maintain their Greek-influenced traditions.[14] Rome would control that which related to commerce and economic matters, including access to the port. Naples benefitted from this arrangement; Rome built roads connecting the city to the capital, and aqueducts to carry fresh water to Naples.[15] As the city thrived, rich Romans built holiday villas on the shorelines near Naples.[16] In fact, its prominence as the city to which Romans came to relax and unwind allowed Naples to exert political, social, and cultural influence.[17]

The Roman decision to allow Naples to remain firmly Hellenic is one that continues to shape the city—or at least, perceptions of the city. The city began to reinvent itself as a resort town after the economy began to stagnate and commerce moved north to Pozzuoli. Naples itself became a city that catered to Romans looking for a leisurely stay along the coast. The Roman aristocracy traveled to Naples to dress like Greeks, behave like Greeks, and relax like Greeks.[18] The Latin verb "*pergraecari*" means "to act like a Greek," implying the individual takes excessive enjoyment in food, wine, and sexual activities. As Romans often lived extremely regimented lives, losing themselves in a city devoted to pleasure had a distinct appeal. Horace called it the ideal beauty spot for romantic and amorous encounters. Virgil's poetry likened Naples to the myth of Arcadia, where the atmosphere invited intellectual discussion and seemed dedicated to happiness.[19] Seneca disagreed, complaining about filthy streets, corrupt judges, and a lifestyle that, on the whole, was immoral.[20] Yet overall, life in Naples was a carefree one—for the Romans, at least—where visitors enjoyed a relaxed indolence. The number of tourists allowed Naples to lift its stumbling economy by catering to the visitors, and the number of jobs for those such as architects, builders, bakers, and goldsmiths rose.[21]

In addition to being known as one of classical Rome's most famous poets, Virgil was also believed to be a magician. He is said to have built a statue of a man, poised with bow and arrow aimed at Vesuvius, to protect the city from its wrath.[22] But in 79 CE, Mount Vesuvius erupted, burying Pompeii, Herculaneum, and Stabiae as well as a number of villas that had been constructed on the slopes of the long-dormant volcano.[23] Due to prevailing winds, Naples managed to escape with just a thin covering of ash.[24] But this explosion has marked Naples and her inhabitants for nearly 2,000 years. One of the more immediate effects was that in the surrounding area, wine, olive, and grain production were severely compromised, a hardship the city endured for around 300 years.[25] Rituals were put in place surrounding the volatility of the volcano, and superstitions abounded, some of which remain today. For example, an erect phallus is visible in Pompeii's lamps, statues, and paintings. According to Jordan Lancaster, "The phallus was a protective charm and is worn by some more superstitious Neapolitans even today to ward off the evil eye."[26] There is also evidence of symbols on walls of these ancient cities that point to communities of early Christians. The new religion quickly took root in Naples, and was strengthened by the visit of Saint Paul to nearby Pozzuoli. Legends tell that the saint also visited Naples itself, and possibly even consecrated the city's first bishop.[27]

Despite the new religion, Naples was still a Hellenic city, and it persisted as one while the Roman Empire declined. Emperor Valentinian III reinforced Naples' city walls, which allowed them to fight off the Goths, Huns, and Vandals ravaging the peninsula, including Rome. He also transformed a villa on Megaride into a fortress, positioned over the Crypta Neapolitana, a tunnel running between Napoli and Poluzzi, allegedly constructed by the magical Virgil in a single night. As another means of protecting the city he adored, Virgil placed a magical egg in the tunnel, warding off human dangers and natural disasters.[28] The final Roman emperor, Romulus Augustus, was exiled to what is now known as the Castel dell'Ovo in 476, amidst a series of explosions from Vesuvius—but neither it nor the invaders could destroy Naples.[29]

Tug of war between East and West

Naples finds its own path

After the fall of Rome and the Western Empire in the fifth century, most of Italy was controlled by Germanic peoples.[30] Naples, on the other hand, was positioned to maintain ties to the still-standing Eastern Empire, centered on Constantinople (now Istanbul). While Europe plunged into the Dark Ages, its citizens marked by illiteracy and ignorance, Naples remained tied to a city dedicated to art, law, culture, and learning.[31] With the sea on one side and steep hills on the other, the city also seemed to be safe from attack, yet it still slipped into a period of decline. As malaria rates increased, the population decreased. And when the Roman state broke down completely, agriculture and trade in Naples suffered. If this weren't enough, fights between the Eastern Empire and invading Goths lasted for over 15 years.

The Byzantine Empire, determined to have Naples for its own, captured it in 536.[32] Soon after, the Lombards, a people with roots in Northern Europe, invaded northeastern Italy. They reached Campania in 570, where they elected to stay, intermarrying and adopting the culture and traditions of the locals.[33] Naples managed to hold out, their nearness to the sea keeping supply lines open. The closeness of the Lombards, aligned with the Catholic Church, was worrisome to the Orthodox Byzantines. The Catholic Church took advantage in other areas, exerting influence over politics and government, while Greek Orthodoxy held out in the few places the Lombards did not manage to conquer.

As the Roman Catholic Church continued to grow in influence, it became vital to the southern peninsula's social and cultural life. The first Western monastic order was established to the north of Naples in the sixth century.[34] The monasteries were essential to enhancing culture in their regions, as each community had a library where texts were translated and transcribed; by the next century, Naples had become a major center for translation and transcription. The stories of local saints were enlivened by the customs of the area. For example, in the story of Saint Restituta, the North African woman was set on fire and pushed out to sea for refusing to give in to a Roman official's sexual demands. A local woman was said to have received a message from an angel to organize those on Ischia to meet and bury the bones of the young martyr. The connection of the sea to saints is common to stories of martyrs in Southern Italy; according to Lancaster, they reflect both the myth of the siren and the birth of Venus.[35]

The Neapolitans achieved their independence peacefully. Under Byzantium, the city grew and prospered. While a small surplus was available for trade—with places as far away as Marseille—for the most part Naples took care of herself first, raising pigs, goats, sheep, and chickens, hunting deer and rabbits, fishing and netting the fruits of the sea. They raised produce in abundant city gardens. Their territory expanded, running from Cumae to Salerno, settlements in Capodichino to the northeast, Capodimonte and Vomero in the west, and Ischia and Capri defending the bay. When Emperor Constans II visited in 661, he appointed the first Duke of Naples, Basil, awarding the city an autonomy that it continued to strengthen, its geographical remoteness from Byzantium making it nearly a sovereign state.[36]

In 763 Naples' autonomy became official. Stephen II, the last duke of Byzantine Naples, recognized the sovereignty of Pope Paul I, defying the emperor; in so doing, he became the first duke of the independent duchy. The new Duchy of Naples was one of the few areas in Italy to remain politically independent of the Lombards.[37] This independence allowed Naples to continue to thrive. The economic structure shifted from subsistence farming to basic urban manufacturing, primarily focusing on the shipbuilding, ceramics, and textiles industries. The city was also able to collect revenues by taxing the ships entering its busy port. This prosperity permitted the establishment of cultural institutions, such as schools of singing and writing as well as private libraries. The importance of culture in Naples, particularly as it relates to music, literature, and the theater, remains to this day.

The Holy Roman Empire

Like the rest of Southern Italy, Naples could not evade the invading Normans. The conquerors eliminated the two dominant foreign powers—the Byzantines in Calabria and the Arabs in Sicily[38]—placing the majority of the land south of Rome under their control. On Christmas Day 1130, Antipope Anacletus II crowned Roger II king of Sicily. The Norman king's opposition to Pope Innocent II set him on course for ten years of warfare; while papal supporters concentrated on Salerno and Bari, Roger moved to Campania, where his successful invasion in 1137 marked the end of the independent Duchy of Naples and its absorption into the kingdom of Sicily. As Palermo was now the capital, and customs emerged from the island, Naples' power and prestige rapidly diminished.

The Normans helped to centralize the government of Southern Italy and turned Sicily into a great maritime power.[39] However, their rule was rarely stable, as they constantly fought off baronial revolts and North African rebellions. Unsurprisingly, the death of King William II, Roger's grandson, produced greater turmoil, as he had named his aunt, Constance, and her husband, the future Henry VI, Holy Roman Emperor, as successors. The people of Sicily, unwilling to be ruled by a German, chose Tancred of Lecce, an illegitimate son of Constance's brother Roger.[40] Henry's first attempt to take the throne occurred just after he was crowned emperor, but his siege of Naples failed due to the persistence of Tancred as well as an epidemic that swept through the imperial army.[41] Pope Celestine later acknowledged Tancred as the king of Sicily, but when his death in 1194 elevated his young son William III to the throne, Henry struck again.[42] This time he met little resistance as he entered Palermo on November 20, and was crowned king on Christmas Day.

More importantly to the cultural formation of Naples, upon the death of Henry in 1197, Frederick II, just three years old, ascended to the Sicilian throne in May 1198. His young age weakened the kingdom's power; Pope Innocent III reasserted authority and the deposed King William III attempted to recapture the throne in 1202. King Frederick's regent was killed in the resulting war, which ended in 1207. Frederick came of age in 1208, became king of Germany in 1212, and was crowned Holy Roman Emperor in November 1220.

Unlike many of his predecessors, Frederick bypassed Germany and spent the majority of his time in the Kingdom of Sicily. Although the court was based in Palermo, Frederick preferred Naples as it was closer to both Germany and to the Papal States.[43] Frederick corresponded with Arab scholars on philosophical and scientific matters, and established a Muslim colony in the town of Lucera, where Islam could be practiced openly.[44] Frederick also ensured the Neapolitans knew the importance of culture and of its development. He established the first university in Naples, which, unlike other contemporary Italian universities, was not affiliated with the Church.[45] Thomas Aquinas, from whose work much of modern Catholic doctrine derives, graduated from the University of Naples, demonstrating that his curiosity about science and philosophy was not a threat to the Church.

Frederick built on the legal reforms of his grandfather Roger II and, in August 1231, issued the Constitutions of Melfi, which contained over 200 laws and proclamations. His goal was to make Sicily a model state in which all rights and obligations were subject to the rule of the king.[46] These laws were of extreme importance; in addition to demonstrating the primacy of written law, they also served as the basis of Sicilian law until the 19th century, and ensured that the kingdom became an absolutist monarchy—one in which the king was not bound by either those same written laws, or customs.

After Frederick's death in 1250, his son Manfred's troops attempted to occupy Naples, but the city asserted its dominance and became a free state under papal protection. The pope, however, wanted a sympathetic leader installed in Naples, as he feared the city's reunion with the Kingdom of Sicily and, by extension, the Holy Roman Empire. On January 6, 1266, Prince Charles of Anjou was crowned King of Sicily in Rome.[47] Frederick's heir, Prince Conradin of Sicily, just 14 years old, began traveling to Italy to claim his birthright. The prince represented the Holy Roman Empire. Charles defended the papacy. When Conradin was taken prisoner, he was sent to Naples to be publicly executed, reminding the city not just of their political allegiances, but of the importance of the Church in everyday life.

Flourishing under French rule
The Angevins
The Angevins, the royal house of France from which Charles was descended ("Angevin" means "from Anjou"), ruled Naples for nearly 200 years. Naples' fortunes continued to increase when the Angevins chose to move the capital of the Kingdom of Sicily to Naples.[48] This decision caused Naples' prestige to grow, placing it on equal footing with other European capitals, a center for trade and diplomacy. Yet, despite this growth in

fortunes and prestige, it was around this time that the North and South divide became undeniably evident. The North may have been occasionally threatened by powers across the Alps, but it was urbanized and extremely rich. In the middle of the peninsula was the religious state, which ruled the international church. Across the dividing line was a powerful Southern kingdom, but much of its money went to support the royal court, with cities and merchants burdened by heavy taxation to appease the royals' ambitions. Soon, the South was reduced to selling food and raw materials to the North, while the North's economy continued to grow. At the same time, King Charles and Queen Beatrice established an elaborate court in Naples, taking advantage of the taxes paid by their citizens. Shortly thereafter, the Kingdom of Sicily began to be referred to as the Kingdom of Naples.[49]

Although Naples rose up in rebellion in 1284, leading to a negotiation that cost Charles the loss of Sicily, it was under the House of Anjou that the city truly began to flourish. The sovereigns expanded the city walls and constructed beautiful churches and palaces. In fact, the Angevins were great patrons of the arts, consistently adhering to a French Gothic style in the buildings constructed around the city, and even the new royal castle. For the residents, the swamps were drained, the streets paved, and the cathedral restored. The population grew from 30,000 to nearly 100,000 during the time of the Angevin reign.[50] In fact, by the beginning of the 14th century, Napoli was proud to be considered one of the most important European cities. It was not only the royal capital of the House of Anjou, but also the principal port and the kingdom's mercantile and financial center.

Close political ties existed between Florence and Naples, while artistic and cultural exchange flowed between the two cities.[51] This is likely due to the interests of Robert, Charles' son, known as Robert the Wise. Under his reign, Naples was highly regarded as a center of learning for scholars, and the court was home to poets, writers, artists, musicians, and theologians. Florence's significant banking families also had offices in the city, which encouraged manufacturing, particularly of cloth and silverware. Still, the desires of most Neapolitans remained the same: a happy, peaceful, and sensual life, not one tethered to their jobs or wealth.

The legend of Saint Gennaro permeates the city, with citizens praying to him for good luck on everything from lottery tickets to calcio matches. This stems from his most famous feat, the liquefaction of his blood, which first occurred in 1338. While this occurrence is now celebrated as a sign of good fortune for the people of Naples, at the time, just five years later, King Robert the Wise died and Naples entered a century of instability, overseen first by a strong-willed queen. Princess Joan, his 16-year-old granddaughter, inherited the throne, but her personal life and political alliances almost immediately threw the kingdom into turmoil. King Robert had arranged for her to marry Prince Andrew of Hungary, which she did two days after her father's death. Almost immediately, Queen Joan began to alienate the Pope by protesting his attempts to work with her husband rather than with her. Joan fought against her husband's coronation and bestowed gifts upon influential supporters. When those on her side discovered the Pope had ordered Andrew to be crowned king, they assassinated him in 1345. Two years later, she married her cousin, Louis of Taranto, with whom she had previously been suspected of having an affair. Andrew's brother, King Louis, sent the Hungarian army to invade Naples to avenge his murder (in which Joan had possibly played a part) and the queen's betrayal, demonstrated by her refusal to marry another of Andrew's brothers.

As the Hungarian Army approached Naples, Joan fled to France in January 1348. But by the summer, she was able to return to her kingdom. Louis had chosen to abandon the city after a ship from Crimea brought the bubonic plague to its shores, killing more than 50,000 people in the city and surrounding areas.[52] The disease would ravage the city four more times over the next century. Joan continued ruling in a similar manner, alternately alienating and winning over both her subjects and various popes, marrying twice more, and losing children to early death or miscarriage. She adopted Louis I of Anjou and supported the antipope during the Western Schism. Ultimately, she was assassinated by her potential heir, Charles of Durazzo, upon whom Pope Urban VI had bestowed the Kingdom of Naples after declaring Joan a heretic.

For all Neapolitans, what she got right was calming the population after a massive tsunami rocked the coast in 1370, causing extensive damage to the Castel dell'Ovo. Queen Joan reassured the terrified citizens that Virgil's egg had survived the collapse of the towers, and Naples would remain secure. It did withstand the short reign of Charles, also assassinated during his pursuit of the Hungarian throne, and his son, Ladislaus, who chased land in Hungary, Croatia, and central Italy. Finally, another strong woman, Joan II, reconquered Rome, sent her abusive husband packing, and assassinated her lover when he attempted to usurp her power. When she died in 1435, Anjou rule effectively ended—she had named her relative René as her successor, but in doing so dismissed Alfonso V of Aragon, who began a siege of Naples in 1441, after which René quickly returned to France.

Sparkling under a Spanish star
Aragonese Naples

When Alphonse took the throne, he began centuries of Spanish rule in Naples. In 1501, Sicily and Southern Italy were again united as one kingdom, its capital located in Naples. The Aragonese, from Catalonia, ushered out the medieval period and began Naples' renaissance.[53] Under Alphonse, new piazzas and fountains were constructed, while old streets and civic buildings were repaired. He also completely refurbished the New Castle, which served as the royal residence, even including a triumphal arch over the entrance. The court there adhered to Spanish customs, traditions, and language. Although the Neapolitans did not consider Alphonse one of their own, this did not deter him from creating a better Naples. His domestic policy stabilized the kingdom, while his trips through the city to see how his people went about their lives enhanced his ability to address their concerns. In short, Naples was soon a great European capital once again.

Under Alphonse, humanism—the belief "that a body of learning, humanistic studies (*studia humanitatis*), consisting of the study and imitation of the classical culture of ancient Rome and Greece, would produce a cultural rebirth,"[54]— flourished in Naples. In the city, humanists were most inspired by the beauty of the area, and creating an idyllic world to match that beauty was the goal. The Accademia Pontaniana remains today as a place where the greatest scholars in many of the humanities gather together.

It was during this time that people began to once again realize the wonder and fascination Naples held and embark anew on trips to visit the city. Jacopo Sannazaro's famous poem, *Arcadia*, was widely read and celebrated not only throughout Italy, but the rest of Europe. It contains a famous description of Naples under Alphonse:

> Naples, as you have all heard several times, is the most fertile and enjoyable part of Italy, placed along the sea coast, a famous and most noble city, happy in both arms and letters, perhaps more than any other city in the world . . . I remember the delights of the bay, the marvelous and great buildings, the pleasant lakes, the enjoyable and beautiful islands, the sulfurous mountains, and the happy coastline of Posillipo with its grotto, inhabited by delightful villas, and softly bathed by the salty waves . . . What can be said of the games, the festivals, the frequent practice of many arts, studies, and laudable activities? Truly, there is no city or province or even the most opulent kingdom which is so greatly blessed.[55]

It is clear from the writings of the day that Naples was at the least a wonderful place to visit, though it is more difficult to discern whether the working class were just as happy. It is known, however, that during this time, local industries, such as cloth making, began to develop. Naples' wool was renowned, and its silk considered to be of the highest quality, so it is unsurprising guilds began to develop around these industries. They created trading standards and offered education to those wanting to join the profession, and provided support to widows and orphans of fellow guild members. While the guilds were primarily for the middle class, the standards established and education offered benefited the workers. The invention of the printing press also helped Naples to flourish, providing jobs and spreading culture. In addition, during this time the vernacular of Naples, *Napulitano*, came to prominence in a way it no longer is today, spoken alongside other languages such as French and Greek at cosmopolitan gatherings.[56] Texts also shifted from being written in ancient French to Napulitano, signaling that the Aragonese were beginning to identify more with their Southern kingdom.[57]

Soon, however, the rivalry between France and Spain for control of the

Kingdom of Naples flared again. The nobility disdained King Ferdinand I (commonly known as Ferrante), who followed Alphonse, while the pope revoked Ferrante's right to the throne due to his illegitimate birth. With Ferrante in Calabria, John of the House of Anjou occupied the capital, and it took Ferrante seven years to get the kingdom back under his control. An intelligent yet calculating, even ruthless figure, Ferrante committed such atrocities as inviting his enemies to a banquet at which he promised peace, after which he proceeded to slaughter many of them (it is said that he fed some to the crocodiles in the moat surrounding his castle; others were mummified, dressed, and displayed in the dungeon).

But there was a brighter side to this dark figure. Although threatened by Venice, the Papal States, France, and the Ottoman Empire, and disdained by the nobility in his own kingdom, Don Ferrante pursued policies intended to elevate Naples, both politically and culturally. Essentially, he laid the foundations for the modern city and the policies that underlie its institutions. He refortified the city by strengthening its walls, completed the redesign of the 13th-century Castel Nuovo and rebuilt the Naples Cathedral, built multiple palazzos, and erected the magnificent city gate, the Porta Capuana. Beyond his oft-nasty reputation, Ferrante was a true Renaissance man, who loved art, music, and literature, founding the first school of music in Naples, bestowing patronage on local artists, and reforming the university (although his subjects were then prohibited from studying outside the kingdom). Napulitano was declared the official language, increasing literacy and encouraging writing and poetry.

While Ferrante and his queen were once fortunate enough to view the liquefaction of San Gennaro's blood, the protection it gave to the city quickly faltered upon his death. Soon Naples was caught up in the Italian wars as the French, Spanish, and German armies fought to establish European hegemony.[58] The nobility called on the King of France, Charles VIII, heir to the Angevin dynasty, to return. Once the French conquered the Aragonese in a naval battle, they entered Naples victorious. The aristocracy refused to back Ferrante's son, Alphonse II. He abdicated in favor of his young son Ferrante II (often called Ferrandino). The young, cultured king was beloved by his subjects, but ruled for fewer than 2 years before dying of malaria at age 29. The crown, already unstable, passed to his uncle, Frederick III. At this time the French realized the weakness of the Aragonese succession and took Naples almost unopposed.

However, the French rule lasted only a few years. Rather than fight, the monarchs of France and Spain decided to divide the territory, with Sicily going to the Spanish and Naples to the French. But as soon as the French returned to Naples, Spain reneged on the pact, and by 1502 the hostilities had simmered over into open warfare.[59] The Aragonese again conquered in 1503, bringing the two kingdoms together once more. Ferdinand II of Aragon and his wife, Isabella I of Castile, ruled their kingdoms jointly until the queen died in 1504, after which Ferdinand expelled the Castilians.[60] The kingdoms were officially separated, and both Naples and Palermo became vice-realms, gaining both status and autonomy.[61] As part of the wide Spanish empire, Naples prospered and grew quickly, from a city of 100,000 to 300,000 a century later.[62] However, the city did not have the infrastructure to support such a population—particularly with agricultural supplies decreasing as farmers moved to the city—nor did they have the right government to help it flourish. The viceroys collected taxes and sent much of the money back to Spain, while the nobility became insolent and allowed corrupt workers to control the land they owned. Although one viceroy, Don Pedro Alvarez de Toledo, helped to make Naples beautiful again with rebuilt walls, the construction of a new palace, and a residential district to house soldiers, what Naples really needed was more residential housing.[63] Then, as now, crowds were often the first thing visitors noticed when arriving in the city.

The city was filled with the rural poor, the provincial elites, and the feudal nobles. It saw traffic from all over the Spanish kingdom, as well as other parts of Italy and Europe, as merchants, administrators, soldiers, and diplomats passed through. It was little wonder, then, that a bureaucrat could describe the city as "the whole world" in the 1630s.[64] However, although supposedly steady under Spanish leadership, Naples being "the whole world" became tiresome quite quickly.

Vesuvius erupted in 1631, and the death and destruction it wrought led to San Gennaro becoming the city's most popular saint.[65] It is said that as the volcanic materials reached the city, the Neapolitans turned to San Gennaro, and the eruption stopped due to his intervention.[66] Next, the city of Naples revolted in 1647–1648. As Spain fell into crisis, so too did Naples. In the midst of a long-term recession, the people were taxed more and endured higher inflation; at the same time, organized crime demanded protection money from artisans, merchants, and shopkeepers, further discouraging small businesses. As taxes increased to stratospheric rates in order to support the Spanish army and pay for royal weddings, the people of Naples grew restless, and began their revolution on Christmas Eve 1646, although fighting didn't truly break out until the summer of 1647, when the King broke his pledge to lower tax rates. Under the leadership of Tommaso Aniello, known as Masaniello, the poor across the city turned to violence, and dreamed about returning to an independent republic. On the cusp of a compromise, the viceroy invited Masaniello to the palace, where he fell victim to a homicidal plot. With his death, the hopes of independence—or, at the least, political representation for all social classes—was extinguished.

Then, in 1656, the bubonic plague returned, likely via a ship from Sardinia, and killed about half of Naples' population. Naples had remained the biggest city in Italy, but also its most crowded, and unsanitary conditions allowed the plague to move quickly through residents for six months. Doctors had no idea what it was or how to treat it, and the kingdom's rulers began to believe it stemmed from "poisonous powders" sprinkled throughout the city by their Spanish enemies. In order to prevent panic, they minimized the disease's threat, avoiding quarantines and lockdowns and permitting trade and other commercial activities to continue. Similar to what occurred during the COVID-19 pandemic, later measures such as health certificates were falsified, and many evaded stay-at-home orders.[67] After six months, the plague began its retreat. To celebrate its eradication in true Neapolitan fashion, a ten-day celebration began on December 1, with a service thanking Saint Francis Xavier, and continued with parades, feasts, and fireworks, all accompanied by joyous music.[68]

This was unsurprising in Baroque Naples, who maintained her role as a major center of art and architecture, and her musical, literary, and intellectual life only continued to grow. The social classes were connected primarily through religion, as festivals and saints' days brought Neapolitans together for feasts and celebrations. The rulers also knew they could keep their subjects happier if they entertained them, a mission often accomplished by music, of which Carolyn Gianturco says, "Of all the adornments of baroque Naples, it might be said that music was the richest and was indulged in the most frequently."[69] But for the most part, the poor and working class were left behind; the small artisanal shops only made enough to meet their families' basic needs, while the unemployed scraped by through the Neapolitan *arte di arrangiarsi*,[70] the art of getting by, through resourcefulness and innovation. Complaints about the city's trade imbalance—exporting food while their own were starving, for example— fell on deaf ears, as most of the money generated by trade went to foreigners. And the government continued to use its funds to appease the higher classes through the arts. One example of the excess obvious in Spanish Naples was *guglie* (rococo obelisks topped with statues of saints), erected in thanks for delivery from the plague.[71]

The House of Bourbon

The wars of succession continued, and in 1707, Austria came to rule Naples. However, their presence was so short they barely left their mark on the kingdom. In 1734, Charles of Bourbon, the son of the king of Spain, conquered the city. Naples remained under Bourbon rule until unification in 1861, although for two short periods they lost the throne, the first time to revolutionary forces, and the second to Napoleon.[72]

A taste of Bourbon rule

Much has been written about Naples under the House of Bourbon, as the city differed from most during the Age of Enlightenment, and because of the revolutions it produced. In most of the Western world, the Enlightenment consisted of a series of reforms that pushed aside religion and tradition in favor of a philosophy that relied on intelligence, and focused on a civil society that came together in a manner in which citizens were bound by the laws defined by the rules of nature, rather than any political power.[73] In Naples, however, a new kingdom had just been born, and the needs of the court and desire to construct new royal buildings to ensure the capital fully embodied its kingly stature outweighed immediate reforms.[74] Essentially, Enlightenment and nation-state building came together in the new kingdom and, as such, the focus was not on civil society but rather how the elites and nobility could construct a city that would attract the highest classes of people to view the theatrical new buildings, listen to prestigious new musical programs and operas, and discuss the new ideas arising in Europe.

This resulted in social and economic reforms in Naples in the second half of the 18th century; however, those reforms were rooted in a small group of progressives, isolated from the majority of Naples' people. Because the reforms were not firmly rooted in the culture of Naples, they ultimately failed to achieve significant, lasting success.[75] While Naples appeared as a vibrant, cultural city to privileged residents and visitors, the majority of citizens saw little benefit; a planned housing welfare project was put on hold with just 20% completed, and the public granaries promised never materialized. This dichotomy is still evident today. Yet Lancaster alleges the period of Bourbon rule was the city's happiest time, as it offered a certain degree of political stability and independence.[76]

The most renowned cultural contribution of Bourbon Naples was the discovery of the lost Roman cities of Pompeii and Herculaneum, attracting the attention of the entire world. The discoveries were one of the defining events of the century, and King Charles capitalized on them, encouraging the excavations and establishing a school of architecture, whose publications circulated widely.[77] However, soon after the discovery of the cities, Charles abdicated the throne, leaving the crown of Naples to his third son, Ferdinand—just eight years old at the time.

The young regent was left in the care of the prime minister, Bernardo Tanucci. Tanucci maintained his hold on power even after King Ferdinand came of age. The undereducated Ferdinand spoke only in Napulitano, established the lottery, spent much of his time hunting, and even set up a stand to sell what he'd caught, although he often gave away much of it to the poor. It is little wonder, then, that the majority of Naples loved him, feeling that he was one of them. This type of adoration, in which Neapolitans embrace a figure they're able to relate to, continues today.

In the winter of 1798, the Bourbons raised up against French soldiers residing in Rome. After the unsuccessful, indeed humiliating, campaign, the French took revenge and surrounded Naples. A bloody battle ensued in which the *lazzaroni* (the poorest of the poor), nearly all of whom were anti-French, took to the streets.[78] The king and his court fled to Sicily at the end of 1798, leaving a power vacuum in Naples. The city was cracking apart. No longer could its charming, seductive atmosphere fool tourists into believing the poor were at peace. It became clear that the kingdom had utterly failed to implement reforms or create a stable, democratic civil society. Yet the lazzaroni remained loyal to the king, even after French republicans took the city and the Parthenopean Republic was declared on January 21, 1799. Between a lack of support for the new government and the punishing war reparations imposed by the French, the Republic lasted just five months. The Republicans surrendered in June and the monarchy was restored.

However, the Bourbons barely had time to celebrate. Napoleon Bonaparte began his sweep of the continent in 1800, displaying military genius and political ambition, and implementing radical social changes. By the time he reached Naples, the royal family had fled once again; this time, the lazzaroni felt no desire to fight against the French, remembering how their allegiance to the king had cost them a great deal in 1799.[79] When Joseph, Napoleon's brother, became the King of Naples in February 1806, the citizens barely took notice.

But Joseph took note of the disarray within the kingdom. The legal system, guided by 12 legal codes, was in such a mess that there were no overarching rules; a decent lawyer could find support for virtually any argument.[80] In addition, the royal family were known to change the laws on a whim, and the courts deferred to them. Even verdicts were reversed by royal will. Taxes imposed were far from equal; the church lands were heavily favored while royal and feudal lands were excused. The nobility paid other, much smaller, tributes, while the king took his revenue not only from the royal domains, but the numerous public offices that were for sale. The majority of the kingdom's taxes went straight into the royal treasury; certain revenues were bestowed as royal favors, but the majority of funds were not redistributed with any sort of fairness.

According to documents of the time, the kingdom was failing by nearly every measure, except perhaps tourism. Land was able to be held by very few, and the abundance of the kingdom's natural resources were of fixed supply determined by local rulers. Few industries and little manufacturing existed. Commerce was hampered by the duties set on all imports and exports, making it difficult to obtain foreign products that might make life easier. Public works were neglected. Even those that would help protect the city such as fortresses were left unbuilt, leaving Naples unprotected and its citizens used to a certain level of servitude.[81]

Neapolitan soldier and historian Pietro Colletta, in his monumental work *History of the Kingdom of Naples*, writes that due to these and other fractures in society:

It was therefore impossible to restore order in the State through its own elements; a new king and a new kingdom were needed, and events of sufficient importance to restore civil strife, and to present one common aim for which to labor and to hope.[82]

Joseph was eager to be well-liked by his new subjects, but at the same time he understood the need for a strong program of reform, modeled on the French Revolution and its aftermath. He deemed the House of Naples incompatible with such ideals, and abolished feudal privileges and taxes. Those who had suffered under the previous king for their liberal ideas were permitted to take office. Joseph nationalized the Church's lands and suppressed monastic orders. To improve the kingdom and provide work for the lazzaroni, the king began a series of public works programs, many of which involved constructing roads and highways. Although his reign lasted only two years, King Joseph demonstrated the best of the Enlightenment.

Napoleon replaced Joseph with cavalry officer Joachim Murat, the husband of his sister Caroline. While he continued Joseph's reforms, the new king governed

Naples for almost eight years, and so was able to leave much more of a mark than his predecessor. He further dismantled the Church and the nobility's privileges through the implementation of the Napoleonic Code, which also transformed the courts and government bureaucracy.[83]

Similarly, taxes were reworked to fall into line with those of France, and organized around a single, direct land tax. In addition to establishing schools for the general population, he merged academies to create the Royal Society. He also established the Botanical Gardens, the Conservatory, the Observatory, and the Bank of Naples. Murat organized the city into municipal districts and a city council was formed, with the new mayor of Naples as its head. What might have been most important to his ability to create change in Naples was his embrace by the city. Once again, a man from humble origins who had worked himself up into a position of power, and who professed to adore Naples and showed it through his enthusiasm for the city, became a beloved figure in Neapolitan history.[84]

But while Naples loved him, Napoleon treated his brother-in-law as subservient to his own rule, rather than an independent king. As such, Murat began to dream of being the future king of a united Italy. He encouraged societies like the *Carbonari* (charcoal burners), groups of which were spread across Italy, agitating for a liberal, unified Italy through the rebellion of the proletariat, led by intellectuals and students.[85] Yet Napoleon's defeat and exile did not lead to an increase in Murat's power. Instead, King Ferdinand returned, uniting his territories as the Kingdom of the Two Sicilies in 1816.

Bourbon Restoration

Under Bourbon rule once more, the middle classes were prepared to continue the reforms of Joseph and Murat, expressing their desire for a representative government and a constitution. Ferdinand, meanwhile, intended to restore an absolutist monarchy. An uprising of the Carbonari forced Ferdinand to relinquish absolute rule and establish a constitution in 1820. Any French reforms were implemented exceedingly quickly, but they did not flourish in Naples and the South, as the laws in the region had been created with a top-down strategy and didn't take into account any Neapolitan customs or culture. This, scholars argue, was the moment when the two halves of Italy truly diverged, and marked the beginning of the "Southern Question," which continues to hamper Italian nationalism today.[86]

The constitution itself was dissolved in 1821, after Ferdinand placed his son, Francis, on the throne while manipulating Austria into overthrowing the Naples government.[87] Reformists found no friend in the next king, Ferdinand II, who took power in 1830. His outdated ideas conflicted with the desires of the liberal bourgeoisie, who were excited about the concessions being granted in Northern and Central Italy. The upper and middle classes desired a constitution, while the lower classes grew tired of famine. The king did grant another constitution, but it was insufficient: word of arguments between the crown and the members of parliament reached the streets, and the people began to rise up in Palermo in 1848. King Ferdinand displayed no weakness, calling on the royal troops to put down the revolution. He became an absolute monarch once more, and adopted a "paternalistic and isolationist" system of policy as a means of protecting the kingdom.[88]

Like the reforms his grandfather Ferdinand I set in motion prior to the arrival of the French, the people realized that the policies of Ferdinand II favored the nobility rather than the middle and lower classes. For example, the first Italian railroad was constructed in the kingdom, but it was created not to transport goods, but to ferry the king's court from Naples to their seaside residence in Portici.[89] Still, Naples did make significant scientific, technological, industrial, and cultural advances in the final 30 years of Bourbon rule. The kingdom was actually designated as third in the world in industrial development (and first in Italy) at the Paris International Exhibition in 1856. The first seismic observatory in the world was constructed on Mount Vesuvius in 1841. Naples boasted the highest number of doctors per capita in Italy and the lowest infant mortality rate. They constructed the first steel suspension bridge in Italy, over the Gagliano River, in 1832. Naples also had the highest number of printing presses of any city in Italy, as well as the longest continually active opera house.[90]

Yet despite the city's development, travelers continued to view Naples as a sensuous destination, flocking to the kingdom to revel in its natural beauty and take in its romantic landscapes. They filled the San Carlo Opera House, where famous composers brought to life songs of the beauty of the setting and the romance that often blossoms in Naples. The School of Posillipo not only painted landscapes, but contributed to the myth of a happy underclass by inserting lazzaroni enjoying themselves into the pictures. The poet Percy Bysshe Shelley captured his adoration of the city's enchantments in *Ode to Naples* while in the midst of a cholera epidemic, Italian poet Giacomo Leopardi concentrated his works on Mount Vesuvius. Only Charles Dickens called out these creatives, and the travelers who admired them, for paying no attention to the difficulties of the city's poor. He alluded to the Brothers Grimm finding inspiration for their bloody fairy tales during the time they stayed in Naples.[91]

The truth was that, as Luigi Settembrini wrote, the foreigners only heard talk of progress in the kingdom, while reveling in the kingdom's beauty, and concluded that the residents possessed an "enviable happiness." Instead, "In the Kingdom of the Two Sicilies, in the country which is said to be the garden of Europe, the people die of hunger, are in a state worse than beasts, and the only law is caprice."[92] However, despite the conditions of the people in most of the kingdom, this misconception lives on today as a mythical point of reference for those who dislike life in a united Italy. After all, prior to the country's founding in 1861, the Southerners were king (literally). They had established their place in the Western world, and a great many respected Europeans traveled to their capital to find a way to live life differently. The kingdom controlled shipping lanes, a maritime fleet, and was in the process of building highways and railroads. King Ferdinand II commanded the largest army and navy in Italy, and was inclined to use it. Naples controlled more gold reserves than the rest of Italy combined.[93]

To tumble from the top of the mountain to the lowest trenches, as the Kingdom of the Two Sicilies did upon unification with the rest of Italy, has left many in the South with the faint hope of the dissolution of Italy and a return to a constitutional monarchy.

The Southern Question

In most political, economic, sociological, and historical literature involving Italy, reference to the "Southern Question" is almost certain to be made. What is more difficult is discerning what, exactly, the Southern Question *is*, and more importantly, how true the majority of the answers actually are.

For the sake of simplicity, the Southern Question refers to the country's severe case of economic dualism, which had existed even before unification, and continues today, as the people of the *Mezzogiorno*—the seven states that made up the Kingdom of the Two Sicilies, as well as Sardinia—in which 32% of the Italian population lives, make only 56% per person of those in North-Central Italy.[94] It is not so much a question as it is an examination of the reasons this inequality has existed for over 200 years.

Yet the Southern Question is about so much more; entire books by experts in this area can be found. This chapter can do no justice to the question in the space available. What must be noted, though, is that its mere existence shapes post-unification discourse and perceptions of the South, both by outsiders and by Southerners themselves. When reviewing the last two centuries of Neapolitan history, it's essential to keep in mind that much of what is revealed about the South is not documented by those in the Mezzogiorno, and so the reader must ask themselves if the author's view corresponds to reality. Readers should also consider what might be missing or ignored when the South is discussed. Finally, another crucial part of the Southern Question is determining how the divide influences the thoughts and actions of those in power— particularly during unification, and after World War II.

Risorgimento

Many find it hard to believe that Italy as we know it has only existed since 1861. In fact, author Mark Doidge argues that Italy is essentially a modern construct that refers to the geographic peninsula south of the Alps.[95] When examining regional differences, and further exploring the Southern Question, it is easy to remember that Italy was once a number of kingdoms and city-states, with their own traditions and even their own languages, that were brought together by Italian nationalists, not always by the will of their people.

"*Risorgimento*" means "rising again," but it refers to the unification of the Italian peninsula under one monarch. As in Naples, a number of Italian states fell under

French rule during Napoleon's reign, and were introduced to the liberal ideas that stemmed from the French Revolution. After Napoleon's defeat in 1815, however, the states fell back under the control of their former, conservative rulers. For Naples, this meant the Bourbons ruled for an additional 41 years, long enough for the monarchy to tamp down most of the reforms and continue its absolutist control. Elsewhere, secret societies, like the nationalist group the Carbonari, began to form, even before the defeat of the French. They did succeed in helping spread liberal ideas—for one, they had a strong base in Naples and provided the main impetus behind the 1820 revolution and demand for a Neapolitan Constitution. Yet one of the main problems with the Carbonari and others is that they had no single goal: some were republicans and others monarchists, some were federalists and others wanted a unified Italy.[96] On the other hand, Young Italy, founded by Giuseppe Mazzini in 1831, was avowedly republican and sought to encourage the masses to revolt against the authoritarian regimes, and to teach them that a unified Italy would be in their best interests.[97]

Neither group was able to succeed in the revolutions of 1848. It was clear from the ruthlessness with which Ferdinando put down the Sicilian drive for independence, and the revolts that spread to Naples, that he would not easily be parted from his crown. Instead, Piedmont, a region in the northwest of the peninsula, took the lead, and with the help of the French liberated themselves in 1859. From there, Piedmont set its sights on the South, the Kingdom of the Two Sicilies. Southern Italy lagged behind the North in a number of categories, including literacy; 87% of those in the South were illiterate. The region was so under-developed that the Piedmontese felt justified in their invasion, believing they brought "civilization" to the South (although there were those who simply looked down on the South, and found that a sufficient reason to invade). Others, trying to explain the differences between the two, theorized that the country contained two separate races that were incompatible with one another.[98] Marxist philosopher Antonio Gramsi, who literally wrote the book, *The Southern Question*, argued that at the time of unification, the North and the South were "absolutely antithetical" to one another.

Yet no matter how incompatible the supposed "races" might be, no matter what disasters might be caused by the unification of two parts of a peninsula anathema to each other, by 1861, most of what is now Italy had come under Piedmontese control, including the Kingdom of the Two Sicilies. Unlike elsewhere, Neapolitans were divided on the issue of unification. When Giuseppe Garibaldi had entered the town, many welcomed him; according to Alexandre Dumas, they followed him from the sea to the cathedral to the palace, singing songs against King Francis and chanting their thanks for Garibaldi.[99] At the same time, the army—unlike elsewhere—had not overwhelmingly rebelled against Naples' King Francis.

To ensure the Pope's sovereignty remained intact, Napoleon made a trade: the King of Sardinia would be released and his forces allowed to move through the Papal States and southward to Naples. King Victor Emmanuel met up with them on October

9, 1960 and, after Garibaldi had given up his dictatorial powers and bowed down to Victor Emmanuel as the "King of Italy," they rode into Naples side by side. There Garibaldi left, his work finished, while the assembled troops fought the Neapolitan forces for three months. Even after Naples was forced to surrender, proud groups of Neapolitans, loyal to their king and kingdom, fought the new Italian government for years. Yet Victor Emmanuel was crowned king of Italy on March 17, 1861, and Rome was declared the capital of the country ten days later—despite the Papal States not being part of the kingdom yet. King Victor Emmanuel still had work to do to unite all of Italy, but 1861 is considered the year Italy officially became a country.

The Bourbons, who died in exile in Rome, remain associated with Naples and the South, given that they had created an independent Neapolitan state. Even with the state of the Kingdom of Naples when the Bourbons fled, their time on the throne is still considered a high point in Naples' history . . . particularly as Naples quickly became bitter about Garibaldi bringing them into the Kingdom of Italy.[100]

Naples after unification

Most who supported the Risorgimento had done so based on promises that the regions would maintain their distinct identities. Instead, despite the Sardinian king Victor Emmanuel II ascending to the throne of a united Italy, it was Piedmont that took over life in the new country: Piedmont's constitution, Piedmont's tax system, Piedmont officials in charge.[101] Since Northern laws were typically liberal, some were in favor of this change. But the people of Naples generally opposed things like anticlerical provisions in the law, and uprisings in Naples continued, a combination of those still supporting (and supported by) the Bourbons, and spontaneous peasant revolts.[102] The Neapolitans—no matter how drab and impoverished their lives may have seemed to those up North—did not want to give up their kingdom or their culture. They did not want to give up their *language*; Napulitano is classified as a distinct language to Standard Italian (which is derived from Dante's Tuscan), and while the two can often be mutually understood, Napulitano is considered essential to the area's history and culture. The city was the site of many protests, against the city's isolation from the richer markets of the North, against the way the local agricultural systems were destroyed, against the fact that the city's textile and engineering industries simply collapsed.

According to Gramsci, the North had many advantages the South did not even come close to approaching when unification occurred. While the North had infrastructure set to move it into the capitalist age, the South remained disconnected, with few roads, ports, and waterways, and an agricultural system that could not even feed the local people. Under Ferdinand II, who had ruled until 1859, nothing was done to put in place a system of public hygiene, and many areas lacked sewer systems and experienced frequent water shortages. Unsurprisingly, one of the first "problems" for the Piedmont parliament to address was the Southern Question: how to govern a land that was considered barbaric and uncivilized? They debated visiting to try and understand Southern society and the problems it faced, but instead determined its people were so far removed from a civilized lifestyle that instead it would establish order by using force, justifying this decision in the name of implementing a "Piedmontese mentality."[103]

The city's resources had been so thoroughly depleted that the economy stagnated and, 20 years after unification, the city was beset by crime, unemployment, and immense overcrowding. In the poorest neighborhoods, it was common for an entire family to live in one small, windowless room.[104] Given the lack of a functioning hygienic system or consistent garbage removal, most trash was thrown on the street. It is no wonder, then, that during the 19th century, Naples experienced eight cholera epidemics as well as four of typhoid, and the poorest neighborhoods were hit the hardest. With the government unwilling to visit Naples to determine what the city needed most, it passed a law to eliminate the slums, which more or less turned into a project of gentrification, with beautiful historic neighborhoods erased, piazzas created, hotels established, all while the former residents of these "slums" waited for new public housing to be built.

The disconnect between Naples' problems and the central Italian government provided an excellent opportunity for the Camorra to put its stamp on the city. When it was clear the state wasn't about to help Naples, organized crime took over many of its duties, such as providing charitable relief to the poor and offering young men jobs. It used racketeering to hide the money it made from collecting taxes and protection money.

As the South grew even poorer, its people began to emigrate, both to the country's North and across the Atlantic. Thousands poured into steamers to New York, both from Naples and nearby provinces. When they arrived in the United States, most men went into jobs in agriculture or construction. Without the cheap labor of the arriving Italians, US economic development may have looked quite different.

Getting by in modern-day Naples

And without unification, Naples' economic development may have looked quite different. The resources poured into the North bypassed Naples, leaving it far less industrialized, but Italy's involvement in World War I ensured the city would suffer at least as much as the rest of the country. In the *biennio rosso* (two red years, 1919–1920), shipbuilding firms went bankrupt, unemployment skyrocketed, and those in the South protested by occupying land belonging to the wealthy. The economic crisis led to a political one, prohibiting parties from building coalitions—and enabling the Fascist Party to come to power.

After three years of a near civil war throughout the country, Benito Mussolini was appointed prime minister in 1922. Like most cities in Italy, Naples suffered under the fascist regime. It was during World War II, however, that Naples' almost-unwilling integration into a unified Italy became its downfall. Seeking an easy route by which to take the peninsula, the Allies began their bombing campaign in 1940, hitting Naples hard in 1943. The near-flattened city was occupied by the Nazis after the Allies left, but Neapolitans rose up against their occupiers in what is known as the *Quattro giornate di Napoli* (Four Days of Naples) and drove the Germans out.

After the war, most of the money Italy received from the Marshall Plan was funneled to the North. Meanwhile, in Naples, the Allies believed the best way to combat communism was to allow the Camorra to fill in the gaps left by the state, and corruption grew. The *Cassa per il Mezzogiorno* (Southern Development Fund), established in 1950, was meant to help industrialize the South; instead, the majority of the plants established employed few local employees and most made little money. The North grew more resentful of the South, as they considered themselves to be subsidizing an area too lazy to commit to work, while Southerners remained distrustful of the North, as projects continued to fail and unemployment skyrocketed.[105]

Like the rest of Italy, Naples had a turbulent time of it during the 1960s and 1970s, with labor strikes, student protests, and domestic terrorism spreading across the peninsula. But, in addition to its economic difficulties, Naples also had to address a cholera outbreak in 1973 and a devastating earthquake in 1980, while a waste management problem that lasted over two decades, and the increasing influence of the Camorra, kept tourist dollars away from the city. In the early 1990s, the European integration project hit Italy hard.[106] With the budget restructuring, the country could no longer support the Mezzogiorno. While the issues of the past 100 years will be interwoven with the history of Napoli's soccer club in the following chapters, it is sufficient to say here that, as opportunities have continued to fade in the early 21st century, more and more young people in the South are simply living "the art of getting by."[107]

---

[1] Lancaster, J. (2005). *In the shadow of Vesuvius: A cultural history of Naples*, 9–10. Palgrave Macmillan.

[2] Austern, L., & Naroditskaya, I. (Eds.). (2006). *Music of the sirens*. Indiana University Press.

[3] Lancaster, 11.

[4] Di Mauro, M. (n.d.) "Myth of Pathenope [sic]: 3 stories of the foundation of Naples." *Visit Naples*. https://www.visitnaples.eu/en/neapolitanity/tales-of-naples/myth-of-pathenope-3-stories-of-the-foundation-of-naples. Accessed August 8, 2020.

[5] Ledeen, M. (2011). *Virgil's golden egg and other Neapolitan miracles: An investigation into the sources of creativity*. Transaction Publishers.

[6] Encyclopedia Britannica. (n.d.) Naples: History. In *Encyclopedia Britannica*. Retrieved August 13, 2020 from https://www.britannica.com/place/Naples-Italy/History

[7] Lancaster, 12.

[8] Naples: History.

[9] Lancaster, 12.

[10] Naples: History.

[11] Lancaster, 13.

[12] Id., 15.

[13] Id., 16.

[14] Id., 17–18.

[15] Naples: History.

[16] Lambert, T. (n.d.). A brief history of Naples, Italy. *Local Histories*. Retrieved September 22, 2020 from http://www.localhistories.org/naples.html

[17] Taylor, R. A. (2021). *Ancient Naples: A Documentary History Origins to c. 350 CE*. Italica. https://doi.org/10.2307/j.ctv1t8q8n3

[18] Lancaster, 18.

[19] Id., 22.

[20] Id., 27.

[21] Id., 20.

[22] The legend of Castel dell'Ovo in Italy. (2021, June 9). *Italian Tribune*. https://italiantribune.com/the-legend-of-castel-dellovo-in-italy/

[23] Taylor.

[24] Lancaster, 29.

[25] Taylor.

[26] Lancaster, 32.

[27] Id., 34–35.

[28] The legend of Castel dell'Ovo.

[29] Lancaster., 35.

[30] Lambert.

[31] Lancaster., 37.

[32] Lambert.

[33] Lancaster, 39.

[34] Id., 40.

[35] Id., 41.

[36] Lambert.

[37] Lancaster, 44.

[38] Id., 49.

[39] Houben, H. (2002). *Roger II of Sicily: A ruler between east and west*. Cambridge University Press.

[40] Encyclopedia Britannica. (last updated 2020, January 1). Henry VI: Holy Roman emperor. In *Encyclopedia Britannica*. Retrieved August 24, 2020 from https://www.britannica.com/biography/Henry-VI-Holy-Roman-emperor

[41] Fuhrman, H. (1986). *Germany in the high middle ages c1050-1200*. Cambridge University Press.

[42] Kaplan, P. H. D. (1987). Black Africans in Hohenstaufen iconography. *Gesta*, *26*(1), 29–36.

[43] Lancaster, 51.

[44] Unfortunately, the town was composed of prisoners and the Muslims regarded as Frederick's personal servants, but it—along with the miniatures, statues, and architectural embellishments, suggested an interest in Black culture and Black participation in the Kingdom of the Two Sicilies, and it demonstrated the further fracturing between Frederick and the papacy. Kaplan, 32.

[45] Lancaster, 54.

[46] Id., 53.

[47] Id., 56.

[48] Id., 57.

[49] Id.

[50] Id., 58.

[51] Id., 62.

[52] Id., 68.

[53] Id., 73.

[54] Nauert, C. G. Jr. (2006). *Humanism and the culture of Renaissance Europe* (2nd ed.). Cambridge University Press.

[55] Lancaster, 78–79.

[56] Gilbert, J., Keen, C., & Williams, E. (2017, May 11). The Italian Angevins: Naples and beyond, 1266–1343. *Italian Studies*, *72*(2), 127.

[57] Lee, C. (2017). Writing history in Angevin Naples. *Italian Studies, 72*(2), 148–156. https://doi.org/10.1080/00751634.2017.1307553

[58] Astarita, T. (Ed.). (2013). *A companion to early modern Naples.* Brill, 3.

[59] Lancaster, 87.

[60] Nowell, E. (1976). Old world origins of the Spanish-American viceregal system. In F. Chiappelli, Michael J.B. Allen, & Robert L. Benson (Eds.), *First images of America.* University of California Press.

[61] Lancaster, 87–88.

[62] Lambert.

[63] Lancaster, 89–90.

[64] Astarita, 3.

[65] Id.

[66] Lancaster, 98.

[67] Bifulco, M., Pisanti, S., & Fusco, I. (2021). Lessons from the 1656 Neapolitan Plague: Something to learn for the current coronavirus pandemic? *Vaccine, 39*(27), 3641–3643. https://doi.org/10.1016/j.vaccine.2021.05.046Bilf

[68] Johnston, K. (2021, March 1). Naples memorialized its 17th century plague with a festival for healing, and so should we after COVID-19. *The Conversation.* https://theconversation.com/naples-memorialized-its-17th-century-plague-with-a-festival-for-healing-and-so-should-we-after-covid-19-154774

[69] Gianturco, C. (1993) Naples: A city of entertainment. In G. J. Buelow (Ed.), *The late Baroque era. Man & music* (pp. 94–128). Palgrave Macmillan, 94. https://doi.org/10.1007/978-1-349-11303-3_4

[70] Lancaster, 91.

[71] Id., 116.

[72] Astarita.

[73] Imbruglia, G. (2000). Enlightenment in eighteenth-century Naples. In G. Imbruglia (Ed.), *Naples in the eighteenth century: The birth and death of a nation state* (pp. 70–94). Cambridge University Press, 70–71.

[74] Maiorini, M. G. (2000). The capital and the provinces. In G. Imbruglia (Ed.), *Naples in the eighteenth century: The birth and death of a nation state* (pp. 4–21). Cambridge University Press.

[75] Davis, J. A. (2009). *Naples and Napoleon: Southern Italy and the European revolutions, 1780-1860.* Oxford University Press.

[76] Lancaster, 126.

[77] Id., 138.

[78] Id., 169–70.

[79] Id., 179.

[80] Colletta, P. (1858). *History of the Kingdom of Naples: 1734–1825, volume 2.* T. Constable and Company.

[81] Id.

[82] Id., 8.

[83] Lancaster, 180.

[84] Id.

[85] Id., 182.

[86] Davis, 2.

[87] Encyclopedia Britannica. (last updated 2020, January 1). Ferdinand I. In *Encyclopedia Britannica.* Retrieved February 10, 2020 from https://www.britannica.com/biography/Ferdinand-I-king-of-the-Two-Sicilies

[88] Lancaster, 186.

[89] Id.

[90] Mendola, L. (2012). Kingdom and house of the Two Sicilies. *Best of Sicily Magazine.* http://www.bestofsicily.com/mag/art425.htm

[91] Lancaster, 192.

[92] Id., 194.

[93] Mendola.

[94] Pescosolido, G. (2019). Italy's Southern Question: Long-standing thorny issues and current problems.

Journal of Modern Italian Studies, 24(3), 441–455. 10.1080/1354571X.2019.1605726

[95] Doidge, M. (2015). *Football Italia: Italian football in an age of globalization*, 46. Bloomsbury, 14–15. http://dx.doi.org/10.5040/9781472519221.0012

[96] Encyclopedia Britannica. (2017, February 09). Carbonari. In *Encyclopedia Britannica*. Retrieved November 17, 2020 from https://www.britannica.com/topic/Carbonari

[97] Encyclopedia Britannica. (2017, February 09). Risorgimento. In *Encyclopedia Britannica*. Retrieved November 17, 2020 from https://www.britannica.com/event/Risorgimento

[98] Niceforo, A. (1901). *Italiani del nord e Italiani del sud*. Fratelli Bocca.

[99] Lancaster, 197.

[100] Id., 198–200.

[101] Clark, M. (2009) *The Italian Risorgimento* (2nd ed.). Routledge.

[102] Hearder, H. (1980). *Italy in the age of the Risorgimento 1790–1870*. Routledge.

[103] Moe, N. (2002). *The view from the Vesuvius: Italian culture and the Southern Question*. University of California Press.

[104] Lancaster, 201.

[105] Encyclopedia Britannica. (last updated 2020, January 1). Italy. In *Encyclopedia Britannica*. Retrieved December 4, 2020 from https://www.britannica.com/place/Italy

[106] Maddaloni, D. (2016). Whatever happened to Italy? The crisis of the Italian pattern of development in the era of globalization. *Athens Journal of Social Sciences, 3*(4), 299–319.

[107] Id.

CHAPTER 2

Founding a Napoli

2

While the official founding of SSC Napoli occurred in 1926, football began to take root in the city 20 years before. Like all things related to the Neapolitan club, their formation is separate and distinct from that of others founded in Italy around the same time. This book begins with a chapter on the history of Naples because it is crucial to understanding the history of, and current state of, SSC Napoli. The poverty of the South, the people's resilience, their belief that their city was superior because it was ruled by generous patrons, the superstitions attached to nearly every aspect of their fandom, even the fact that the Greeks began to attach more importance to Neapolis, the settlement on the Bay of Naples, than to Cumae, high on the hill and reaching down to the coastline. All of this influences the way that Neapolitans view themselves, often as a separate and distinct entity from the rest of Italy, produced by royalty and still not entirely separate from it—which in turn shapes the way that the Napoli football club have and do conduct themselves, how the Napoli fans support their team and react to league decisions, and how the rest of the *calcio* world views the club and their supporters.

As John Foot writes:[1]

> Most serious Italian fans are well aware of the date of foundation of their club, its record, its founders and its historic players, managers, and even the various stadiums where the club has played. All these historic features are a strong part of a civic religion—adherence to which is a crucial aspect of fan identity. Founding myths, legends, and stories permeate this self-styled football history, as tales are handed down from generation to generation.

Yet football has changed, even in the fewer than two decades since Foot's *Winning at All Costs: A Scandalous History of Italian Soccer* (distributed as *Calcio: A History of Italian Football* elsewhere) was published. No longer does one need to be part of this "civic religion" to be considered a "serious Italian fan." In most countries, people can watch calcio on TV or through an app. Via websites, blogs, forums, and Twitter, fans outside Italy can learn about the team that they have adopted, rather than one passed down through an Italian family. However, internet searches can take one only so far into the history of their club. By writing this book on Napoli history, both new and seasoned fans will have the ability to grow their fan identity, learning both the facts and the legends about this tremendous club.

Football grows Neapolitan roots

What was crucial to the development of football in the city was its location on a port. In Italy, the first games that resembled today's football took place in port towns—Genoa, Livorno, Palermo, and Naples—places where British sailors could hold kickabouts.[2] In 1896, the first game in Naples featured Circolo Italia, a team composed of sailing team members and boat racers, facing off against foreign sailors from the port.[3] The game was held in Campo di Marte, near where the Naples airport is today.

These foreign sailors—from Britain, mostly, but also from other countries such as France and Switzerland—came from countries where football was already popular. The local bourgeoisie soon became "infected" with the love of the game, due to their consistent dealings with foreign traders and maritime agents.[4] Although now football might be looked down upon as a working-class game, a means to an end for the poor, it was the opposite at the turn of the 20th century. The game was played by what some might call the "idle rich," while the poorer classes stuck with their sports clubs, which were dominated by individual sports such as cycling, gymnastics, and maritime activities. It would only be later that the *lazzaroni,* the outcast class comprising the city's poorest, would take up the ball for kickabouts in the narrow streets and hope it might be a way out of poverty. At the time, though, they were too busy dealing with overcrowding, cholera epidemics, and an economic structure that failed all but the rich.

The interest football stirred up demanded a club from the city of Naples, but the new institution—Naples Foot-Ball & Cricket Club—was formed by foreigners, English sailor William Poths and partner Hector M. Bayon, in 1904. Wisely, Poths also took on a Neapolitan partner, Ernesto Bruschini. Naples engineer Amedeo Salsi assumed the role of the club's first president, although he was helped along by Poths and Bayon, as well as two amateur players, Catterina and Conforti, whom even the Napoli website does not mention by first name. Poths, an employee of Cunard Shipping, had emigrated to Italy just the year before, and wanted to found his own club, despite multiple clubs already putting down roots in Naples.[5] According to the history section of Napoli's own website:

> There were already several teams in Naples: the aristocratic Open Air team, founded by the Marquis Ruffo, the Costa brothers, Verusio, D'Andria, the Panaria brothers, Alfonso Parise and Alfredo Reiclin; the Helios team, founded by Matteo Giovinetti, with black and white checked tunics; and the Audace team, with green and white colors, founded by Gustavo Romano, the De Giuli brothers and Pepèn Cangiullo, a goalkeeper very much admired for his diving technique.

Yet the club founded by Poths was considered "the first true Neapolitan football club to represent the city."[6] In the first match played by Naples Foot-Ball &

Cricket Club, from which the current SSC Napoli would emerge, the players trotted out in a sky blue and navy striped shirt and black shorts to face the English crew of the boat Arabik, a friendly game which Naples FC won 3–2 via goals from William MacPherson, Michele Scafoglio, and Léon Chaudoir—a surprise as Arabik had just beaten Genoa, Italy's oldest and most illustrious club, 3–0.[7]

While the Italians still generally preferred cycling, they began taking an increased interest in this new sport. In Napoli, street games, similar to those that take place today all over the globe, often took place near the *mandracchio* (a port reserved specifically for small boats and fishing vessels).[8] Naples has no such port now, but it played an important part in the city's football history. Although today's Serie A tends to hide its origins, paying lip service to the years from 1898 to 1922 by saying the clubs were "organized into regional groups," it was first limited to teams from the North—a prejudice that filters down to teams and players to this day. Without a league to play in, the team now known simply as Naples Foot-Ball Club competed against sailors from the port. Thomas Lipton created the Lipton's Challenge Cup, in which the teams of Southern Italy competed for a trophy. The cup lasted just six years, from 1909 until war broke out in 1914, and Palermo and Naples each won three titles.[9]

In October 1911, prior to winning their final Lipton's Cup, the foreign section of what had become Naples FBC broke away.[10] After a falling out with their Italian teammates, Hector Bayon and Steinnegger (known as both Walther and Emil in the history books[11]) formed their own rival club called US Internazionale Napoli. By then, in 1912–1913, a Campania section of the Italian Championship existed, and Naples FBC finished top before losing to Lazio in the next round.

The next season, the results were reversed, but Lazio still managed to beat Internazionale Napoli, with a rather embarrassing 9–0 scoreline. In 1914–1915, the 1st round of the two-legged competition, played in April—which saw Internazionale win 5–2 on aggregate—was voided due to irregularities. The rematch took place in May with Internazionale winning the first leg 3–0. However, the second leg was never played due to World War I. That year Genoa won the title, having come 1st in the Northern competition, although the trophy is disputed by Lazio, who placed 1st in Central Italy, and would have likely faced Genoa had the war not stopped the competition.

As teams such as Puteolana, Bagnolese, and Savoia became competitive in the Southern League, both Internazionale and Naples FBC began to face financial hardship. That the clubs had money troubles was hardly surprising; the war years had been particularly hard on Naples. In addition to the deaths that touched almost everyone in the city, the factories that had sustained the population during wartime closed, causing the people to become even more entrenched in poverty. In 1922, the two merged, becoming Foot-Ball Club Internazionale-Naples, abbreviated as FBC Internaples. Soon, however, the name would change again, as the world continued to shift. Benito Mussolini, founder and leader of Italy's Fascist Party, was installed as the country's prime minister after a coup that ousted Italy's monarchy. The next year, 1923, he became the leader of the now fascist state, taking the title *il Duce*.

Mussolini makes his mark

At the time, football was not the country's most popular sport. Individual feats of athleticism, such as cycling, were more appealing to Italians. The country's current love of football is inseparable from the building of the fascist state. The entire peninsula, fighting on the side of the Allies, greatly suffered during World War I, and struggled to industrialize after the war.[12] The country, unified only on March 17, 1861, still felt more like a collection of independent regions than a nation. It didn't help that the South had been hit the hardest during the war, and felt the state was doing little to assist it in developing. Resentments that had existed since the time of the *Risorgimento*, or Italy's unification, deepened, with the North feeling as though the South was holding Italy back from economic success, and the South believing they might be in a better position if they hadn't been drawn into a united Italy.

Mussolini quickly surmised that football was a sport that would translate easily into a means of growing nationalism, helping the people identify as "Italian" rather than refer to the region where they were from.[13] The first step was to separate it from its English roots. Rather than adopt an Italianized version of the word "football," as many other countries had done (i.e., *fútbol, fussball)*, Mussolini chose calcio, to suggest a connection with *calcio storico Fiorentino*.[14] Although the new sport had little in common with the centuries-old game that originated in Florence, the goal was to encourage people to believe that the sport had existed before the British sailors began their kickabouts. While Mussolini may have not been able to define exactly what he was doing, history reflects that adopted practices must undergo a process of nationalism before they appeal to the majority of the public.[15]

Mussolini next realized that the world would take notice of the power of fascism if his country displayed excellent football. However, the 1925–26 season was rather embarrassing for the country, with a number of games postponed due to crowd trouble as well as a strike by the match officials, who were fed up with the fact that clubs could submit names of referees they did not want officiating their games.[16] Given that the fascist regime was not a fan of protests, and that the overseeing agency was in crisis, Mussolini took the opportunity to step in and further influence the game.[17] The fascist regime sought both nationalist and centralized control of the sport, doing so by handing down specific orders that would apply to all clubs in the "new" nation.[18] On August 2, 1926, the *Carta di Viareggio* (Viareggio Charter) was published.

The charter introduced radical changes to the way calcio would be played throughout Italy. It divided the players between amateurs and non-amateurs, the latter to circumvent promises made to the International Association of Association Football (*Fédération Internationale de Football Association* or FIFA) that the Italian Football Federation (*Federazione Italiana Giuoco Calcio* or FIGC) would be an association for amateurs. At the same time, it acknowledged and permitted the previously clandestine practices of clubs who paid for players in the transfer market or provided salaries for top players. In line with fascist ideology, the charter prohibited teams from having more than two foreign players on their books. Finally, it declared that big cities with clubs that were not performing at the level Mussolini wanted—in other words, those that could not compete with the big clubs from the North—should consolidate their cities' teams to create stronger clubs. This order was directed at Florence, Rome, and Naples.

According to club mythology, Associazione Calcio Napoli were formed on August 1, 1926. An article from *il Mezzogiorno* contradicts this, stating that a meeting on August 25 involved a serious discussion of changing the name from "Internaples" and founding a new society.[19] The information about the incorporation of the new team is also mixed, with some saying that Internaples simply changed their name, while the club's official history hints it also absorbed Audace, Open Air, and Juventus del Vasto.[20] While the club mention nothing about fascist dictates, Rory Smith argues the regime put pressure on Internaples to absorb these clubs and change their name, as occurred in Rome and Florence at the same time in accordance with the Charter.[21] One consequence of this change was that rather than unifying the nation through the sport, it divided cities and the provinces that surrounded them; for example, those in Campania's countryside would often choose to support big Northern teams like Juventus or Internazionale in order to differentiate themselves from the residents of the city.[22]

Just as Mussolini recognized he needed strong teams to reflect the glory of fascism, he also understood that having two leagues, each separated into regions, would continue to divide the nation. He appointed the fascist president of the Italian Olympic Committee, Lando Ferretti, head of the FIGC, and tasked him with the responsibility of bringing the different leagues and regions together to form a united national league.[23] Of course, Mussolini often communicated his thoughts about the structure to ensure it would advance his plan to have football help unify the nation. As impressive new stadia were built to signal the now-unified clubs' strength, Mussolini set about changing the names of English-sounding squads: Genoa and Milan took on their Italian names, Genova and Milano, while Inter Milan, after merging with Unione Sportiva Milanese, was forced to change to Ambrosiana.

Mussolini did his best to mold the game in such a way that it fit the regime's needs as a tool of propaganda and national identity.[24] The name Associazione Calcio Napoli aligned with the edicts handed down from above, and it is likely the reason it was changed from Internaples in 1926. The club was even granted the Arenaccia, a military ground, as their home turf.[25] However, the team did not exhibit the strength that Mussolini was seeking in order to bring teams from the South into the first national tournament. Their first season as A.C. Napoli, 1926–1927, was the first to feature a truly national league, with three teams from the South added, although the teams were still divided into groups for the first phase. It was a horrible season for the new Napoli, one in which they secured only one point, a draw against Brescia, and just seven goals, leaving them with a goal difference of -54. The bright prodigy Attila Sallustro, who had scored 10 goals in 13 games and lifted Napoli to promotion, scored just once, in a 9–2 defeat to Ambrosiana. On the back of this embarrassing finish, the team earned the nickname *I ciucciarelli* (the little donkeys). Legend has it that a well-known local nicknamed Fichella could often be seen in the company of his old donkey, bent from carrying things for his master for years. One day, after another defeat, an older fan looked at the horse on Napoli's badge and announced to fans gathered in a Bar Brasiliano, "This doesn't look to me like a horse. It has a crooked back, and looks weak, like Fichella's donkey."[26] The joke quickly spread across the city. In true Neapolitan fashion, rather than hang their heads in shame, Napoli embraced the nickname, even replacing the horse on its crest with a donkey.

Despite their last-place finish, which should have seen them relegated, Napoli and Roma (which had absorbed all the Roman teams save one, Lazio), who also should have been sent down, were readmitted to the 1927–1928 Divisione Nazionale by order of the Fascist Party, who were still trying to build a truly national league, and therefore required teams from the South. Napoli recorded their first-ever league win, 4–0 against Reggiana, but took just four points in the *andata* (first half of the season). By the time the *ritorno* (second half of the season) ended, they had also beaten Pro Vercelli and Genova. With 15 points, Napoli finished ahead of Lazio and Reggina but still in the relegation zone. However, this season saw no teams relegated—again Lazio and Napoli were readmitted due to needing teams from the South, while retaining Livorno and La Dominante (now Sampdoria) ensured the groups would be even. The FIGC also rescued the other two relegated clubs, Hellas Verona and Reggiana, to have greater representation from the territories annexed during World War I. Finally, they brought in Fiorentina and three other minor clubs, leading to a 32-team championship that would see 14 clubs relegated at the end of the season, establishing the Serie A and Serie B divisions and one national league for all. All of this was done, according to *La Stampa*, for political as well as sporting reasons.

Surprisingly for Napoli, they finished 8th in their group in the 1928–1929 season, a single point above Biellese and relegation. However, they very nearly missed out on the inaugural Serie A season. Originally, only the top eight teams were set to be admitted, which would have led to a playoff between Lazio and Napoli, who were tied on points in Group B. However, it was crucial to have Triestina, the 9th-place finishers in Group A, in the top division due to nationalistic reasons—the team plays in Trieste, an area Yugoslavia claimed to be part of their territory. By admitting Triestina from Group A, the two Southern teams from Group B could also enter Serie A's 1st season—now composed of 18 teams—consistent with the regime's goal that teams from the South would play in a unified league, thus reinforcing the country's budding nationalism. Yet Mussolini also wanted to be sure teams from the South could "hold their own" in such a league (in fact, they were told they needed to improve if they wanted to stay in the division).[27] The previous owner, Giorgio Ascarelli, who would return for the 1929–1930 season,[28] gave his assurances that Napoli had the funds to compete.

Napoli managed to woo William Garbutt from Roma, who under his guidance had finished 3rd in Group A the previous season after barely missing out on relegation in 1927–1928, and Antonio Vojak, who had won the 1926 championship with Juventus and had narrowly missed out on honors the next two seasons.

Garbutt was an intriguing character. He had played for Reading and Woolwich Arsenal before he sustained a horrific injury with Blackburn Rovers.[29] He moved to Genoa, the most English of Italian cities, finding work as a dockhand to support his family. Then, despite his complete lack of experience at coaching, Genoa CFC appointed him as manager on July 30, 1912. According to his memoirs, Vittorio Pozzo, later to guide the Italian men's national team to two World Cup victories, was present for that match;[30] many believe he is the reason Genoa took a chance on the young man with no management background. At age 29, Garbutt began to change the face of Italian football forever.

Prior to his arrival, teams prepared for games with a few kickabouts. Garbutt introduced modern training techniques and professional behavior, emphasizing physical fitness and tactical training. In his first year in charge, Genoa finished 2nd in the Liguria–Lombardy division, and then 2nd in the final round; therefore, it was Pro Vercelli who advanced to the National Final, where they thrashed Lazio 6–0 in the first championship that allowed teams from the South to participate. Clearly, those teams needed to up their game, adopting the same professional training models and being willing to pay for players from other clubs. The same happened the next year, although Casale took the place of Pro Vercelli, facing the Lazio team that thrashed Internazionale Napoli 9–0 in the Southern Italy final and beating them 9–1.

Genoa finally won the title in 1914–1915, although the title is disputed. The season was suspended as Italy entered World War I, and Genoa were declared champions due to being top of the table in the Northern League with one game remaining. Lazio asked the FIGC to award a joint *scudetto*, arguing that they had won the Central Italy semifinal, and the war had prevented the Central and Southern Italy final from being played. Only Lazio had qualified, but in Campania, either

Internazionale Napoli or Naples would have faced Lazio in the final. The first game was won 3–0 by Internazionale Napoli and the second 4–1 by Naples, but the result of the second leg was never validated because Italy entered the war the next day. The last time Lazio made a serious push for the 1915 scudetto, club president Claudio Lotito said that because Genoa and Lazio were due to meet in an end-of-season final, a game that was never played due to the war, the *Biancocelesti* had a legitimate claim to the title, but "it was taken away from us by the war."[31] However, the original decision to award Genoa the title occurred because it was generally accepted that they were the best side in Italy.[32] That decision was upheld, but the *Laziali* continue to press the FIGC for a change.[33]

During the war, Garbutt returned to England, but a significant wage increase tempted him back to Genoa, the club he had led to the 1922–1923 title.[34] Genoa was awarded the scudetto the next season as well, although the team once again faced controversy. During the second leg of the Northern League final in Bologna, riots broke out, the game was called off, and Genoa were awarded a 2–0 victory;[35] although Bologna had won the first match 1–0, Genoa had now beaten them on aggregate and moved on to the National Final. There, they beat Savoia 4–2 over two legs. Meanwhile, Internaples could not advance from the semifinals, while the next year they couldn't even make it out of qualification. Clearly, something had to change.

First, though, Garbutt was drawn to the new AS Roma team created by a merger, orchestrated by Mussolini, between a number of teams in the capital. It is said that it was Il Duce's request that Garbutt step up and help give Rome a team it would take pride in.[36] However, he remained there only two seasons. After taking them from a relegation battle up to 3rd in Group A, he moved on to Naples—another team created by the fascist regime's dictate. It's not clear whether this move was arranged by Mussolini, given Napoli had a Jewish president, Giorgio Ascarelli, who would have faced persecution.[37] Instead, it is said that Ascarelli called him to ask him to coach the Napoli side, and Garbutt was intrigued by a new challenge[38] (given Napoli's place in the table hadn't exactly been stable prior to the league switching to its new format in 1929). What is known is that in Naples he is celebrated as the first man to lift them up from the trenches and consistently achieve a top 10 finish, a feat they would not achieve again for another 30 years. In 1932–1933 and 1933–1934, the team even finished 3rd.

He also captured the hearts of Neapolitans forever after adopting one of their own. He took in a young orphan girl, Concettina Ciletti, endearing him further to the fans.[39] As will be evident throughout Napoli's history, the way players, coaches, and even owners become legends is not only through the goals they score or the players they buy, but by becoming part of the city itself, a character that stands out amidst hundreds of bold personalities. To embrace Naples as one's home is to earn the respect of Napoli fans for life.

– HEROES –
Attila Sallustro (1925–1937)

Napoli appearances: 271
Goals: 114

Atilla Sallustro was likely the first footballing hero of Naples, no matter which name the city's football club went by. The Napoli fans love a transplant, as they love to see a player embrace the city and become "*Napulitano.*" Sallustro may have had an unfair advantage here, given his move from Paraguay to Italy when he was 12. His wealthy father took every advantage he could to ensure his son could play football on the peninsula, and he was almost immediately accepted into the youth team of Internazionale Napoli. He entered what was now FCB Internapoli before turning 18. The next year, the club—conforming to fascist dictates and joining with other teams in the city—became A.C. Napoli, and Sallustro became Austrian coach Anton Kreutzer's first-choice striker. Unfortunately, that first season was a dismal one, in which the club earned just a point. It is difficult to become a hero when your side is sitting last, and your team only remains in the first division because the league needs competitors from the South.

Fortunately for Sallustro, the arrival of better players the next season, such as central midfielder Oreste Tosini, and the better link-up play between fellow forward Ernesto Ghisi, started to make him look like a star and enabled him to score 22 goals in 1927–1928. The various managerial teams that took over in 1927 and 1928 were content to let him take the ball from midfield and dribble it to the net. But when *il Mister* William Garbutt arrived in 1929, he felt Sallustro too selfish and individualistic. He surrounded him with veteran poacher Antonio Vojak from Juventus and attacker Marcello Mihalich from Fiumana, the former who scored 20 goals that season, the latter 10, with *il Veltro* (the greyhound) claiming 13. Sallustro's quickness meant he could quickly turn the ball from defense to attack, moving the entire team forward more quickly.

As good as Sallustro was, he appeared for the national team only twice, as the Italy team coach Vittorio Pozzo preferred Giuseppe Meazza for the striker role. The Napoli faithful barely cared, especially as he and Mihalich were the first Napoli players called up for the *Azzurri* (the nickname for the Italian national team). Sallustro is said to be the city's first football idol—fans would often gather under his balcony and beg until he made an appearance. It likely helped, too, that he was a good-looking man from a wealthy background, and he loved their city. Because of his family's riches, his father refused to let him take a salary from the club, declaring it "unseemly."[40] On the other hand, gifts from the club were acceptable, and the nicest he received was a black Fiat Balilla 521. A reckless driver, like so many in Naples, Sallustro hit a pedestrian; when the man recognized him, he actually apologized to Sallustro, telling him he could do whatever he liked.[41]

While it is often difficult to pinpoint exact stats from a century ago, the most trustworthy source shows Sallustro played 12 seasons with Internaples/A.C. Napoli, notching 271 appearances and 114 goals.[42] He abruptly retired at age 31. He returned to coach the last two games of the 1960–1961 season, after which he became the stadium director of the Stadio San Paolo, responsible for its safety and cleanliness, a role he held for 20 years. He died shortly after, in Rome rather than his beloved Naples, as he was sent there when he became ill. In an interesting twist, Diego Maradona, after whom the current stadium is named, wanted the San Paolo stadium to be named after Sallustro.

Attacker Antonio Vojak came to Napoli the same year Garbutt arrived. Vojak, born in an area of Croatia ceded to Italy in 1918, began his career with Lazio but played just ten games with the team. He scored an impressive seven goals in those ten games, catching the eye of Juventus. The Old Lady had finished just 3rd in the Group B table the season before, failing to even make the semifinals. With Vojak linking up with Ferenc Hirzer, the top scorer that season (he held the record as Juventus' top single-season scorer until 2019–2020), Juventus won the title for the first time in 20 years. Vojak himself went on to score 46 goals in 102 appearances with the club, but Juventus added no additional hardware to the trophy cabinet during those years. Still, they were stronger than Napoli, and Garbutt insisted he needed Vojak in his squad. Therefore, Ascarelli offered a sizable amount of money in order to "snatch him" for Napoli.[43]

Vojak was certainly a quality addition to the squad, but the fans' personal favorite was Attila Sallustro, who was a son of the city—something that can be born or made; either will convince the fans to take the player into their hearts and then into their collective memories. Born in Asunción, Paraguay, he and his Italian parents moved to Naples in 1920. That same year, he became part of the Napoli youth system. Given the time he spent in the city and his long association with the club, the Neapolitans accepted him as one of their own. In fact, he was the first in a long line of Napoli idols. In 1926, when the merging of multiple clubs within the city formed A.C. Napoli, Sallustro took on the role of first-choice striker in the new squad. He was just 17 years old at the time.

When Garbutt arrived for the 1929–1930 season, Sallustro's role was suddenly in jeopardy. Although he'd played well enough for the struggling side, especially considering the players he was surrounded with, the new coach believed he was more concerned with his own scoring record than with the team as a unit. Yet as Garbutt took the team in hand, bringing in new, higher quality players like Vojak, and as the team as a whole started playing as though they belonged in the top division, Sallustro blossomed.

Because of his quickness, he was called il Veltro by the journalists of the time. He could take the ball from midfield, dribble past two or three opposing players, then more or less simply walk it into the net. His primary rival, Giuseppe Meazza, played for Inter, or Ambrosiana as it was known for much of his career. The Northern striker was almost always selected for the Azzurri, and scored 282 goals for the *Nerazzurri* in his 13 seasons there.

Anticipating great things under Garbutt, work began on a new stadium in Naples during the 1929–1930 season (when Napoli finished a respectable 6th). Unlike other clubs upon whom Mussolini bestowed stadia, such as Florence's Stadio Giovanni Berta (named after the Florentine fascist but now called the Stadio Artemio Franchi), the regime offered no help to Napoli. Again, this is likely due to the fact that one of the founders of the club, Ascarelli, was Jewish. As such, Ascarelli set about building his own ground, on land that he owned, using his own money. The ground was called "Vesuvio" and inaugurated in February 1930, when Napoli beat Triestina 4–1. Unfortunately, Ascarelli died shortly after the ground opened, falling victim to appendicitis at only 34 years of age. After his death, the fans demanded the stadium's name be changed to "Stadio Ascarelli."[44]

At the time of writing, the Ascarelli is the only stadium Napoli owned and operated themselves, although they are currently working through red tape to finally construct another, nearly a century later. Even their ownership of the Ascarelli did not mean they had full control over their grounds; the fascist regime would continue to interfere with clubs and their related institutions until the end of World War II.

For Mussolini, the primary reason for the establishment of a national domestic league was to identify and train the players who could best represent Italy on the world stage.[45] The league had helped unify the Italian people and encouraged a sense of nationalism, but to truly demonstrate the benefits of fascism, its cherished symbols needed to be shown to the rest of the world.[46] Il Duce would need to put on a magnificent show in which his team emerged victorious. While pursuing the opportunity to host the 1934 World Cup, he built new stadiums and refurbished old ones. His show gave him a reason to change the name of the Napoli stadium, the largest in the city, from that of its former Jewish owner to the more Italian "Stadio Partenopeo."

Mussolini likely also wanted to capitalize on Naples' popularity, one that was attributed to his own rule. An article from 1932 hailed Naples as the "Most Perfect Tourist Port," thanks to its "resurrection" created by the Fascist Party.[47] Naples itself did not crow about its association with the reigning party (on the other hand, it did not form organized groups ready to resist fascism, either). However, the Association of Italians Abroad sent 100 volunteers to Naples, ready to help Mussolini in the Second Italo-Ethiopian War, and the arrival of those individuals was said to have looked like a massive family reunion, although it was unclear whether the volunteers were actually Neapolitan.[48]

Given the political situation in Italy at the time, and the obvious knowledge that Mussolini would use the tournament as a propaganda vehicle for fascism, it's almost unbelievable that the country would be chosen as hosts—but it's a good reminder that FIFA has been under scrutiny for the means by which they choose World Cup hosts for nearly 100 years. Rumors of intimidation and allegations of payments made to FIFA officials continue to circulate, but it is likely that the assurances made that the Italian government would underwrite any losses incurred during the tournament was what swayed FIFA.[49] Unsurprisingly, even before the trophy was awarded, accusations were leveled at Mussolini that he had directly interfered in the tournament to further promote the benefits of fascism. Italian writer Marco Impiglia told the Associated Press, "The fascist regime made a political abuse of the event. It was a questionable win and it raised many doubts at the time."[50] One specific accusation is that the dictator personally selected the referees for the Italian national team matches. Prior to the semifinal against Austria, Mussolini allegedly took the referee, Ivan Eklind, to an exclusive dinner.[51] During the match, Enrico Guaita scored before 20 minutes were up, but the Austrians complained the match was fixed, with the referee ignoring decisions they believed should have gone their way.[52]

Given the fact that a critical mission of the 1934 World Cup was to promote and enhance an awareness of "Italianness," it may be surprising to learn that Guaita was born in Argentina. So were essential squad members Luis Monti and Raimundo Orsi, both of whom played for Juventus. *Oriundi*, foreign-born players (typically from Argentina or Brazil) with native Italian ancestry, were in fact welcomed into the side, as Mussolini wanted to demonstrate that an "expansive, colonial Italy" would naturally include persons with Italian blood living overseas.[53] Napoli, of course, had known the importance of oriundi since 1920, when Sallustro joined the side.

It is also said that Mussolini overrode instructions from FIFA, and changed the venues where certain games would be held. As will be discussed later, Italian officials should have considered taking a page from the dictator's handbook, and at least discussed with FIFA where specific matches would be held for Italia 90. A San

Paolo stuffed with Napoli fans hosting a Diego Maradona-led Argentina side in a semifinal with Italy may have harmed the host country's chances all those years later.

Nevertheless, during Mussolini's time in power, Italy's wins in the 1934 World Cup, the 1936 Olympics, and the 1938 World Cup went quite some way to uniting the country's people under the banner of "Italianness." The dictator's ability to grow nationalism through the sport is one that Italian officials have yet to replicate. Much of this has to do with the "Southern Question," the discussion over whether those from the Southern end of the peninsula should be included as part of the country. This question was briefly presented in "A Short History of Naples" and will be examined throughout the book.

Ironically, just as Mussolini was using the league to strengthen the national side in preparation to show off fascism to the world, Napoli was still having fun with football. On January 29, 1931, 11 chorus girls from the Teatro Nuovo dancing company lined up against a side composed of Napoli supporters who worked in the textile mill formerly owned by
Giorgio Ascarelli. The large crowd that attended saw the women win 5–3, with all of the women's goals coming, rather surprisingly, from their center-back. William Garbutt, in particular, liked the "*arte calcistica*" he saw of two especially high-spirited dancers, the center-forward and one of the half-backs.[54]

Beyond the excitement of the chorus girls, the 1930–1931 season was a bit of a roller coaster for Napoli, one that ended on a rather slow and familiar track. Very little changed from the season before. Garbutt continued on, and Mihalich, Sallustro, and Vojak were still up front—the latter scoring 20 goals in the league, Mihalich contributing 12, and Sallustro still willing to play in a less flashy role, scoring 10 as he provided for his partners. The team finished 6th, again without reason to worry about relegation. The next season Garbutt stuck with the same plan; the problem was that his strikers' ability and accuracy had decreased. Sallustro had the most goals that season with 12, while Mihalich and Vojak scored only 9 each. They finished 9th in 1931–1932, and it seemed clear that a change was needed for 1932–1933.

Garbutt remained. So did Sallustro and Vojak. Mihalich made no appearances that season, as Napoli had acquired another attacker, Pietro Ferris, from Pro Vercelli. He would later go on to win the scudetto with Ambrosiana Inter and the 1938 World Cup with the Azzurri as well as play a significant role in the *Grande Torino* squads. For Napoli, he turned provider, allowing Vojak (22) and Sallustro (19) to rack up impressive goal tallies. The defense, anchored by the veteran Paolo Innocenti, who spent 11 seasons with the club, had played together long enough to prevent many goals from sneaking in. For the first time, Napoli finished 3rd in the league, level on points with Bologna. They repeated the feat the next season, conceding the second-fewest number of goals and celebrating Vojak scoring 21. The only major change was that Gino Rossetti joined the attack from Torino, where he would return after four seasons with Napoli. He contributed seven goals and appeared in all but one match, while captain Sallustro sat out nine matches and scored five goals.

Although Napoli finished 3rd in the 1933–1934 season, only one player received a call-up to the 1934 World Cup Italy squad. Antonio Vojak had made one appearance for the national team, in 1932, although he had to use the Italianized "Vogliani" rather than "Vojak" due to fascist anti-Slav laws. Vojak was Napoli's top scorer and assist provider in the previous season, so it is likely prejudice that kept him out. Meanwhile, manager Vittorio Pozzo favored Giuseppe Meazza over Napoli's idol, Attila Sallustro. The lone inclusion, Giuseppe Cavanna, Napoli's starting goalkeeper, never touched the pitch, as he served as backup to the squad's captain, Juventus' Gianpiero Combi. The next season, Cavanna would go on to record the lowest goals conceded per game average (0.722) for Napoli, a record that stood until 1970–1971. Despite the lack of hometown players, the Stadio Partenopeo hosted a Round of 16 match in which Hungary beat Egypt 4–2, and the third-place match, a thriller in which Germany edged out Austria 3–2.

After the World Cup, Napoli's 1934–1935 squad changed little from the line-up they'd used to capture 3rd the season before. Garbutt remained, as did Sallustro and Vojak. However, the team came in 7th that season, most likely because they focused on the Mitropa Cup. This was their first time playing in a major international competition, conducted among the successor states of the former Austro-Hungarian empire. Napoli's 1st round saw them facing off against Admira Vienna. In Vienna, they left with a 0–0 draw. In the next leg at the Stadio Partenopeo, Sallustro and Vojak both scored, the latter from the spot. Unfortunately for the hosts, Admira Vienna came back late in the second half to end the game 2–2. Unlike in the years that were to follow, where in many tournaments the number of away goals scored by each team decided the winner in the event of a tie over two legs, the teams were sent to a neutral venue in Switzerland for a playoff. There, the Viennese side dominated, scoring their first goal within two minutes, and ultimately recording a decisive 5–0 victory.

The following seasons saw Napoli finishing 8th, then 13th. In 1935–1936, Károly Csapkay replaced William Garbutt, while Antonio Vojak left for a season with newly promoted Genova 1893 (now Genoa). Giovanni Busoni, from the relegated club Livorno, stepped in for a season to score 12, and then left for Bologna, that season's champions. A mercenary often receives vitriol from even his own fans in Naples, and so that season it seemed there was little to cheer for. In the next, the new owner replaced Csapkay with Angelo Mattea, and while players like Pietro Ferris, Giuseppe Cavanna, and Paolo Innocenti left, few were replaced, and the team struggled. Placing 13th in 1936–1937 failed to impress their new owner, Achille Lauro—and given the prestige with which he was held in Naples, and the comparative power bestowed upon him by the Fascist Party, he was a man most people wanted to impress.

1 Foot, J. (2006). *Winning at all costs: A scandalous history of Italian soccer*, 28.

2 Foot.

3 Signorelli, A. (Ed.). (2002). *Cultura popolare a Napoli e in Campania nel Novecento*. Guida Editori.

4 Id.

5 SSC Napoli. (n.d.). *Football History*. Retrieved December 8, 2020, from https://www.footballhistory. org/club/napoli.html

6 SSC Napoli. (n.d.). From Naples Football Club to Internaples. *SSC Napoli*. Retrieved August 19, 2020 from https://www.sscnapoli.it/static/content/From-1904-to-1921-81.aspx

7 FIFA. (2009, August 3). Napoli back among the big boys. *FIFA.com*. https://www.fifa.com/news/ napoli-back-among-the-big-boys-1045169

8 SSC Napoli. (n.d.) La Storia del Calcio Napoli diventa un'opera d'Arte. *SSC Napoli*. Retrieved August 10, 2020 from https://www.sscnapoli.it/ilnapolinelmito/guidabreve.pdf

9 Quartarone, R. (Updated 2016, August 25). Lipton Challenge Cup. *Rec.Sport.Soccer Statistics Foundation*. http://www.rsssf.com/tablesl/lipton-sicily.html

10 Salvi, S., & Savorelli, A. (2008). *Tutti i colori del calcio*. Le Lettere.

11 Naples Football Club—Stagione 1910–11—Storia. (2017, August 8). *Calcioantico*. https:// calcioantico.altervista.org/naples-football-club-stagione-1910-11-storia/

12 Lea, G. (2015). The relationship between Mussolini and calcio. *These Football Times*. https:// thesefootballtimes.co/2015/07/20/the-relationship-between-mussolini-and-calcio/

13 Doidge, M. (2015). *Football Italia: Italian football in an age of globalization*. Bloomsbury Academic, 46. http://dx.doi.org/10.5040/9781472519221.0012

14 Archambault, F. (2018). "Italy." In J-M. De Waele, R. Spaaij, S. Gibril, and E. Gloriozova (Eds.), *The Palgrave International Handbook of Football and Politics* (pp. 105–124). Palgrave Macmillan.

15 Id., 106.

16 *Annuario italiano giuoco del calcio v. 1 1926–27*. (1927). Societa tipografica modenese.

17 Martin, S. (2004). Football and fascism: The national game under Mussolini. Berg.

18 Doidge, 46.

19 L'Internaples muta il nome in A.C. Napoli e prepara le basi per le prossime battaglie. (1926, August 26–27). *il Mezzogiorno*. https://upload.wikimedia.org/wikipedia/commons/a/a1/Il-Mezzogiorno-26-27-agosto-1926.png

20 SSC Napoli. (n.d.). The history from 1926 to 1962. *SSC Napoli*. Retrieved August 19, 2020 from https://www.sscnapoli.it/static/content/From-1926-to-1962-82.aspx

21 Smith, R. (2016). *Mister: The men who gave the world the game*. Simon & Schuster.

22 Archambault, 107.

23 Lea.

24 Martin, S. (2018). Football, fascism and fandom in modern Italy. *Revista Crítica de Ciências Sociais*, *116*, 111–134.

25 SSC Napoli. (n.d.). The stadiums where SSC Napoli teams have played. *SSC Napoli*. https://www. sscnapoli.it/static/content/Playgrounds-90.aspx

26 Materazzo.

27 Archambault, 106–07.

28 SSC Napoli. (n.d.). All the presidents from Ascarelli to De Laurentiis. *SSC Napoli*. https://www. sscnapoli.it/static/content/Presidents-88.aspx

29 Foot, 25.

30 Id.

31 La Lazio e uno scudetto di 100 anni fa ultima magia dell'imprevedibile Lotito. (2016, July 21). *La Repubblica*. https://ricerca.repubblica.it/repubblica/archivio/repubblica/2016/07/21/la-lazio-e-uno-scudetto-di-100-anni-fa-lotito41.html

32 Lazio: Serie A side could be awarded the 1914–15 championship. (2016, July 21). *BBC Sport*. https:// www.bbc.com/sport/football/36858549

[33] Larsson, A. (2019, May 30). 2015 scudetto: New proof sent to FIGC. *The Laziali*. https://thelaziali.com/2019/05/30/lazio-scudetto-1915-new-proof-figc/

[34] Foot, 26.

[35] Acerbi, E. (last updated June 6, 2004). Genoa Cricket & Football Club—Short Historical Overview 1893–1960. *Rec.Sport.Soccer Statistics Foundation*. http://www.rsssf.com/tablesg/genoa.html

[36] Smith.

[37] Amato, M. (2011, November 15). Napoli onora Ascarelli, il presidente ebreo che il Duce tentò di cancellare. *L'Unità*, 47.

[38] Materazzo, G. (2013). *1001 storie e curiosità sul grande Napoli che dovresti conoscere*. Newton Compton.

[39] Foot, 26.

[40] Carratelli, M. (2008). Attila Sallustro, centravanti e divo. *La Repubblica*. https://ricerca.repubblica.it/repubblica/archivio/repubblica/2008/12/14/attila-sallustro-centravanti-divo.html

[41] Attila Sallustro, il Veltro e quella guida spericolata sulla Balilla 521. (2021, March 22). *Corriere dello Sport*. https://www.corrieredellosport.it/news/motori/news-motori/2021/03/22-80161530/attila_sallustro_il_veltro_e_quella_guida_spericolata_sulla_balilla_521

[42] Sappino, M. (2000). *Dizionario biografico enciclopedico di un secolo del calcio italiano—volume 2*. Baldini & Castoldi.

[43] Materazzo, G., & Sarnataro, D. (2014). *I campioni che hanno fatto grande il Napoli*. Newton Compton.

[44] Amato, 47.

[45] Martin, Football, fascism and fandom.

[46] Doidge, 46.

[47] Guidi, A. F. (1932, June 4). Napoli, Il più perfetto porto turistico (Naples, the most perfect tourist port). *Il Progresso Italo-Americano*.

[48] Lee, J. H. (2016). *"To the seventh generation:" Italians and the creation of an American political identity, 1921–1948* [Unpublished doctoral dissertation]. Columbia University.

[49] Hart, J. (2016, July 27). When the World Cup rolled into fascist Italy in 1934. *These Football Times*. https://thesefootballtimes.co/2016/07/27/when-the-world-cup-rolled-into-fascist-italy-in-1934/

[50] Did dictators fix World Cup matches? (2013, April 25). *ESPN*. https://www.espn.com/soccer/news/story/_/id/1425310/researchers-say-dictators-fixed-world-cup-matches-1934-1978

[51] Lea.

[52] Weiner, M. (2010, June 8). When worlds collide: Soccer vs. politics. *CNN*. http://edition.cnn.com/2010/SPORT/football/06/08/world.cup.soccer.politics/index.html

[53] Lea.

[54] Giani, M. (2020). Playing football with the chorus girls: Vaudeville women's football in Naples [1931]. *Playing Pasta*. https://www.playingpasts.co.uk/articles/football/playing-football-with-the-chorus-girls-vaudeville-womens-football-in-naples-1931/

CHAPTER 3

Playing through Pain

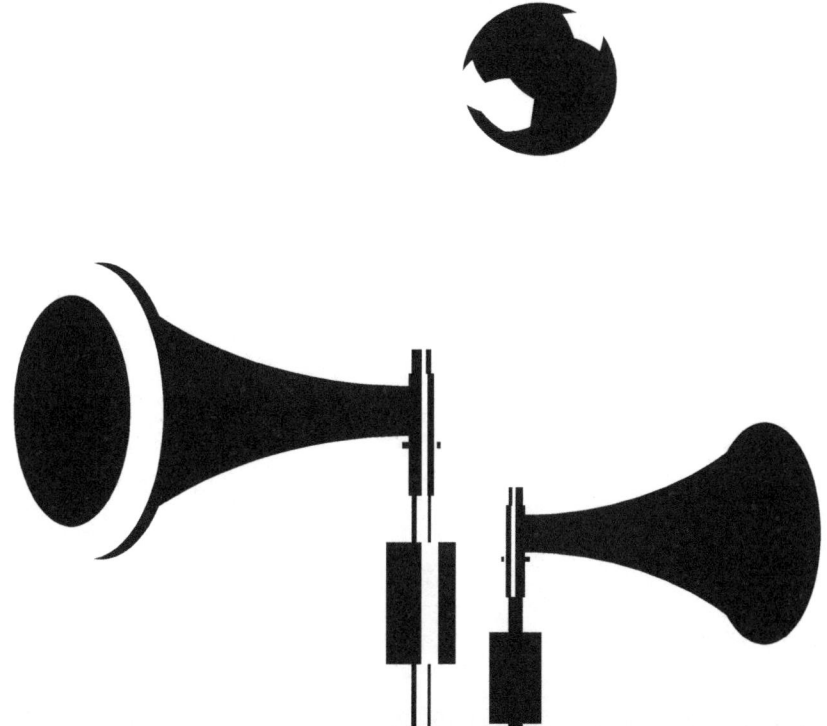

3

In 1936, Benito Mussolini's army was mired in the Second Italo-Ethiopian War, an aggressive, expansionist act that signaled the country's clear alliance with the Axis powers. Mussolini sent troops to fight in the Spanish Civil War in support of fellow fascist Francesco Franco, issued a set of anti-Semitic laws that closely resembled those of Nazi Germany, and entered into the Pact of Steel with Hitler in May 1939. They were hoping for territorial spoils. Instead, they wound up with spoiled territory. By November 1942, British forces had defeated Italian fascists in North Africa. The Allies began their invasion in July 1943, in Sicily. While most major Italian cities suffered due to the Allied bomb strikes, Naples was hit hardest; after the war ended on May 2, 1945, little money was put into efforts to rebuild the South.

At the same time, Achille Lauro was stepping into his first stint as president of SSC Napoli. A man whose "extensive and ramified system of interests"[1] led to a populist movement dubbed "Laurism" in honor of its leader. Lauro was the owner and founder of the Lauro Fleet, one of the most important companies in the South. He had joined the *Partito Nazionale Fascista* (PNF) in 1933,[2] and within five years risen to membership in the Chamber of Fasces and Corporations, Italy's lower house of legislature. After one of Napoli's officers begged for his help, *O Comandante* took the reins and accepted the club's mountain of debt.[3] However, despite his riches, and his connections with the ruling party that helped him increase his wealth, allowing him to funnel more money into the club, Napoli failed to impress. By the end of the decade, they'd been sent down to Serie B.

In the shadow of war

As the machinations leading to what would inevitably be a wide-ranging war became more obvious, Mussolini grew more determined to unite his country behind the glory of Fascism. By going against British demands, *il Duce* found himself at the height of his popularity in 1936, when most thought the Ethiopian War had ended with the capture of the capital Addis Ababa and the fleeing of Emperor Haile Selassie.[4]

Despite sanctions placed on Rome by the League of Nations, the war was incredibly popular with Italians, who appreciated Mussolini's defiance of global leadership.[5] True victory over Ethiopia in 1938 led to an even greater surge in nationalism as Italy celebrated growing its empire, just as the major world powers had done. Mussolini had gotten what he wanted—an expansion of the Italian state beyond the territories of Italian Eritrea and Somalia. However, that same year, his "volunteers" fighting in the Spanish Civil War were defeated. By the end of the war in 1939, approximately 4,000 Italian troops had been killed and 11,000 injured.[6] Meanwhile, after a long campaign of racist propaganda by the Fascist Party, Mussolini ordered the implementation of a set of harsh race laws meant to hurt the Jewish population of Italy; despite the propaganda, most citizens disagreed with the anti-Semitic regulations. Italy was about to tip over the edge, and the only reassurance its population had was the number of anti-Fascist individuals and groups that were willing to stand up to the ruling party.

The fall occurred in 1940, when Italy officially entered World War II. Despite the Pact of Steel between Mussolini and Hitler, in which both Italy and Germany committed to notifying one another before engaging in military action, as well as taking part in any conflict involving the other, Italy did not step in to help when the Nazis invaded Poland in 1939. It was only when it appeared as though France would be taken and the war would shortly end that Mussolini directed his troops to enter the conflict, despite the majority of the population having no interest in tying themselves to Hitler.[7]

An increasingly desperate Mussolini felt sport was the way to keep the people of the peninsula together. It is likely he knew he would need their unified support should Italy be involved in a full-scale world war. However, clubs still clung to their cities' and regions' pre-unification identities and traditions; after all, Italy was still a rather young state. Therefore, Mussolini felt it best to capitalize on the Italians' 1934 World Cup win and push on to 1938, weaving myths and creating symbols around the national team to encourage a sense of national identity.[8] While Italy's win in the previous World Cup, played on home turf, drew accusations of bribery and undue influence, the same could not be said about the French tournament. Italy dominated, beating the hosts 3–1 in the quarterfinal, and ultimately beating Hungary 4–2 in the final.

That's not to say fascism played no part in the 1938 World Cup. Prior to June and the tournament's kickoff, Mussolini had already sent his troops to back General Franco in the Spanish Civil War and had signed a peace accord with England, causing France to fear a coming war. Coach Vittorio Pozzo, back for his second World Cup, believed part of the reason Italy triumphed was due to their ability to win the "battle of intimidation."[9] The team was not well received by the crowds in France, and when they first lined up in their black shirts, giving the Fascist salutes, the stadium filled with deafening insults and shrill whistles. When the ruckus finally began to die down, the team lowered their arms. At once the noise started again, and at once the team lifted their hands in salute. They outlasted the crowd, and thus—in Pozzo's eyes—intimidated each opponent before the game began.

To bring his team together under the banner of nationalism, Pozzo had players of rival teams room with one another, secluding the squad in *ritiro* (retreat) and, while there, forcing upon them militaristic maneuvers such as marches through the woods.[10] Fortunately for Pozzo, he had few rivals to keep apart, having selected players from just 9 of the 16 Serie A clubs, and none from winners Ambrosiana-Inter's crosstown rivals, AC Milan, despite the club's finish in 3rd. Napoli, having finished 10th in the 1937–1938 season, had no players in the 22-man squad, despite Pozzo calling up a goalkeeper from nearly-relegated Lucchese, and a forward from Serie B side Pisa. At the time of the 1938 World Cup, Attila Sallustro had finally aged out, heading to Serie C side Salernitana while his rival, Giuseppe Meazza, favored under Pozzo, captained the squad.

Although it has not been expressly stated, it is possible that Pozzo, at Mussolini's urging or through his own beliefs, specifically excluded players from the South from his World Cup squads. Included in the 1934 squad was only one player from a club from the *Mezzogiorno*, Napoli goalkeeper Giuseppe Cavanna, who never saw the pitch—and none were involved in the 1938 squad. Cavanna hailed from Vercelli in Piedmont, where unification first arose. In fact, of the forty men called to the 1934 and 1938 squads (four players were called up to both), just one, Mario Pizziolo, could be said to be of "Southern" origin, and only because his region, the Abruzzo, was historically tied to the Kingdom of the Two Sicilies.

Most today would likely place his province of Pescara in Central Italy; in addition, he'd entered the Livorno (Tuscany) youth system early on. Given that one of the fascist regime's first tasks was to squash rising communism in the South, perhaps Mussolini, and by extension Pozzo, did not feel players from those regions could adequately represent Italy on the world stage, given that a major part of the campaigns was to spread the ideals of fascism.

Of course, it could also be as simple as the fact that the South simply did not have enough players to choose from. In the 1933–1934 season, only Napoli and Palermo represented the Mezzogiorno, although given Napoli had placed 3rd, it would seem at least one of their players would be chosen for a starting spot in 1934. In 1937–1938, only 10th-placed Napoli and 13th-placed Bari hailed from the South; their teams' positions could have reasonably kept their players out of contention. Although one might still wonder—were there truly no talented players of Southern origin playing for Northern clubs? After all, the players selected from the Roman clubs had been brought in from the Northern provinces. Perhaps Mussolini knew Pozzo would bring him victory again, and the photograph of him with the cup-winning team would not serve as such strong propaganda had anti-fascists been allowed to participate.

Despite having no players in the World Cup-winning team of 1938, Napoli did not escape the tentacles of fascism that seemed to extend to all aspects of Italian life.

– STADIUM –
Stadio Arturo Collana

In use: 1929–present
Alternate names: Stadio del Vomero, Stadio XXVIII Ottobre, Stadio del Littorio, Stadio della Liberazione
Tenants: Internaples, SSC Napoli (1933–1959), Internapoli

Despite being re-christened the Stadio della Liberazione in 2012, the stadium, which has stood for nearly 100 years, is most commonly called the "Collana." The name refers to famous Neapolitan journalist Arturo Collana, who founded Società Sportiva Napoli in May 1944 to give the city a bit of respite from the hell it endured during World War II and the liberation campaign.

In 1930, Napoli had opened their first (and to date only) privately owned stadium, the Stadio Vesuvio. Shortly after, it was renamed the Stadio Ascarelli, after the club's recently deceased Jewish owner. In preparation for the 1934 World Cup, the stadium underwent massive renovations so that it could comfortably hold 40,000 supporters—double the number it had been designed for. Benito Mussolini refused to allow a Jewish surname on one of his host stadiums, so it was renamed the Stadio Parthenope.

While the reconstruction project was underway, Napoli played their home matches at the Collana, which held just 15,000 people. The Ciucciarelli moved back to the Parthenope after the tournament, never imagining their stay would be so short. Sadly, in 1942, Napoli's Stadio Parthenope was reduced to rubble during an Allied bombing campaign. As more and more damage accrued in the area, the decision was made to raze the stadium; the only sign of its existence now is in the name of the neighborhood rebuilt around the grounds: the Rione Ascarelli.

Napoli returned to the Stadio Collana for the 1942–1943 season, after which the league halted for two years. In September 1943, the Kingdom of Italy signed an armistice with the United Kingdom and the United States; in response, German forces attacked and occupied Italian cities, including Naples. When all men from 18–33 years of age were ordered to perform compulsory labor, and learned they would be deported to Northern Italy and Germany, the civil uprising truly began. During the *Quattro giornate* di Napoli, the people of the city rose up to defend themselves, just four days after the Germans had arrived. The Neapolitan men who refused the order were rounded up and brought to the Collana.

Shortly after league play resumed, another tragedy struck at the stadium. On January 27, 1946, during a "Southern Derby" between Napoli and Bari, Napoli's forward, Albanian Riza Lustha, scored a goal in the 3rd minute. The crowd rushed forward, the railing gave way, one of the terraces collapsed, and 114 people were injured seriously enough to be sent to the hospital.[11]

With such a small capacity and such a propensity toward danger, the City of Naples broke ground on a new stadium for Napoli in 1948, to be called the Stadio del Sole. The stadium would hold over 90,000 and, as such, took quite some time to build. Therefore, Napoli were still playing at the Collana on November 6, 1955, when Bologna came to town. After the visitors scored the equalizer in injury time, the referee blew the whistle, infuriating the supporters. Around 300 fans ran onto the pitch, chasing the referee, while those who remained in the stands threw various objects, such as glass bottles. Both fans and police were injured, approximately 160 in all.[12]

The Stadio del Sole opened in December 1959, quickly changed its name to the Stadio San Paolo, and has been called the Stadio Diego Armando Maradona since December 2020, a month after the God of Naples died. Napoli, with the 3rd-highest average attendance in Italian football, have truly grown out of the tiny Stadio Collana. According to the stadium's website, its primary use today is as a youth center, where prospective athletes are trained in disciplines from volleyball to martial arts.

Napoli's turbulent war years

Achille Lauro first took charge of the club in 1936, when Vincenzo Savarese stepped down after four years as president.[13] After his studies, Lauro followed in his shipowner father's footsteps, founding the Flotta Lauro, one of the most powerful Italian shipping fleets of all time. At the time he assumed the role of Napoli president, Lauro had been a member of the National Fascist Party for three years. By 1938, when he became a member of the lower house of the legislature of the Kingdom of Italy, Lauro had made his first millions by participating in the arms trade that helped the Fascist Party move ahead with their colonial plans in Africa.[14] Fairly quickly, the size of his fleet grew to 21 ships. With help from the PNF, Lauro would eventually construct a true financial empire around his business. But at the moment, he'd been called upon to help save his city's team from financial ruin.

Despite the powerful Achille Lauro taking the helm, Napoli experienced a decade of turmoil, both before and after the war. In fact, the new president oversaw a dip in Napoli's fortunes. The club had finished 8th in 1936, and had boasted one of the league's best scorers in Giovanni Busoni, who knocked in 12 in 27 games. However, the club needed to sell Busoni to pay off their debts (the total of which is not known); he went to title-holders Bologna, where he failed to make the list of top scorers. They also sold off fellow attacker Pietro Ferraris to Ambrosiana-Inter. Together the players brought in 190,000 lire to the club, and Lauro added 60,000 of his own.[15] Finally, Lauro appointed Angelo Mattea, most recently manager of Serie B side Messina and the vice-coach to the Italian national team, to replace Károly Csapkay. Unsurprisingly, the weaker squad fell to 13th in 1936–1937.

Lauro's second season, and the second for coach Mattea, saw Attila Sallustro finally depart. In response, the side picked up forwards that again looked like they would produce plenty of goals, but no center-forward was strong enough to take the place of Napoli's first hero. The newly transferred 25-year-old Giuseppe Gerbi, scooped up from Serie B side Messina, scored just eight, Germano Mian six, Fillipo Prato four, and Nicolò Nicolosi just two. To compare, the league's *Capocannoniere* (player who scores the most Serie A goals in a season), Giuseppe Meazza, scored 20 goals. Lauro also purchased Nereo Rocco, a winger from Triestina, who hadn't delivered much for the *Alabardati* the season before. However, as he'd scored 11 goals in 1935–1936, he seemed a decent bet, especially when Napoli bought his teammate Germano Mian at the same time. Mian, too, was coming off a poor year, but had scored 10 the season before. Both players stayed at Napoli for three seasons, departing in the summer of 1940. Neither produced anything close to what they had in their glory days at Triestina. But while his playing career at Napoli was relatively unremarkable, with 7 goals in 52 appearances across three seasons, Rocco would go on to become one of the greatest Italian managers of all time, winning two league titles and several European trophies with Milan.

Mattea had nowhere near the talent Rocco would develop, but managed to survive the worst of the 1937–1938 season. Then a 2–0 loss against Atalanta saw Napoli slip to 13th in the Serie A table. Lauro's patience proved finite and the disappointing manager was sacked halfway through the season, after a 3–1 away loss to Bari. In came Hungarian coach Eugen Payer from Serie C side Fiumana, about as untested a manager as Napoli could find. The club didn't make any sort of miraculous comeback under Payer, but they picked up a number of points from draws, most significantly against Juventus, Bologna, and Ambrosiana-Inter, the eventual *scudetto* winners. It was enough to bring Napoli to 10th, 12 points above the relegation zone.

In 1938–1939, Napoli could breathe a little when they managed to finish 7th— tied on points with 5th-placed Roma and 6th-placed Liguria—and some credit should be given to the Napoli fans. The squad's season record is distinctly lopsided: 9 wins and just 2 losses at home, where they scored 21 of their 30 goals of the season. On the road they did well to close down and play for a draw; overall, seven of the draws that season ended 0–0. It was a common theme in Italy to play for a draw, particularly away, rather than really take the game to the other team. Even after Payer was fired at the beginning of February, thanks to a goalless draw at Novara that saw Napoli drop down to 13th, the *Commissione Tecnica*—the technical committee that consisted of Amedeo D'Albora on the bench, Paolo Jodice as the fitness director (who ordered all the players to move to Vomero so he could more easily control them),[16] and three players, one for each area of the pitch—seemed committed to playing cautiously, even at home.

On the road, Napoli won just once that season, a 2–1 victory over eventual 3rd-place finishers Ambrosiana-Inter. Of greater cause for celebration was their home victory over Juventus in Round 19, shortly after Payer was sacked. The season hadn't been particularly fruitful for Juve, but any time Southerners could bring the Old Lady to her knees was celebrated; it was even better when the win resulted in Napoli finishing higher in the table. After all, Juventus had already won seven *scudetti*, and had been *the* team to beat in the early 1930s, when they won five straight championships. That success is almost tangential to Napoli's rivalry with Juventus. Some go as far to say it dates back to the feud between the House of Savoy, which led the drive for Italian unification and ruled from 1861, and the Kingdom of the Two Sicilies, which was subsumed by Savoy. What is certain is that it is related to the difference between the industrialized North and the less-developed South.[17] Since the *Risorgimento* in 1861, the South had been left behind, with many of its citizens moving to the North or even overseas in search of better jobs, as Naples and the areas around it remained mired in agriculture, shipbuilding, and tourism. The fact that Southerners continued to work the land after unification led to them being labeled *terroni*, a classist term that remains a xenophobic slur, tainting its target with a brush wide enough to encompass poverty, bad manners, poor hygiene, a lack of education, and more. "Terroni" continues to be used to insult Southerners, particularly in stadiums, where many clubs are still not punished when the word is displayed on banners and used in chants, despite the fact that Italian courts recognize it as a discriminatory insult.[18]

In this game, rather than play cautiously, it seems the Commissione Tecnica decided to hit Juventus hard. The first goal came in the 28th minute from defender Achille Piccini, his first of the season. Just before the break Nereo Rocco slotted in his fourth of the season. In the second half it was all about Germano Mian, who had been purchased to pair with Rocco, but had not produced the goals Napoli had expected. His two in the second half, bracketed around the one Juventus goal, were his first of the season.[19] Yet Napoli fans were apt to forget his earlier lack of production when he sent in a last-minute goal that resulted in a final scoreline of 4–1.

Despite the decent finish that season, it was becoming clearer that the players Lauro was bringing in weren't quite up to task. Carlo Biagi did well enough in midfield at Pisa to be called up to the 1936 Olympic squad, which eventually won gold.[20] That is likely the reason Napoli snapped him up. But he never truly made an impact on the Napoli squad before retiring in 1940. Natale Masera and Nicola Ferrara looked to be forwards on the cusps of their best years, but neither started more than half the matches, and only Ferrara managed to score—twice, against Novara in a 4–0 home win, and against Bologna in a 2–0 away loss. Both were sold at the end of the year. Attacking midfielder Dino Poggi, just 21, stuck around until 1939, but that may have been because he was forgotten; he played just five games with the club. Still, he fared better than defensive midfielder Dino Da Caprile, with four matches in two seasons. Mario Pretto, a right-back, achieved the most of the first players bought during the first Reign of Lauro, becoming a mainstay in the squad by the time he was 22, and staying with Napoli until he retired 13 years later.

Given the lack of true talent up top, the team scored just 30 goals in the 1938–1939 season. They finished 7th, tied for points with Roma and Livorno, meaning that the league relied on goal average (goals scored divided by goals conceded) to determine their final position. It mattered little as only the top two in Serie A had something to celebrate, and the bottom two something to mourn. Yet Napoli still finished just seven points from the bottom, and both relegated teams had scored more goals. An offensive threat was needed for the 1939–1940 season.

In retrospect, it feels odd that the season was played, given that Hitler had invaded Poland less than three weeks before. But despite Italy's alliance with Germany, Mussolini was not willing to commit militarily to the conflict, and so the season marched on. The threat of war was real, but in Naples, the threat of relegation seemed more pressing to the fans. The club, now managed by Adolfo Baloncieri, brought in from Liguria (who had finished 6th the season before) kept sinking lower and lower in the rankings. Again, a lack of production was a problem; Carlo Alberto Quario, on loan from Ambrosiana-Inter, scored approximately a third of their 26 goals. He was meant to link up well with Rocco, but the latter failed to do his part, scoring exactly zero. Fortunately, 26 goals turned out to be sufficient; with Fiorentina, Napoli, and Liguria all tied on 24 points, goal average was used to determine who would be sent down with Modena. Napoli had scored one more goal and conceded three fewer than Liguria, just barely saving themselves with a 14th-place finish. Napoli's skin-of-their-teeth survival is one reason Lauro left his post at the helm of the club, disillusioned as he was by their inability to succeed.[21] In addition, his fleet of ships was called into action when Italy entered the war on June 10, 1940; five days later, he asked Gaetano Del Pezzo to take over Napoli, as he had to attend to other matters. While Lauro had failed to reverse the club's fortunes on the pitch, he had done what he was brought in to do, which was balance the books.[22]

"After" Lauro

Del Pezzo, a famous name in the city as his father and grandfather had both held the mayoral position (in addition, he was also the Marquis of Campodisola and ninth Duke of Caianello) assumed the Napoli presidency during the 1940 transfer market. He had played for Naples Foot-Ball Club from 1910 to 1921[23] and became the president of Napoli Internazionale when it merged to become A.C. Napoli in 1922.[24] A smart man, he handed over the presidency to Count Tommaso Leonetti before the 1940–1941 season began.

Antonio Vojak returned to the club, this time as the manager. Vojak did his best to stamp his mark on the squad, getting rid of deadweight like Rocco, Carlo Biagi, and Filippo Prato. Luigi Rosellini, in his final season at the club, grew into his own and scored the most goals for Napoli, 13, only to be snatched away by Milan the next season. Two new faces, Evaristo Barrera and Umberto Busani, both brought in from Lazio, were able to help their new side, with nine goals and eight goals respectively. Perhaps one of the best choices—although certainly not the most flashy—was the purchase of Luigi Milano from Lazio. The central midfielder had been a steady presence in the *Biancocelesti* squad and quickly became one at Napoli, playing every league game that season, combining well with old squadmates Barrera and Busani.[25] Finally, Arnaldo Sentimenti stuck around to captain the side and occasionally take over between the sticks, but the 30 goals he'd conceded in 20 appearances the season before were substantially improved upon by yet another former Lazio player, the new starting goalkeeper Giacomo Blason, who let in 22 in 18 appearances. Napoli managed to bring themselves up to 7th in Serie A, tied on points with Torino.

A small comfort in the city of Naples, which almost immediately felt the effects of the country joining the Axis powers—the first bombs dropped on November 1, 1940, just three days after Mussolini manipulated his puppet state, Albania, into launching an attack on Greece. The British used their base on the island of Malta to attack the city in the early hours, targeting industrial areas in the east and neighborhoods around the central train station.[26] With the war encroaching, Neapolitans refused to accept the destruction of their city, peppering its walls with words like "obey," "believe," and "fight." The fans applied the latter two to the club as well, pushing thoughts of war out of their minds as the bombs slowed and they rallied behind their side.

Just as it seemed that Napoli might be starting to settle, that the team had moved on from the horrific 1939–1940 season and embraced a year in which they were steady and safe, that a new manager and smart buys in the transfer market were going to pay off—that Napoli could, at the least, remain a midtable team, if not improve their standing—everything went south for the *Azzurri*. Napoli lost their most talented player when Rosellini went to Milan. Count Leonetti left after 13 months, in October 1941, and ownership of the club changed yet again. Luigi Piscitelli, a local bureaucrat, took over. Vojak remained as *il Mister*. A month later the bombings intensified as the Allies targeted Naples' port, factories, and central train station, killing a number of civilians. In such an atmosphere, it was unsurprising that the 1941–1942 season marked the first time Napoli were relegated from Serie A.

Going down

While Napoli had two decent seasons under President Lauro, it's difficult to deny that his approach to the transfer market—sell high, buy low, balance the books—caused his club to falter on the field. As players constantly moved to and from the club, and as coaches were hired and fired, any success on the pitch seemed to come down to luck more than a solid plan from those in charge of Napoli.

The fallout from Lauro's actions continued to haunt the squad, while the rotating presidency helped little. Piscitelli appeared to have no real transfer strategy either (except, perhaps, to keep the club financially stable and thus compliant with FIGC rules). He bought two forwards, Umberto Menti from Milan and Carlo Paoletti from Genoa. Menti had played approximately half of the season before being benched toward the end, having scored only three goals. He then scored just two in his first season in Naples. Meanwhile, Paoletti had made just seven appearances in two seasons for Genoa, and had never scored in Serie A. Carlo Alberto Quario, a regular in the squad, returned to Ambrosiana-Inter, taking his eight goals with him. Yet Napoli really didn't need forwards so much as they needed a creative player in the middle of the park to replace Rosellini, a fan favorite for his technique and vision. He scored a respectable number of goals for a midfielder, but more than that, he had a deadly shot from distance, and most of the goals he scored were fantastical. He was, in short, one of those players that everyone loves to watch. Napoli retained their top-scoring forward, Evaristo Barrera, who had scored nine the previous season, and their proficient attacker Umberto Busani, who, with his nose for goal, was one of the best wingers Napoli had had in their early history.[27]

On the other side of the pitch, Piscitelli made another series of strange changes. Defender Maximiliano Faotto, who'd missed just two games the season before, was sold to Lazio. Napoli had given up 48 goals the past season, yet that was about average for the league (exempting Bari). So getting rid of Faotto, a defender who'd played nearly every match that season, made little sense. Neither did selling Renato Cappellini, a central midfielder in his prime who started much of the season, to Roma. It was as though each president chose players at random, with an eye nearly solely on cost, rather than basing decisions on the talent of the players brought in, or how the squad would work together. More than anything, it was stability that Napoli needed; that stability was further shaken when Piscitelli sent young defender Aldo Monsellato on loan to Salernitana and sold Alberto Rossi outright to the same club.

Yet all in all, the team put together by Piscitelli and coached by Vojak was fine on paper—certainly strong enough to finish midtable. Napoli beat both Juventus and Milan, who may have finished only 6th and 10th, respectively, but wins over such Serie A stalwarts would have made the fans' hopes rise. In addition, they took points from the eventual 3rd-place team, Venezia, the 2nd-place team, Torino, and the scudetto winners, Roma, by holding them to draws at home.[28] But in the end, Napoli just couldn't produce the goals—and couldn't stop enough of them—to save themselves from relegation. Just one more win would have seen them tied for safety with Ambrosia-Inter; instead, they finished a point behind Atalanta and Livorno. Had they managed to hold on to Rosellini, or even Faotto to help strengthen the defense, it's likely they could have kept their heads above water.

However, choices had been made, and Napoli entered the 1942–1943 season determined to fight their way back to the first division. Surprisingly, Luigi Piscitelli stuck around for another year, although Antonio Vojak left his former squad. In his place came Paulo Innocenti, who as a player had won the title with Bologna in 1925, only to go on and stay with Napoli for 11 seasons, 1926–1937. He was just 41 years old when he stepped up as il Mister for Napoli's first Serie B campaign, which turned out to be his only season on the bench (to clarify, this refers to a manager rather than a substitute hoping to take the field; this terminology is common when discussing *calcio*). Perhaps it was pure frustration that led him to a different career. He went on to become a trusted associate of Achille Lauro and wound up opening a centrally located bar that became popular with Napoli fans.[29] But prior to that, Innocenti fought a hard battle and just barely lost. A defender himself, it's little surprise that Innocenti devised an organized squad that set out to shut other teams down; of the 18 teams in Serie B that season, only Pro Patria conceded fewer goals. Napoli also bought Vinicio Viani, also known as Viani II, from 14th-placed Livorno, realizing that a man who'd scored 10 goals in 19 Serie A appearances, including 5 in his last 6 games, might be a worthy investment. Viani II managed 16 that Serie B season, proving calculated risks are worth taking . . . even if they don't quite pay off as hoped; after all, Viani II had scored 35 goals in 31 Serie B appearances 2 seasons previously.

Wisely, the club managed to hold on to Umberto Busani, such an integral part of the squad who contributed ten goals that season. Overall, Napoli's 46 goals scored was about average in the league that year. Combined with the low number of goals conceded, they looked destined to go straight back up. Unfortunately, the final day of the season pitted them against Modena, who had a two-point advantage over Napoli. The game had to be played at the Stadio Collana, as British bombs had destroyed Napoli's Stadio Ascarelli in July 1941. With two minutes left to go, Modena's Alberto Eliani lobbed a ball over the head of captain and goalkeeper Arnaldo Sentimenti. Napoli's loss not only handed Modena the title, but also allowed Bresia to sneak into 2nd thanks to their win on the final day. Dreams of returning right back to Serie A were smashed.

Momentum might have propelled Napoli back up the next year, but the FIGC canceled Serie B for two seasons due to World War II.

Naples under siege

It's quite surprising that Serie B kept playing until 1943—and Serie A until 1944—given the destruction inflicted by the Allies on the peninsula, Italy's woeful military maneuvers, and the loss of life that occurred from both. Italy officially entered World

War II in June 1940, when Italy attacked French troops in the Alps, a battle cut short by France's surrender and the Franco-German Armistice on June 22. This furthered Italy's belief that the war would soon be over and they would emerge on the side of the victors.

After the battle against the French, Italian forces regrouped under Mussolini's direction and, in October 1940, attacked Greece from Albania, a maneuver so hastily planned and inadequately enacted that the Germans had to come in and not only take the country themselves, but rescue the Italian soldiers. Despite further German assistance, Italy lost its East African empire and then suffered the indignity of having to surrender in May 1943 after their unsuccessful battles in North Africa. Italy was simply weak, militarily, and it showed. Many soldiers felt no attachment to the battles fought on land far from home. Clothing, food, fuel, and vehicles were all scarce due to the lack of safe supply routes. Raw materials were in extremely short supply, and when available, it mattered little, as the factories of Northern Italy were subject to heavy bombing.[30]

Naples had the misfortune of experiencing the greatest amount of bombing of any Italian city between 1940–1944.[31] The city suffered approximately 200 air raids, many coming before the city had adequate air-raid shelters or ships capable of anti-aircraft maneuvers.[32] In portions of the city's underground, formed by the Greeks for their city of Neapolis and converted by the Romans into aqueducts to provide water to their citizens, messages from those who hid in the tunnels during the bombings are still visible. The majority of the attacks were directed at the harbor area, as the Port of Naples was one of the main industrial areas of the country as well as the primary route to Africa. The Allies sought to destroy the area's factories and disrupt communication lines. The combination of the frequent attacks, a decimated military with few weapons up to the challenge of fighting back, the spread of typhus and other deadly diseases, and the lack of safe shelters for residents meant death and destruction came earlier for Naples than the rest of Italy.

The worst began in December of 1942; on the 4th, the 14th-century Basilica of Santa Chiara was destroyed. Even as football continued, so did the bombings, which became a daily Naples occurrence in January of 1943. Citizens quickly became accustomed to sprinting for shelter at the sound of the air siren. Approximately 20,000–25,000 civilians were killed, most in 1943, when the bombing intensified during the Italian Campaign or the "Liberation of Italy." Neapolitans were forced to barter or outright beg for food, while many women braved the black market to buy or sell goods such as coffee or flour. The Allied forces believed (rightly, it turned out) that Italy was the weakest link of the Axis powers, and an invasion might cause Italy to drop out of the war.[33] The best means by which to occupy Italy began by capturing Sicily before shifting to Naples and its surrounding territory.

Allied forces landed on the mainland on September 3, the same day the armistice between Italy and the Allies was signed. However, Naples was left unprotected. Once the armistice was announced on September 8, the Allies thought the German forces would retreat to the North, as Hitler was convinced Southern Italy was unimportant. Instead, chaos reigned as the Germans took over the city.

Unlike elsewhere on the peninsula, Neapolitans, barely removed from the days in which they had their own kingdom, were prepared to defend their city themselves. In what is known as the Quattro giornate di Napoli the people of the city rose up to defend themselves against the German occupiers.

From the start, the situation in Naples, where bombs were still falling, was pure pandemonium. The officials in charge collaborated with the Nazis or, like the military generals in charge of keeping orders, fled the city, with many Italian troops following. On September 12, the same day Colonel Walter Schöll took command of the military occupation of Naples, 4,000 Neapolitans were deported, most of whom became forced laborers. Those who resisted were executed. That same day, the German military commander declared the city was in a state of siege, and gave permission to execute up to 100 Neapolitans for every one German killed.

When all men from 18–33 years of age were ordered to perform compulsory labor, and learned they would be deported to Northern Italy and Germany, the civil uprising truly began. Of the Neapolitan men who refused to go (only 150 of an estimated 3,000 had shown up on September 26), 8,000 were arrested and brought to the Stadio Vomero.[34] This was the same stadium where Napoli had played the final game of the 1942–1943 season, and where they would play after the war.

After years of bombing, the German control of the city, their indiscriminate public executions, and now the further looting and destruction of their town, the people of Naples did not need an external force or organization to spur them to action. Throughout much of Italy, pockets of resistance had formed or strengthened as the bombs began to fall, leading to an agreement among the antifascism parties to work together to overthrow the government and the Axis powers. But Naples, one of Italy's poorest cities even before the war, had watched the North industrialize while they experienced chronic underinvestment from the government in Rome, and were not about to wait for their fellow Italians.

The citizens of Naples stole ammunition and rifles, resisted forced deportation, and, on September 26, began to riot, pushing back against the deportations and freeing the resisters. The first of the "four days" began the next night. The German soldiers had captured around 800 Neapolitans, and riots broke out in various neighborhoods around the city. On the 28th, the fighting intensified. Again the future Napoli stadium was involved, with an assault on the Stadio Vomero; this time, partisans forcefully dragged Schöll and other troops to the stadium and forced them to release the prisoners held there.[35] On the 29th, the insurgents became more organized at the local level, although they remained separate from other Italian groups. One incredible example of the Neapolitans' bravery involved 11-year-old Gennarino Capuozzo, who stood in front of a German tank, holding a hand grenade, until the soldiers surrendered.[36] Finally, on

September 30, even while the fighting continued in some areas, the Germans began to flee, executing people as they went and setting fires behind them. On October 1, the Allies officially liberated Naples. The Germans had not done what Hitler expressly had told them: organize resistance to the Allies in Naples to head off the offensive they knew was coming.

Naples played no role in the Italian Social Republic, the puppet state formed under a rescued Benito Mussolini but dependent on the Germans for control. Unlike in other Italian cities, where the new government prompted Italians to fight one another in a civil war, in Naples the citizens—despite being loosely organized and having no real leaders—were almost completely united, fighting against the Germans rather than Italian fascists.[37] On the other hand, the Allied campaign had wreaked havoc on the city, and the Germans had barely held back as they departed. The fleeing Germans dismantled or removed all communications, transportation, water, and power grids. They collapsed bridges, tore up railroad tracks, and sunk many of the ships that were still in the harbor. Communication lines, railroad tracks, bridges, hotels, buildings . . . the only things the Germans left untouched were the churches and monasteries. There was no electricity, no gas. People had to line up at the fountains to get water. While the Port of Naples was reopened a week after capture, the people of Naples were forced to depend upon the Allies for basic survival items for many months.[38] And when the Allies left, there was no social order for the people of Naples (or the rest of the Mezzogiorno) to fall back on.

Yet football pressed on, a way to make life feel a bit more normal. Famous Neapolitan journalist Arturo Collana formed Società Sportiva Napoli in May 1944. A month later, Societa Polisportiva Napoli was founded by Luigi Scuotto. After a long negotiation, the two clubs merged into Associazione Polisportiva Napoli on January 19, 1945. Less than ten days later, on January 28, the team played its first match. As the FIGC kept football on hold, many clubs in the South formed a tournament in Campania, consisting of teams from the top three divisions and one team of the military police. The new Napoli club finished 3rd, behind Stabia and Salernitana.[39] The team then merged back into A.C. Napoli for the 1945–1946 season.

Italian partisans executed Benito Mussolini on April 28, 1945, and took his body to Milan, hanging it upside down for all to see. Prior to the dictator's rise, many in the South, tired of the divide between the haves and the have-nots, had begun to engage in communist activities. Yet in the 1946 referendum, Naples was the biggest city to cast their votes for a continued monarchy; more than three-quarters of the residents voted for Italy to remain a kingdom. To the citizens, it made the most sense. Throughout the push-and-pull, back-and-forth among the empires that had maintained a loose hold over Naples, the city had been one of the most prestigious in the world. The South had floundered after the 1861 Risorgimento and under the rule of King Victor Emmanuel II. Mussolini had reneged on his promise to eradicate the Mafia and forced the South to remain an agrarian economy.[40] What the citizens of

Naples remembered was that, prior to unification, a true monarchy had produced a dazzling city of art, music, literature, and beauty.

However, the majority elsewhere in Italy voted otherwise, and their vote established the Italian Republic on June 2, 1946.[41] Again, the Mezzogiorno found itself overlooked. The Allies turned away from the region's poverty and permitted the Italian police to crack down on protests against inflation and food shortages.[42] In addition, wary of the spread of communism in the South, they permitted the mafias to return to rule the cities and towns. In Naples, that meant the Camorra, who forced out communist organizers and leaders to gain near complete control of the city. The United States even made deals with the Camorra, asking for intelligence and protection of the shipyards. In return, the Camorra constructed a far-reaching black market to help the Neapolitans survive after the war.[43] Yet the effects were long-lasting; the corruption and political influence of the Camorra continues to linger in Naples.[44]

Meanwhile, the United States, who occupied the North, helped it rebuild its infrastructure. Alone, Italy could have taken decades to rebuild, thus creating a greater economic gap between it and the more advanced nations; however, the peninsula's geography made it a fortunate recipient of generous funds from the Marshall Plan. Italy, with its strong Communist Party, was considered a valuable "hinge state" between the Iron Curtain and the Free World, and it was assumed that the 1.5 billion US dollars poured into the country would keep Italy onside.[45]

"Normal," or as close as could be

Achille Lauro, the former Napoli president, deserves some of the credit for ensuring Naples did not simply wither away, cast out into the bay. The years between the time he left, in 1940, and when he officially returned to the helm of the club, in 1952, were both cruel and kind to the shipowner. Given the government had requisitioned 52 of his 57 ships during the war, Mussolini had granted him, in partial compensation for the loss, an interest in several Neapolitan newspapers as well as the Banco di Napoli. However, in 1943, the Allies charged him with aiding and abetting fascism. He spent 22 months in a prison camp before being acquitted of all charges and released. Lauro then began to build up his company once again, this time in the passenger market, sending emigrants to Australia and New Zealand and bringing tourists from there back to Europe. It was this business that made him a multimillionaire.[46]

In order for football to resume in 1945, the *Divisione Nazionale*, Italy's first division, would need to be split into two sections, North and Central/South Italy, as routes between the two still needed to be reconstructed. Because the South had always had far fewer strong teams, the Central/South region would need to include multiple teams from Serie B to even things out. Napoli, who'd been 3rd in Serie B before the league shut down, were called up, along with five other Serie B teams. The section had just five teams whose rightful position was in Serie A. The reason for the odd number of clubs, 11, in the regional competition was because Pisa's stadium had been bombed during the war and the club did not have the funds to repair it.

In the first local derby against Salernitana, a club in Salerno near Naples, the score was 1–1 in the 35th minute. The referee awarded a dubious penalty to Napoli. The home fans were so upset that the police struggled to control them. Napoli missed the penalty, but tensions remained high both on and off the pitch, with many small fights breaking out when the referee called a foul or when fans weren't happy with his decision. Referee Stampacchia kept halting the game due to fouls, unsportsmanlike behavior, and even punches being thrown . . . on the pitch. In the stands, fight after fight broke out. Then all hell broke loose after the sound of gunshots echoed through the stadium. Stampacchia fell on the pitch, pretending to be dead, perhaps trying to calm the situation by doing so. He was brought to the locker room, both sets of players followed, and the game was suspended. The committee in charge of the regional grouping declared no more games would be played at the stadium in Salerno; moreover, they decided to stop the championship due to the many incidents occurring in the league. The Central/South tournament was suspended for four weeks and began again on May 24.[47]

Napoli won the top spot in their section, automatically ensuring they would enter the playoff for a Serie A spot. Oddly enough, their new coach, Raffaele Sansone, played the first game of the season for Napoli before retiring, at which point he stepped into the role of il Mister (this happens a few times in Napoli history—after playing one game, a player would step in and become a player-coach—but it rarely occurs today). Sansone had a rather conservative style, like much of calcio during that time, but he also presided over an excellent goalkeeper and defense. The Azzurri had 11 wins to 2nd-placed Bari's 13, but they also let in just 10 goals to Bari's 21. Arnaldo Sentimenti, or Sentimenti II, who had stuck with the team through relegation, had given one last truly great season to the squad. But some of the credit must go to new president Pasquale Russo. In desperate need of an infusion of cash, the association elected the lawyer, industrialist, and mayor of Solofra, an inland town about 70 miles to the east. As soon as Russo got the job, he and sporting director Luigi Scuotto jumped into an old car and drove around Italy to find players.[48] They returned with eight new players, including Carlo Barbieri, who scored 15 goals that season, the most in the squad.

Next, the top four from each section of the two played out a mini-season of 14 matches. Napoli did manage to find the net this time, with 19 goals. Barbieri, who'd scored 7 of Napoli's 10 goals in the first part of the season, scored 8 this time, making the midfielder the second-best goalscorer in the Divisione Nazionale. Napoli went on to finish 5th, leading the pack for the South; poor Bari, in 8th, had just 5 points to champions Torino's 22. It barely mattered, however, as all eight teams in the playoff went through to next season's top division.

In 1946–1947, Serie A returned to its normal format, with the addition of two teams, making it a 20-team tournament, as it remains today. This included the teams that had made the Divisione Nazionale (Torino, Juventus, Milan, Inter, Napoli,

Roma, Pro Livorno, and Bari), the teams that would have been in the top flight after the 1942–1943 season (this added Genoa, Bologna, Fiorentina, Lazio, Atalanta, Vicenza, Triestina, and Venezia) along with newly promoted Modena and Brescia), and Alessandria, who had won the Serie B–C championship of 1946. The final team that ascended was the newly created Sampdoria, a team with a rather ironic beginning. In 1927, the Fascist Party forced Sampierdarenese and Andrea Doria to merge, becoming first La Dominante, then Liguria, before splitting after two seasons. Sampierdarenese then absorbed two smaller Genoa clubs to become Liguria once more, in 1937. Due to this interference, Serie A admitted Sampierdarenese, although Andrea Doria were the better team in 1945–1946. After these machinations, the two clubs threw up their hands and made their own decision to combine, becoming UC Sampdoria.

The season was an average one for Napoli. For once, they remained under the same president, Pasquale Russo, and Sansone remained as il Mister. However, one interesting name was added to the club sheet—Attila Sallustro, who returned to his team to take over from Scuotto as sporting director. While other sides were taking advantage of the fall of fascist dictates that prohibited them from buying foreign players, Napoli had but one player born outside Italy, Michele Andreolo, a Uruguayan who'd become a naturalized Italian citizen.

A devastating accident occurred midseason. Riza Lustha had signed from Juventus, where he had scored 46 goals in 85 games. He was expected to be a huge success at Napoli, but fans spent the first half of the season waiting for him to score. Noted Neapolitan journalist, Carlo Di Nanni, famously wrote in the *Sport del Mezzogiorno* newspaper, "*Quando Lustha segnerà, se ne cadrà lo stadio*" (When Lustha scores, the stadium will collapse).[49] Horrifyingly, exactly that occurred when, on January 27, 1946, Lustha scored his first goal of the season against Bari at the Stadio Vomero. When fans celebrated a little too much, one of the guard rails on the terrace collapsed and 114 people were injured.[50]

Besides Umberto Busani recovering his Serie B scoring boots, about the only thing Napoli were good for that season was putting a wrench in other teams' spokes in the second half of the year, where they beat 3rd-place Modena and drew with 2nd-place Juventus. But given that Torino won the title with ten points to spare, Napoli's help was likely not required.

Yet after the conclusion of the following season, Napoli would have longed for a midtable finish. Instead, they wound up dead last, in 21st—fascinating for a league that only had 20 teams. But no, Napoli were not so bad that they deserved a place worse than last; at least, not on the pitch. It was just that kind of year in Serie A, a kind of year that would almost become a normal occurrence every decade or so.

First, Serie A had 21 teams because Triestina, who had shown nearly unbearable weakness the season before, finishing at the bottom of the table with just 18 points, 32 goals scored, and 79 conceded—by far the worst team in the league—were readmitted to the top division, forcing Brescia (who were apparently very polite about it all) down to Serie B. This was due to political plotting of the highest order. In Trieste, negotiations between the Allies and Yugoslavia had resulted in the region becoming an autonomous state, the Free Territory of Trieste. The Allies administered Zone 1, the Yugoslavs Zone 2. Both had their own football club, funded by their government, playing in their respective countries' top division. Until Triestina were relegated in 1947, that is. Two politicians from Bologna asked the prime minister, Alcide de Gasperi, to interfere with the sport's decision, successfully arguing the need for Triestina to compete at the highest level. The other Trieste club, Amatori Ponziana, was playing in another country's top division, and fans would likely leave Triestina to support the Yugoslav side. Thanks to funding by the Italian government, Triestina could buy the most talented players, and woo former Napoli player Nereo Rocco to manage the side. No wonder they finished 2nd, tied on points with Juventus and Milan, at the end of 1947–1948.

As for Napoli, more was expected. The side had remained much the same; although Umberto Busani lost his scoring boots once more, Dante Di Benedetti stepped in to fill them. Their defense was just as steady. But when the results didn't come, the fans grew restless, and began to demand that Achille Lauro return to the helm. Instead, in January 1948, the entire managerial side resigned, including Rosso. Though he technically did not take over as Napoli president, Vincenzo Savarese returned to delegate power to the appropriate persons.[51] Even amidst all this mess, the squad was strong enough that Napoli should have been able to escape relegation, even if it was just by a hair. Instead, they were embroiled in one of Italy's famous calcio scandals.

Caso Napoli, the imaginatively named "Napoli Case," involved accusations of bribery between Napoli and Bologna. Just days after the June 6, 1948 match between the two sides ended 1–0 to the visiting Napoli—thanks to a 90th-minute goal by Naim Kryeziu—Renato Dall'Ara, president of the *Rossoblù*, accused them of corruption. He had received an anonymous letter stating that Napoli midfielder Luigi Ganelli had convinced his friend, Bologna forward Bruno Arcari, to go easy on the visitors. The family connections helped; Arcari was poised to marry Ganelli's sister-in-law. Therefore, such a meeting would not have been suspicious, but Napoli president Muscariello and the former captain, Paolo Innocenti, were there, as well as Bologna technical director Hermann Felsner.[52] The result would not have mattered much to Bologna, who were a solid midtable team at worst, but the Partenopei were trying to escape relegation. Due to the league situation, it seemed reasonable that Napoli had been willing to pay around 3 million lire to fix the game, and the plot extended past those present in the meeting to encompass almost everyone on the pitch. Dall'Ara asked for Napoli to be relegated and all those involved to be handed lifetime suspensions.[53]

Regardless of the accusations, Napoli's performance on the pitch in 1947–1948 saw them relegated to Serie B after finishing 18th. The club appealed to the FIGC to allow them to stay in Serie A, as the association had allowed Triestina to do the year before. The Azzurri compared themselves to the team in Trieste, saying they should be allowed to remain due to "Southern pride," and asking for a 21-team league the following season as well. Then Napoli began to strike out at the North, speaking of a cabal that controlled the National League, as well as a subset of a "Milanese caste"[54] who were angry at the allocation of funds that had been granted to Napoli to construct a much-needed new, stable stadium. The League, annoyed by the tangential nature of Napoli's arguments, dismissed them, relegated Napoli to last place, and banned Ganelli as well as Napoli president Giuseppe Muscariello for life.

Napoli, clearly not realizing when enough was enough, fought back, again stating a Northern cartel wanted to keep the Naples club in its place rather than provide funds for stadium construction. They stated such allegations by a Northern club were ridiculous; such a meeting may have occurred, but Ganelli was clearly joking when he asked if Arcari would allow the Partenopei this crucial win.[55] Napoli then called a meeting of the Southern clubs participating in Serie A, B, and C to discuss moving FIGC headquarters from Milan to Florence. The move did eventually happen, but at the time the leadership accused Napoli of slander, after which Napoli fought back with their own complaint; because this complaint violated the club's arbitration clause, the league threatened them with further relegation.[56] Finally Napoli backed down and they were readmitted to the league, although in Serie B, where they'd originally been before all the fighting began.

Down in Serie B for the second time in their sporting lives, Napoli faced losing some of their brightest players, as almost always occurs when a team is relegated. Dante Di Benedetti, Andrea Verrina, and Sentimenti II were snapped up by Serie A clubs. Umberto Busani, by that point 33 years old, had little to offer any club and consequently actually moved *down* in the calcio league hierarchy. On paper, the new club president, industrialist Egidio Musollino, looked to have created a squad strong enough to challenge in the second division, primarily purchasing younger players from Serie A teams. The team had plenty of money—Musollino, with capital to spend, had joined the club staff in 1945, at the same time as Alfonso Cuomo (a manufacturer), Giuseppe Muscariello (politician and entrepreneur), and Pasquale Russo (lawyer and industrialist).[57] Russo had had the first shot at the presidency, but was part of the management team that resigned in January 1948. Not implicated in the events of Caso Napoli, Musollino took the reins in August of that year. He appointed defender Felice Borel to the bench as player-coach, followed by Luigi De Manes, who had coached Napoli in 1944–1945, and finally Vittorio Mosele, who led Napoli to a 5th-place finish and another year in Serie B.

While Napoli fans continued to support their team while in Serie B, the events occurring in Serie A dominated 1949 in the worst way possible. On May 6, 1949, a small aircraft carrying Torino FC players and staff, who were on course to win their fifth straight Serie A title, crashed into the retaining wall at the back of the Basilica of Superga, near Turin. The plane was coming from Lisbon, where *il Grande Torino* had just played a testimonial match against Benfica in honor of the Italian side's captain, Francisco Ferreira. The weather was awful—windy, rainy, with a thick fog that severely limited visibility. It is unknown whether the strong wind sent the plane drifting, leading it straight into the church wall, or if faulty equipment caused the pilots to believe that they were flying at a higher altitude, easily able to bypass the 2,195-foot hill that Superga stood upon.

While it was clear from the suitcases found by the searchers that Torino players had been on the plane, it was up to the former manager of the Italy national team, Vittorio Pozzo, to identify the bodies; nearly the entire national team had been composed of Torino players. All 31 people, including the 18 active Torino players, the manager, the coach, four other club officials, three journalists, and four crew members were killed. Sauro Toma, an injured defender who did not travel with the team, was mobbed in the streets by crying fans. Toma lived until age 92, a rather difficult fate for a man who said he was "condemned to survive, while my brothers perished."[58]

Two days later, on May 6, at the request of other clubs in Serie A, the FIGC awarded Torino the 1948–1949 scudetto. They had been leading Inter by four points at the top of the table when the crash occurred. Because Torino could only feature their *primavera* (youth) players during the final four games of the season, their opponents also fielded their youth teams. For the 1949–1950 season, each team in the top flight was asked to donate one of their players to Torino, in order to help them rebuild.[59] That season, crosstown rivals Juventus won the title, while Torino finished 6th.

It turned out there were far worse things than a team's relegation.

[1] Sircana, G. (2005). LAURO, Achille. In the *Biographical Dictionary of Italians* , *LXIV*. Institute of the Italian Encyclopedia Treccani.

[2] Liccardo, G. (2017). *Storia irriverente di eroi, santi e tiranni di Napoli*. Newton Compton Editori.

[3] Acampora, R. (n.d.). Un romanzo lungo cent'anni. *Riccardocassero.it*. http://www.riccardocassero.it/lalungaavventura.htm#1

[4] Smith, M. D. (1983). The Italians were prevented from knowing the war continued by government censorship. In *Mussolini*. Granada.

[5] Kallis, A. (2000). *Fascist ideology territory and expansionism in Italy and Germany, 1922–1945*. Routledge.

[6] Encyclopedia Britannica. (last updated 2020, January 1). Italy. In *Encyclopedia Britannica*. Retrieved December 4, 2020 from https://www.britannica.com/place/Italy

[7] Encyclopedia Britannica. (last updated 2020, January 1). Italy—World War II. In *Encyclopedia Britannica*. Retrieved December 4, 2020 from https://www.britannica.com/place/Italy/World-War-II

[8] Martin, S. (2004) *Football and fascism: The national game under Mussolini*. Berg.

[9] Martin, S. (2018, April 5). World Cup stunning moments: Mussolini's blackshirts' 1938 win. *Guardian*. https://www.theguardian.com/football/blog/2014/apr/01/world-cup-moments-1938-italy-benito-mussolini

[10] Id.

[11] Nardacchione, P. (1946, January 28). Napoli–Bari 2–1; Graci incident per crollo d'una balaustra. *Corriere dello Sport*, 1.

[12] Guarino, C. (1955, November 7). Tumultuoso epilogo dell'incontro Napoli–Bologna. *Stampa Sera*, 1.

[13] Tutti i presidenti da Ascarelli a De Laurentiis. (n.d.). *SSC Napoli*. Retrieved November 19, 2020 from https://www.sscnapoli.it/static/content/Gli-altri-Presidenti-21.aspx

[14] Forlenza, R. (2017). Power, politics and soccer in postwar Italy: The case of Naples. In B. Elsey and S. Pugliese (Eds.), *Football and the boundaries of history: Critical studies in soccer* (pp. 249–266). Palgrave Macmillan.

[15] Acampora.

[16] Zontini, A. (2015). *Storia del Napoli*. EUS.

[17] Formato, A. (2013, February 27). Napoli–Juventus, le tappe di una rivalità più sociale che sportiva. *International Business Times*. https://web.archive.org/web/20150526162330/http://it.ibtimes.com/napoli-juventus-le-tappe-di-una-rivalita-piu-sociale-che-sportiva-1329287

[18] Lo chiamavano «terrone», sarà risarcito (2005, April 20). *Corrie della Sera*. https://www.corriere.it/Primo_Piano/Cronache/2005/04_Aprile/19/terrone.shtml

[19] AC Napoli–Juventus FC, Serie A match sheet. (1939, February 19). *Transfermarkt*. https://www.transfermarkt.com/spielbericht/index/spielbericht/2399510

[20] All statistics involving games played, number of goals scored, and transfer dates are from Transfermarkt unless otherwise noted.

[21] Forlenza.

[22] Acampora.

[23] Il Giuoco del Calcio in Naples. (1912). *La Stampa Sportiva*. https://www-byterfly-eu.translate.goog/islandora/object/libria:43665?_x_tr_sl=it&_x_tr_tl=en&_x_tr_hl=en-US&_x_tr_pto=ajax,se#page/15/mode/2up

[24] Carratelli, M. (2007). *La grande storia del Napoli*. Gianni Marchesini Editore.

[25] Id.

[26] Gribaudi, G. (2005). *Guerra totale: Tra bombe alleate e violenze naziste Napoli e il fronte meridionale 1940–1944*. Bollati Boringhieri.

[27] Id.

[28] Id.

[29] Carratelli, M. (2009, January 18). Storia del centravanti che diventò un idolo e non se andò mai più. *La Repubblica*. https://ricerca.repubblica.it/repubblica/archivio/repubblica/2009/01/18/storia-del-centravanti-che-divento-un-idolo.html

[30] Encyclopedia Britannica. (last updated 2020, January 1). Italy—World War II. In *Encyclopedia Britannica*. Retrieved December 4, 2020 from https://www.britannica.com/place/Italy/World-War-II

[31] Gribaudi.

[32] Monda, L. (2005, March 5). Napoli durante la II guerra mondiale ovvero: i 100 bombardamenti di Napoli. Relazione convegno I.S.S.E.S.

[33] Blumenson, M. (1969). *Salerno to Cassino. United States Army in World War II, Mediterranean theater of operations* (Vol. 3). Washington, D.C.: Office of the Chief of Military History, U.S. Army.

[34] Varriale, A. (2014). The myth of the Italian Resistance Movement (1943–1945): The case of Naples. *Mythen—nationale Grenzen—Religionen/Myths—National Borders—Religions, 27*(2), 390.

[35] Id.

[36] Napolitano, S. (n.d.). Four days of Naples. *HistoriaPage*. https://www.historypage.it/four-days-of-naples/

[37] Zuccotti, S., & Colombo, F. (1996). *The Italians and the Holocaust: Persecution, rescue, and survival*. University of Nebraska Press.

[38] Naples—Foggia: The US Army campaigns of World War II [Military brochure]. (n.d.). Retrieved December 8, 2020 from https://history.army.mil/brochures/naples/72-17.htm

[39] Tramontano, E. (1984). Da Sallustro a Maradona 90 anni di storia del Napoli. Edizioni Meridionali.

[40] Varriale, 383–93.

[41] Forlenza.

[42] Varriale.

[43] Nugent, C. (2018, November 16) Organized crime lurks everywhere in *My Brilliant Friend*. Here's the real story of the rise of the Naples underworld. *Time*. https://time.com/5435772/elena-ferrante-hbo-mafia-camorra/

[44] Saviano, R. (2007). *Gomorrah: Italy's other mafia*. Macmillan.

[45] Hogan, M. J. (1987). *The Marshall Plan: America, Britain, and the reconstruction of Western Europe, 1947–1952*. Cambridge University Press.

[46] Forlenza.

[47] Pezzullo, P. (2004). *70 anni di storia della Frattese Calcio 1928–2004.* Istituto di Studi Atellani.

[48] Tramontano.

[49] Carratelli, M. (2010, August 11). L'anatema Di Nanni. *La Repubblica.* https://ricerca.repubblica.it/repubblica/archivio/repubblica/2010/08/11/anatema-di-nanni.html

[50] Zontini.

[51] Id.

[52] Di Franco, O. (2002). Gli scandali del mondo del calcio. *Treccani.* https://www.treccani.it/enciclopedia/gli-scandali-del-mondo-del-calcio_%28Enciclopedia-dello-Sport%29/

[53] Gli "scandali" del calcio vengono al pettine. (1948, July 9). *La Stampa,* 4.

[54] Melillo, G. (1948, August 4). Il Napoli e il suo diritto alla riabilitazione. *Corriere dello Sport,* 1–2.

[55] Id.

[56] L'aut aut al Napoli—La C.A.F. ha respinto i reclami del dr. Muscariello e del giocatore Luigi Ganelli. (1948, August 20). *Corriere dello Sport,* 2.

[57] Tramontano.

[58] Jennings, P. (2019, May 4). The plane crash that killed Serie A's champions and their English coach. *BBC Sport.* https://www.bbc.com/sport/football/46788983

[59] Id.

CHAPTER 4

Slips, Stars, and Silverware

4

Calcio became the prominent sport in Italy following World War II, despite the fact that people had little-to-no disposable income and match tickets were expensive,[1] particularly in Southern Italy, where wages were lower than in the North.[2] While the Marshall Plan, implemented in April 1948, aided reconstruction in the North and expanded its industrial sector, the majority of the aid to the South was spent on increased agricultural production (which *decreased* the number of workers needed) and ultimately widened the economic gap between North and South.[3] When aid ceased in 1950, it coincided with the beginning of the Korean War, which required metal as well as manufactured products that Italy could readily provide—specifically, that the *North* could provide. Per capita income in the South was half that of the North. In 1957, the Treaty of Rome established the European Economic Community, of which Italy was a founding member, allowing for greater ease of export and investment. Around nine million people are estimated to have migrated internally during this "economic miracle,"[4] most moving from South to North, where economic expansion moved quickly, particularly in Milan, Turin, and Genoa.[5]

This massive internal migration in the 1950s and 1960s contributed to an increase in calcio fandom in the large industrial cities, to the point that the first supporters groups began to emerge. It helped that Milan, Inter, and Juventus all benefited not only from their geographical locations but also from industrial patronage and the money it brought in, enabling them to buy the best players and consolidate their positions as the best clubs in Italy.[6] Down in Naples, SSC Napoli simply did not have the funds to compete for the best players. These two decades would be tumultuous, but they also brought with them charismatic characters, new fan favorites, and a shiny piece of silverware.

A decade of ups and downs
In January 1950, Napoli entered the new decade in Serie B, but in a strong position, with the fans still backing their club, even when it wasn't possible for them to attend matches in person. In Naples, the price of a ticket at the Vomero Stadium varied from 100 lire for soldiers and children, and 500 lire for the central terrace. That amounts to approximately 4–20 US dollars in today's money, and was a considerable amount for those living in an area requiring extensive rebuilding, which had also lost its main income source when the factories that had provided wartime materials closed.[7]

Post-Marshall Plan, Italy had implemented its own plan for rebuilding the South. The *Cassa per il Mezzogiorno* (Southern Development Fund) provided aid for roads, schools, electrification, water provision, and land reclamation. When it moved into industrialization, the South stagnated. The government built ridiculously expensive factories in selected areas, yet the goods produced were primarily for export or were simply non-consumer-oriented. In addition, these factories relied on very little human labor. Meanwhile, small firms received few of the benefits poured into these new factories. The South saw little good come from this development, while the North was angry about having to pay for it.

Despite the ticket prices, football had become a shared national culture throughout the country, just as Mussolini had hoped. Much of the reason was down to a surprising source: the Catholic Church. The Church saw football both as a means of ministry and/or conversion and a way to implement political projects that coincided with a Christian social order. The parishes, with their football pitches, often offered the only real entertainment in town, especially in the suburbs and new developments that began to appear after the war. However, the number of these welcoming parishes decreased when moving south, to the point where not one parish in the city of Naples had a football field.[8] Yet Naples would, over time, provide Italy with some of its best players.

Meanwhile, in the South, unrest grew. The Southerners wanted land reform and so organized large land occupations in 1949 and 1950, involving hundreds and thousands of landless citizens. Because the wealthy landowners had plenty of political connections, the government intervened—sometimes lethally—to oust the occupiers. In the end, very little land was transferred. The South's inequalities continued to deepen; in 1950, only a small amount of the aid from the Marshall Plan had reached the South, and Southern Italy, as a whole, lacked the development and thus the money and industrial patronage the fortunate clubs of the North received. But the South was not without individuals able to inject cash into their clubs, and Napoli president Egidio Musollino was one such individual.

After a disappointing previous year saw the team finish 6th, Musollino was even more determined to bring the club back to Serie A in 1949–1950. He put on the bench successful former defender Eraldo Monzeglio, who had won the 1928–1929 title with Bologna and the two 1930s World Cups with Italy, despite his inability to bring Pro Sesto up from Serie B to A in two seasons (and despite his former close friendship with Benito Mussolini[9]). His wartime allegiances may have attracted controversy, but Monzeglio was said to be the steadiest hand at Napoli's helm since William Garbutt left in 1935. He would remain the Napoli manager for seven seasons.

To help Monzeglio succeed, Musollino brought in a slew of new players: Mario Astorri, Vittorio Dagianti, Alberto Delfrati, Costantino De Andreis, Bruno Gramaglia, Paolo Todeschini, and Luigi Vultaggio. By the end of the season,

Napoli were regularly scoring two or more goals per game, and finished with a seven-game unbeaten run. De Andreis and Astorri helped significantly, scoring 12 and 8 goals respectively. Ivo Šuprina, a Yugoslav national who had arrived the year before, grabbed the *Capocannoniere* title with 15. Delfrati was a solid defender, contributing to a backline that conceded just 34 goals in 42 games. Midfielder Gramaglia, who had also played for the *Azzurri* in 1938–1943, became a franchise player, celebrated like Attila Sallustro, making 273 appearances in the Napoli shirt.[10]

Even with such a quality squad, it was a tough fight, with Udinese holding on to 1st place until the last round when Legnano beat them 6–1. With a 2–1 win against Catania, Napoli grabbed one more point than the *Friulani*, as well as the Serie B title. A fuzzy newsreel reveals plenty of fans at the Vomero for the decisive match, including supporters crowded on rooftops overlooking the stadium, eager to welcome their club back to Serie A with a plethora of cheers and flares.

The 1950–1951 Serie A season made clear the widening gap in talent between Milan, Inter, and Juventus and the rest of the league. Having won the title and played quite well at the end of the previous season, Napoli retained coach Eraldo Monzeglio and turned their eye toward acquiring attacking talent. New attacker Amedeo Amadei provided 11 goals, Antonio Bacchetti 10, and Silvio Formentin 7, all in their first year at the club, numbers that would've delighted fans in the previous decade. But that year's top scorer, AC Milan's Gunnar Nordahl, bagged 34, leading his team to the *scudetto*. Milan and Inter both scored 107 goals that season, while Juventus, in 3rd, posted 103. Then, as though a thick black marker had been drawn across the table, came Lazio in 4th, with just 64 goals. Napoli ended the season in 6th with 57 goals, an entirely respectable result for a team that had just been promoted, particularly given the difference in resources between the Big Three and the rest of the league.

When looking at the next season's table and statistics, one can be forgiven for believing that Napoli just continued on in the same way, even improving as more players found their scoring boots. In 1951–1952, thanks to a breakout season from Mario Astorii (13), increased awareness from Amedeo Amadei (12), and a single season from István Mike Mayer in which he scored 8 in just 21 appearances, Napoli were 4th in goals scored with 64, and finished 6th again. But the team was hit by tragedy that winter. Musollino, their owner and patron, watched from his balcony as a restaurant atop a hill in Vomero, in front of his home, burned down. He went back to bed safely, but it was just one more trauma for a person who was already plagued by worries about Napoli.[11] He died of a heart attack in his sleep.[12] After his death, Alfonso Cuomo, a canned food manufacturer, and Luigi Scuotto both helped preside over the club. The two often disagreed on the club's management, and were unable to keep Napoli's finances afloat.[13]

Although it meant little to Napoli, it's worth noting that this season brought yet another change in Italian calcio: the *Reforma Barassi* reduced Serie A to

18 teams and made the top three divisions professional, as well as adding a fourth division rather than relying on regional groupings. That meant the two bottom-placed teams, Padova and Legnano, were automatically relegated. That should have been enough to reduce Serie A to just 18 teams, but the Italians needed to add a little drama—the team that stood above the two automatically relegated clubs would need to play a tie-breaker against the 2nd-place team in Serie B . . . except Triestina and Lucchese were tied for the third-worst in Serie A, which meant they had to play a tie-breaker to reach the tie-breaker. The teams first met in Bergamo, where they drew 3–3. Another game was arranged in Milan, in which a 1–0 victory by Triestina took them to the playoff with Serie B side Brescia. Triestina again triumphed 1–0 in a match played in Valdagno, allowing them to stay in Serie A. The next season, the dramatics died down, and the top two in Serie B simply replaced the bottom two in Serie A.

Yet Napoli had plenty of drama to keep themselves occupied in 1952–1953, which is one of the true joys of following the club (or the league)—hardly will a season pass in which something terrible, or fascinating, or scandalous does not occur. Napoli adhered to this formula nearly every season; whether this is a telling detail of the Neapolitan personality, a reflection of the traits of the current Mister, or simply the result of a football club that frequently changed hands between often-eccentric owners depends on who takes part in the debate. For Napoli, the fun began when Achille Lauro returned to head up the club.

Lauro's true ambition was politics. President of Napoli from 1936 to 1940, he had lost nearly his entire shipping fleet during World War II, but had maintained his riches, good business sense, and political connections. Whether he could leverage the last was questionable, as he had served as a Fascist parliamentarian and been sent to prison during the war. Yet despite his fascist roots, he was liked, even loved, from the start of his new campaign, winning the people's affection by doing things like giving away pasta or even dollar bills.[14] Aware of just how many people in Naples wanted the return of the monarchy (and likely the stability that came with it), Lauro joined the *Partito Nazionale Monarchico* (National Monarchic Party), creating a coalition that easily won power in 1952, then tripled their votes in 1956, making Lauro the mayor of Naples for nearly all of the 1950s.

The "King of Naples," capitalized on the boom years of the 1950s, revitalizing Naples—a city that had lost its port and industrial zones during World War II—by pouring money into businesses, building beautiful fountains, most of the 80,000 new dwellings the city required, a new train station and, of course, the Stadio San Paolo.[15] His "liberal" approach to Italy's bureaucratic processes meant that he often overlooked, ignored, or bypassed the rules to obtain what he needed, enriching several Neapolitan families in the process. His ownership of the city's club only heightened the population's affection for Lauro.

Comandante Lauro was one of the first Italian businessmen to advance his populist politics through the purchase of a calcio club when he returned to Napoli as the "honorary president" in August 1951.[16] His investment in the club was one means of winning the affection of Neapolitan citizens, as a stronger Napoli meant more votes for Lauro. Adopting the slogan "A great Napoli for a great Naples,"[17] he even convinced Napoli players to campaign for him.[18] This later became a strategy of many Italian politicians—piggybacking on calcio's popularity in order to increase their own support.[19] But in Lauro's case, his opponents used his team to downplay his abilities, noting that because SSC Napoli were unable to form a winning team, there was no way Lauro could successfully govern the city.[20]

Determined to prove his detractors wrong, while rebuilding the city, Lauro also rebuilt the club. In April 1952, his presidential post became official[21] and, after paying down Napoli's debt, he put the club up for liquidation. He then formed a new sporting club with the same name, of which he became the largest shareholder.[22]

Under Lauro, one of the first cases of egregious spending by Italian clubs transpired in the transfer of Hans "Hasse" Jeppson, a name most calcio fans don't remember, given the number of record-setting transfers that have occurred since that time.[23] The striker, called "Hasse Goldenfoot" due to his scoring 58 times in 51 matches in three seasons with Swedish side Djurgårdens, had signed with Atalanta in 1951, where his 22 goals in 27 matches (having missed the first seven rounds of the season) helped the struggling side escape relegation. Lauro, knowing he needed to bankroll the club to maintain and grow his popularity throughout the city, had already brought in right winger Giancarlo Vitali from Fiorentina and left winger Bruno Pesaola from Novara for 1952–1953. But he needed to do something special to impress the people of Naples—such as setting the record transfer fee to bring an international star to the city. Napoli ended up in a bidding war with Inter, out of which Lauro (and Jeppson, if we're honest) emerged triumphant. Lauro had to cough up 105 million lire, 30 million of which went straight to the player,[24] but the president believed him to be worth it, that he could make Napoli a title contender.

Unfortunately for Napoli fans, the pressure of going from Serie B to fighting for the scudetto in just three seasons proved to be too much. The squad were undefeated in their first three matches, but Jeppson couldn't connect with his partner up top, veteran Napoli player Amadei. He didn't kick off his goalscoring campaign with his new club until October, four rounds in, in a match his side ended up losing 5–1 to Inter. After the next round, a 2–1 loss to Lazio in which Jeppson also scored, Lauro was irate. His side was meant to be challenging for the title, and he hadn't paid a record fee for a striker who wasn't getting the job done fast enough. Fortunately Eraldo Monzeglio, still Napoli's coach, was one of the few who could control Lauro when he lost his temper, and was usually able to keep him and his aides from entering the locker room.

Ultimately Napoli had one of their most successful seasons ever, finishing in 4th, six points behind scudetto winners Inter, scoring seven more goals than the champions. But while Jeppson was the team's joint-top scorer with Vitali, notching 14, Lauro had expected much more from the player known as " 'O Banco 'e Napule" (The Bank of Napoli), perhaps enough goals to lead them to the title. Napoli owners almost always dream big.

Jeppson improved in the 1953–1954 season, scoring 20 goals, 2nd behind his fellow countryman Gunnar Nordahl of Milan. In the 3rd round, the striker scored four against his old club Atalanta in Napoli's 6–3 drubbing of *La Dea*. Despite their excellent start to the season, Napoli were once again burdened by the weight of expectations, and their results became rather inconsistent. While they finished 5th in Serie A, they came nowhere near as close to the title as they had the year before, ending up 13 points behind Inter, who lifted the trophy once again.

In 1954–1955, Napoli slid down again, to 6th in the final table. The title went to the other side of town, Milan, whose striker Nordahl continued to be unstoppable. Jeppson, on the other hand, managed exactly half of what he'd scored the season before, and no Napoli player appeared on the list of the top 15 Serie A goalscorers that season. The forward did miss much of the season due to injury, but he also fell out with Lauro, something that was never a good idea. The president accused him of being lazy, the same trait that, decades later, would often be attributed to another star Swedish forward, Zlatan Ibrahimović. The problem was that Jeppson, while good in the air and technically gifted, lacked Zlatan's—and Nordahl's, for that matter—ability to penetrate inside the area or to offer up a neat finish. In fact, even though he was able to create stunning individual goals, he was curiously capable of consistently missing easy chances. Still, he helped set the mold for Ibra's type of charisma, his flair and good looks making him a fan favorite.

Given just how much Napoli had paid for their record signing, Lauro wasn't ready to give up on Jeppson—despite how irritated he had become with him—and so decided to bring in the Brazilian attacker Vinício for the 1955–1956 season, to partner with the Swede up top. (It's crucial to note that although Alfonso Cuomo is technically listed as Napoli's president here and in other seasons, he was essentially Lauro's right-hand man, and it was Lauro who remained in charge of the big decisions, even though his title was only "honorary.") However, Jeppson, coming back from a discussion with Inter about the possibility of signing there, was involved in a tragic car wreck that killed his driver and left him injured for the start of the season. He returned in time for the 8–1 smashing of Pro Patria, in which he scored a brace and Vinício a hattrick, which led the fans to believe that this pairing was going to work. Alas, Jeppson really wasn't pleased to remain at Napoli, the pairing never truly felt in sync, and the 30-year-old produced just 7 goals. Vinício, Napoli's top scorer, netted 14, coincidentally the place they finished that season.

– HEROES –

Luís Vinício (1955–60)

Napoli appearances: 154
Goals: 70

Luís Vinícius de Menezes, Italianized to "Luís Vinício" and primarily known as simply "Vinício" (as with most Brazilian football players, he went by one name), visited Rome in 1955 when Botafogo, on tour in Europe, played a friendly there. At the time he'd been with the Brazilian club, his first as a professional, for four years. Achille Lauro, the "honorary chairman" but true power behind Napoli, insisted that other members of his management team, including coach Eraldo Monzeglio, travel to the city to convince Vinício to come to Naples. The problem was, when they arrived, he was in the office of Mario Vaselli, Lazio's vice-president. Lauro picked up the phone and told Vaselli that he needed a favor—he needed Vinício. It's said that Lauro, Naples' mayor at the time, promised Vaselli a lucrative contract to rebuild the Piazza Municipio in Naples.

Vinício's five years at Napoli were filled with quirks and anecdotes such as this one. For example, when the Brazilian was signed, Napoli already had three foreign players, the maximum permitted in the squad in Italy at the time. Therefore, they scrambled to find a solution that allowed them to get Vinício into the squad. The management team decided they needed to find him an Italian parent. A local parish priest found a woman in the city of Caserta by the name of Amarante, Vinício's mother's surname, and concocted a story about how her mother, Vinício's grandmother, had emigrated to Brazil.[25] The Italian authorities, likely wiping away tears of laughter, informed Napoli that no documents existed to connect the Caserta woman to Vinício's family, and so his citizenship could not be changed. Vinício was stuck being called "uncle," "nephew," or "cousin" by locals. Napoli ended up selling Jenő Eugen Vinyei to make room for their new dream attacker.

The behind-the-scenes machinations quickly proved their worth in Vinício's first season, 1955–1956, when he scored 16 goals. A moment most older Napoli fans often talk about when they get excited about the old days occurred in his very first game. Within 40 seconds he sent a thunderous goal into the top corner of the Torino goal from the edge of the penalty area. The game ended 2–2.

In the middle of his first season, Vinício became O *lione* (the lion). He had injured his leg during the match against Milan, but with no substitutes he was forced to continue. He went on to score two goals to defeat the *Rossoneri*, even though it was clear to all that he had difficulties running after the injury.[26] He, along with Hasse Jeppson, was nicknamed "V2," after the German rocket. The media must have been quite disappointed that the nickname never caught on, but putting these two at the front of the attack together never produced results, save for an 8–1 win over Pro Patria.[27] The jealousy between the two was obvious, but it mattered little as Jeppson was sold the next season.

Although Brazilian, Vinício was excluded from his national team due to the fact that he played the majority of his professional career in Italy. Like José Altafini, who would soon become a Napoli star, he could have chosen to play for Italy. Instead, he dismissed the idea of playing international football, and missed out on Brazil's first World Cup win in 1958. Perhaps he did not pursue a national side because Italy failed to qualify for that particular tournament.

Vinício was a player who exuded charisma and who attracted the eye of the spectator wherever he was on the pitch. His height and pace certainly helped, but the way he rolled his socks down and wore no shin guards made it seem that he was invincible. He put his trust in his physicality and athleticism.[28] It appeared to work; blessed with a powerful right foot, Vinício scored 16 goals in his first season at Napoli, 18 goals in his second, and 21 goals in his third and best year. He was the top scorer at Napoli in all three seasons, although it wasn't until he moved to Vicenza that he was crowned the Capocannoniere of Serie A, an award which came in 1965–1966, five years after he left Napoli. The problem was Napoli coach Amedeo Amedei, with whom Vinício had a poor relationship. Amedei wanted Achille Lauro to sell both Vinício and Bruno Pesaloa, whom he felt added little value to the squad.[29] The fans, aware of the dispute, hung a banner that read "*Vendetevi l'anima, ma non Vinício*" ("Sell your soul, but don't sell Vinício").[30] However, in 1958 Napoli had purchased another Brazilian, Emanuele Del Vecchio, from Hellas Verona. Del Vecchio impressed immediately, scoring 15 in all competitions in his first season at the club; Vinício, just 7.

Ironically, Amedei got his wish around the same time he

would leave the club forever; both Vinício and Pesaloa were sold before the 1960–1961 season, while Amedei was first joined on the bench by technical director Renato Cesarini, then Attila Sallustro for the final two matches. It was too late; the three players brought in from Bologna to help compensate for the loss of Vinício simply weren't up to the task, and Napoli ended up finishing 17th, in the relegation zone.

After ending his playing career in 1968 with Vincenza, Vinício immediately turned to coaching, and helped a nearby Campania club, Puteolana, get promoted to Serie D. After bouncing around the lower leagues for a few seasons, he returned to Napoli as head coach in 1973–1976 and again in 1978–1980, bringing exciting, fast-paced football to the club.

One game is indicative of the Napoli fans' growing resentment, evident after just eight rounds of the 1955–1956 season. These fans had been reassured by the return to Serie A, the purchase of Jeppson, that 4th-place finish, the bringing in of the Brazilian to advance the team further. However, they no longer remembered the 8–1 win three weeks prior. On November 6, 1955, Bologna came to town. The Northern side had finished in the top six for the last four seasons, having successfully rebuilt the team that were nearly relegated in 1952, and so it would be no real embarrassment to drop points against the visitors. However, Napoli were winning 3–0 when Bologna came back to make it 3–3. The decisive goal for the *Rossoblù* was a penalty, converted at the last minute, after which the referee, Mario Maurelli, immediately blew the whistle. Already angered by the Bologna players' aggressive approach to the game—Vinício had been fouled so often, he needed to leave the pitch for a few minutes—and the fact that the *Ciucciarelli* had been denied their own penalty, after Jeppson had been fouled inside the area, Napoli *tifosi* invaded the pitch.[31] The supporters chased Maurelli into his dressing room, where he locked the door behind him. One light-hearted moment in this incident, that resulted in 160 injuries, is the replays that show two police officers trying to take down one fan, in what looks remarkably like a Warner Brothers cartoon. Their despicable actions earned the Napoli fans a two-game ban, but Achille Lauro, still the Naples mayor and still the "honorary" president of the *Partenopei*, praised the supporters, saying "Neapolitans are a good and generous people who tend to rise up against injustice and arrogance."[32] Naturally, when the fans returned to the Arturo Collana, they engaged in the same sort of behavior, and earned themselves another home ban.

As the season went on, it seemed incredible that anyone could attend the games in person. Napoli, for all it had undergone in the war, continued to suffer. The

Marshall Plan injected cash into the government coffers, but almost all of it went to the North. As the cities further industrialized, those involved in such industries grew richer, and could buy up football clubs as a bit of a hobby. They used their extra earnings to support the club, as Enrico Befani did for Fiorentina in 1955–1956, when the *Viola* won their first title. Meanwhile, in Naples, although Mayor Lauro held only the title of honorary Napoli president, his money still supported the club, while Alfonso Cuomo supposedly ran its day-to-day operations. However, a 14th-place finish certainly didn't instill much confidence. Lauro continued to pull the reins behind Cuomo for the 1956–1957 season (many sources list Lauro as president, but the official Napoli history states that it was Cuomo), while Amedeo Amadei, who had taken over the bench in January 1956, remained. Rid of Jeppson, who was shipped off to Torino, the club did extraordinarily well in the first part of the season, beginning with an eight-game undefeated run and ending the *andata* in 4th place. They had even beaten Milan, the eventual title winners, 5–3, with all of Napoli's goals coming in the first half. But in the second half of the season, they lost 12 times, something their 2–2 draw with Milan could not compensate for. Despite Vinício's impressive performance throughout the season—18 goals, including the brace that secured the draw against Milan—Napoli finished in 11th.

As was their lot in life, Napoli fans were disappointed by the results that year, yet quickly became optimistic at the start of the 1957–1958 season, when the club brought in Beniamino Di Giacomo, a small yet aggressive forward who radiated charisma, to partner with Vinício up front. Napoli once again started the season quite well, remaining near the top of the table and scoring 23 goals by the end of the 6th round, including a 6–0 drubbing of Verona, in which Di Giacomo scored a hattrick. Then Napoli started to slip a little, but managed to beat Juventus in Turin 3–1, keeping John Charles, the eventual Capocannoniere that season, to just one goal. But again, it was the second half of the season that brought Napoli down from their lofty heights; it began with a 2–1 loss at Genoa and ended with a 1–0 loss to Sampdoria, a 7–0 humiliation at Udinese, a fortunate 1–0 win over Inter, and, finally, a sad 4–0 loss to Lanerossi Vicenza. If (exempting the Inter game) Napoli had managed to win those games, or had overcome Verona or Atalanta, who were both relegated that season, it could have been the year Napoli managed to bring home the scudetto.

But even without the trophy, Napoli had reason to celebrate: they'd beaten Juventus twice that season. When Juventus came to town in April, the crowd noise at the Vomero was tremendous, even before Vinício scored the first goal four minutes in, off a pass by Beniamino Di Giacomo. Unfortunately, Elia Greco quickly tied it up for Juventus with his own goal, although many insist it was Juventus' Gino Stacchini who had the last touch. Then Vinício found Luigi Brugola alone in the penalty area to put Napoli in the lead again in the 24th minute, and the crowd continued to make its presence felt as their side went into the second half 2–1 up. Around an hour in, Stacchini redeemed himself with a left-footed shot from what appeared to be an impossible angle, yet it brought Juventus level. From that point, the visitors threatened Ottavio Bugatti's goal consistently, but Vinício, Napoli's top scorer that season, again came to the rescue with a header in the 77th minute. Then, the horror—an 86th minute equalizer from Antonio Montico, who pounced upon the ball Bugatti had inadvertently pushed his way, and volleyed in the 3–3 goal. The crowd remained in full force, however, and their loyalty was rewarded when Gino Bertucco more or less replicated Juve's equalizer to score the winner two minutes later. When the referee blew the whistle, Napoli fans invaded the pitch. It wasn't a scudetto win, but it was almost as good. Napoli had beaten their rivals twice that season, and that was worth a party, so much so that they carried the players, and even Comandante Lauro, around on their shoulders.

In search of miracles

Dropping down a division is rarely good for a club, no matter where in the world they might be, as they typically miss out on revenue from ticket sales and, once television started to play a role, extra money from games shown on TV. And, of course, many players leave so they can continue playing in the higher division, while at the same time it's more difficult to buy players who would help the side return. But for Napoli, falling into Serie B at the end of the 1960–1961 season may have been the best thing to happen to them since before the war. It was the culmination of an agonizing slide, one that would ultimately result in rewards that were barely short of miraculous.

It's not that Napoli quickly stopped being a threat after 1958, it's more that they were on the downward slope of the roller coaster they'd been on for the past decade. In 1958–1959, they were perfectly respectable—7th place, 13 goals by Emanuele Del Vecchio (whom they'd snatched from Verona when they went down), 50 goals altogether, and the ability to draw against almost any side, often coming from behind to do so (they racked up 16 draws in 34 games). However, over the next few seasons, Napoli were set to experience something more akin to a haunted house, one they wouldn't be able to exit until 1966, and as they wandered, the fans wiped their sweaty palms, jumped in horror, tried to calm their racing hearts, and—to make it all worthwhile—at times clutched one another and wept with joy, so relieved to have bested the bogeymen.

Surprisingly—or perhaps not, if Napoli fans are particularly good at scrubbing records of their club's most difficult seasons—there is little information available about the 1960–1961 season. What we do know is this: Napoli went through a good many changes between 1959–1960, when they'd finished just two points above the drop thanks to a 1–0 win over Roma in the penultimate round with a penalty converted by Del Vecchio, and 1960–1961, when they couldn't manage to claw their way back. During the latter season, Napoli had been hampered by bringing in former Genoa coach Annibale Frossi, who'd led the Rossoblù to 3rd place in Coppa Italia the season before but he oversaw four losses in his first four games before Alfonso Cuomo brought back the adored Amedeo Amadei.

Despite the Napoli hero's inability to keep the ship upright throughout the next 30 games, he did oversee the first match at the new grounds, the Stadio del Sole, on December 6, 1959, where Alessandro Vitali and Vinício led the hosts to a 2–1 victory over eventual Serie A champions Juventus. Napoli had moved into the Stadio Arturo Collana out of necessity after the Stadio Partenopeo was bombed in 1942, and since that time the team had been seeking a more permanent solution, given the capacity at the Collana was approximately half what it had been before. The Naples government had broken ground on the new stadium in April 1952, delivering a brutalist monstrosity in the western suburb of Fuorigrotta, one whose nearly 100,000 capacity could bring in plenty of money for the municipality. Fuorigrotta was one of many suburbs that saw its population boom after World War II, and today it remains the most populated suburb, although its numbers tend to double in size each matchday (the stadium is the 3rd-largest in Italy). Approximately 90,000 attended that first match, celebrating that eternally desired win over their Juve rivals.

Unfortunately, in 1960–1961, neither Del Vecchio nor Amadei could save the Partenopei. Not even San Gennaro, whose blood did liquify in September 1960, could work miracles with this squad.[33] The Brazilian, the team's top scorer the previous season, netted just four in his rare appearances. Technical director Renato Cesarini, who had shaped the mighty River Plate team that dominated Argentina in the early 1940s, joined Amadei on the bench. Renato would later establish a lasting legacy at the club with the creation of the *Zona Cesarini*, the final minutes of the match in which Napoli would manage to score seemingly impossible goals, grabbing points that left their opponents stunned (the tradition was revived in 2009, when Walter Mazzarri was appointed head coach, and Napoli fans kept the faith in last-minute salvation during "Mazzarri Time" throughout his four years at the helm). But at the time, left without starting defenders Rodolfo Beltrandi and Luciano Comaschi, midfielder Bruce Pesaola, and, most disheartenedly, Vinício, the coaching team failed to find ways to get their squad to gel, winning just seven games.

Hoping for a miracle, Cuomo brought club legend Attila Sallustro to the bench for the final two matches, after Napoli lost 4–0 to Juventus at home. But even the returning hero could not help his side conquer Inter at home, and could only stand by helplessly as the season ended with an appropriate whimper, an 86th-minute goal from Vincenzo Di Mauro that was too little, too late as Padova beat Napoli 2–1 at the San Paolo. Picking up points in those last two games *may* have held back the tides of relegation for a few more weeks, as that season three Serie A sides were forced into a rare playoff to decide which team would go down, but the drop in talent and corresponding nose-dive in performance on the pitch seemed to indicate this team was always destined to go down.

Which may have been what saved the club.

As expected, before the 1961–1962 season in Serie B, Napoli did lose important players, such as Ottavio Bugatti, the starting goalkeeper; Emanuele Del Vecchio, the Brazilian midfielder who had scored 27 goals in 68 appearances; and forward Beniamino Di Giacomo, who'd notched 32 goals in his 4 years at Napoli, among others. The staff more or less rebuilt the entire squad to suit Serie B. They purchased well—goalkeeper Walter Pontel came in from Catania for 75 million lire,[34] and defensive midfielder Pierluigi Ronzon turned out to be an excellent buy from Milan. After the crowding of the bench last season, both Amedeo Amadei and Attila Sallustro took their leave, replaced by Fioravante Baldi. The new coach had spent the strange year that calcio had resumed after the war playing as a midfielder for Napoli, and most recently had been sacked by Palermo *in the middle of the game* on the very day they'd secured promotion to Serie A. (In the 21st century, Palermo owner Maurizio Zamparini would become infamous for hiring unknown names to guide the club and firing them—sometimes before they'd even led the team in a single match—then bringing back coaches almost immediately . . . but in his 16 years with the *Rosanero*, even he never sacked a manager at halftime.)

Beyond a 1st-round 1–0 win over Novara, the 1961–1962 side, which had seemed so carefully assembled, were unable to maintain any consistency in the first half of the season. When Napoli then lost 2–1 to Novara in the reverse fixture it looked as though they might be relegated down to Serie C, and so the club wasted no time in trading in Baldi for former forward Bruno Pesaola. New director of sport Roberto Fiore suggested the beloved club legend, despite the fact that he had only served as player-coach for Scafatese, in Serie D (the fourth level of the Italian calcio system), for half that season.[35] Yet somehow, he worked a miracle on this patched-together team of youngsters—the average age was under 24 years old—and the team pulled off a major comeback during the *ritorno*, finishing 2nd after a seven-game unbeaten run and securing automatic promotion back to Serie A.

That incredible comeback facilitated by Pesaola was not the most impressive Napoli feat that season. What astonished calcio fans is that Napoli won the Coppa Italia, a domestic competition involving the top four divisions. The tournament began in 1922 and was meant to be held every 4 years, but the second edition in 1926 was canceled during the Round of 32. In 1935–1936, it was decided the competition would be played annually, but due to the war it was not played in 1942–1943, and did not resume again until 1958. Four years later, Napoli would become the only team outside Serie A to lift the Coppa Italia.

Napoli struggled through the first two rounds of the competition, meeting fellow Serie B side Alessandria first, eventually beating the visitors 6–5 on penalties. Then it was Sampdoria, in Serie A, and another goalless draw that led to penalties. Napoli again triumphed, 7–6. Then, six months later, at the end of April, when Pesaola had been at the club long enough for his squad to have become comfortable with his methods and tactics, Napoli traveled to Torino. The *Granata* had made it to the semifinals of the Coppa the year before, and had finished 6th in Serie A in 1961–1962.

The Partenopei gradually broke the hosts down, with Glauco Gilardoni scoring in the 66th minute, and again in the 78th. A week later, at Roma, it was Gianni Corelli who scored in the 66th, but one goal was enough in that match to send Napoli to the semifinal, to be played at home.

Next, the Ciucciarelli were up against a side a bit less talented than Roma or Torino, but one that would ultimately finish in the top half of the table. Mantova were likely a bit nervous to be playing the giant-killers from Serie B, especially at the San Paolo. Sure enough, Ugo Tomeazzi, who in two seasons with Napoli never scored a league goal, put his side up 1–0 before ten minutes were up. But this time Napoli made a mistake and paid for it; Italo Mazzero converted a penalty to bring the visitors level. There must've been quite a few fingernails being chewed at the Sole that day, wondering if this would be the time they tripped up, the time they'd take the game to penalties but lose, with the heartache even greater now that they had come so close to the final. When Giovanni Fanello scored in the 67th minute, the exhalation of breath likely felt like a strong wind racing down the streets of Fuorigrotta, followed by a roar and the sound of yelling and clapping, drifting into the other quarters of Naples just as those stuck at home or in the bars got word of the goal and added their own screams. And then, as every football fan knows, came more nail-biting, prayers to the calcio gods, shouts to hold on. They did. Napoli, the little Serie B side that could, were headed to the Coppa Italia final.

Unfortunately, the Coppa final is traditionally held at the Stadio Olimpico in Rome, so Napoli were unable to rely upon one of their most powerful weapons: the packed-out San Paolo. The fans with the money and the means to make it to Rome for the final made as much noise as they possibly could, particularly after a SPAL player fouled Napoli forward Amos Mariani.

The pressure fell heavily on Gianni Corelli, who'd been playing with SPAL just the season before. He smashed his free kick over the wall and directly down the center of the goal, a strike that just barely grazed Edo Patregnani's fingertips before crashing into the net. But just four minutes later, Dante Micheli, who had been at Fiorentina the season before when the Viola won the European Cup Winners' Cup, flew up the right wing and threaded his shot between the Napoli defense. With goalkeeper Walter Pontel out of position, Micheli's equalizer seemed far too easy.

Nerves were strung so tightly that they could be plucked like a banjo, as SPAL had at least two chances that could've given them the lead, but screwed them up mightily. Napoli missed a terrific chance to go ahead when a SPAL defender blatantly tripped Ugo Tomeazzi inside the area. However, fans were forced to wait much longer, chewing down the ragged bits of nails that had regrown since the semifinal and their dramatic 2nd-place Serie B finish. Corielli aimed low and to the left, but Patregnani guessed correctly, and this time did manage to get a hand to the ball, tipping it wide. The game stretched on, with the *Spallini* failing to do much of anything, but without Napoli managing to take advantage of SPAL's increasing tiredness.

Finally, far too far into the second half for most supporters, Napoli pushed forward and found a crack in the SPAL defense. A Tomeazzi cross from the left put the ball in front of Pierluigi Ronzon, who stuck out a toe to send a volleyed shot over Patregnani's sprawled form. Immediately the Napoli players who'd rushed forward in the attack embraced, while fans threw flares onto the pitch, helped on by the fact that the stands practically touched the field. After squad captain Mariani hoisted the trophy above his head, he turned and lifted it to the fans, who had been there as Napoli faded away, and deserved to witness their moment of triumph. In the stadium, the Partenopei faithful celebrated wildly, setting off fireworks and other pyrotechnics. The subsequent party in the streets of Rome lasted for hours; someone even stumbled upon a live donkey they invited to the festivities, the Neapolitan sense of humor shining through that reference to their first horrible season and the "Ciucciarelli" nickname.

Then came the 1962–1963 Serie A season, and Napoli plummeted back down to earth. Bruno Pesaola remained on the bench, where he was joined by former Napoli defender Eraldo Monzeglio, who'd managed Napoli from 1949–1956, but now assumed the role of technical director. Few changes were made from the former Serie B side, now in Serie A. Perhaps it was felt that the Coppa Italia win suggested they already had a squad who could compete in the higher division. Three new faces were added: Humberto Rosa, an experienced midfielder who'd been at Juventus the year before; Cané, a 22-year-old Brazilian winger; and Antonio Juliano, who was promoted from the youth team and even younger than the Brazilian, at just 19.

Before league play began, Napoli got their first taste of European play (and perhaps their first real inkling that this season would be anything but easy). In the 1st round of the Cup Winners' Cup, the Neapolitans faced Welsh side Bangor City. They lost 2–0 in Wales, but managed to turn it around in Italy with a 3–1 win. However, that put the sides level at 3–3; the means of resolving this was a playoff match. This time Napoli won 2–1 and moved on to the round of 16. The playoff matches in the Cup Winners' Cup were to become something of a trend that season, and the tiredness it created (and the travel involved) may have contributed to their performance in the league.

They certainly weren't worn out by Coppa Italia matches. This tournament started even earlier than the Cup Winners' Cup, and Napoli, last season's champions, were turfed out by Messina in the 1st round, on September 1. Messina, a lowly Serie B team, beat Napoli 2–1. And they likely went into European play with a cloud of foreboding hovering over their heads.

Whether it was their frustration at being knocked out of the Coppa, their irritation with needing to play multiple games against Bangor City, the burden of underperforming new squad members, or simply bad luck, Napoli had a terrible time of it in the first three games. In the first match of the season, they lost at Roma 3–1. The next week Milan came to visit. It looked like they'd be able to get a point; after Gianni Rivera scored the opener, Juan Carlos Tacchi responded after five minutes, and the score stood at 1–1 until Bruno Mora struck in the 74th, and Paolo Barison in the 76th. Then came another from Rivera in the 81st, and finally a goal from José Altafini in the 88th. Milan walked away from Naples the 5–1 winners. The next week, almost exactly the same thing happened at SPAL. Gianni Bui scored for the home side in the 21st minute, and Napoli responded 2 minutes later with a goal from Gianni Corelli. Then came the SPAL breakthrough in the 59th, after which they scored 4 more for another 5–1 victory against the Partenopei.

In Europe, Napoli were able to keep the goals down in the Round of 16. Against Újpest of Hungary, the first two games finished 1–1. In the playoff game, Napoli shook off some of the cobwebs and came away with a 3–1 victory, taking them to the quarterfinals. There they met OFK Belgrade, a team in the middle of their golden era. The *Romantičari* beat Napoli 2–0 at home, then Napoli beat them 3–1 at home for an aggregate score of 3–3. This time, though, Napoli couldn't finish their opponents off in the third game. OFK won 3–1 in Belgrade and Napoli were out of Europe.

Back in the league, after the aggregate 13–3 score over the first three games, Napoli played excellently at home throughout the first part of the season, dropping points only in a draw with Torino. However, just to even things out, they only earned a point on the road, in a draw at Venezia. Napoli finally won an away game on January 27, 1–0 over Milan. But then they started losing at home, too, and generally performing quite unremarkably, such as a 5–1 home loss to Inter, followed by a 5–1 loss at Fiorentina. Napoli could have used more firepower—Corelli had the most goals, with 9—but their main problem was at the back, where they leaked goals; for example, in 3 seasons with Napoli, Walter Pontel made 82 appearances and let in 101 goals. Given that Pontel had spent most of his time on the bench over the past two years, at Catania and then at Inter, it's likely Napoli didn't make a great choice when investing in a keeper who couldn't cut it in Serie A.

In the final game of the season, the Partenopei, a side that lacked the skills to both net goals and keep them out, were away at Atalanta, that season's Coppa Italia winners. Before 20 minutes were up, Napoli knew they needed to win—in Genoa, the Rossoblù striker Giancarlo Galli had struck in the 18th minute, and if the side held on for all three points, Genoa would finish above Napoli. By the 82nd minute, Napoli's hopes were dead in the water as Flemming Nielsen celebrated scoring La Dea's second goal. Although Corelli knocked one in three minutes later, no one could find a second, much less the third goal that would have kept them up. When the whistle blew signaling Napoli's loss, the visiting fans were so upset that they took to the pitch, chased off the referee, and dismantled one of the goals. The *carabinieri* had to be called to put a stop to the madness. It was a humiliating end for last season's Coppa Italia winners, the comeback kids who had managed to get themselves promoted to Serie A for one haunting season.

[1] Archambault, F. (2018). "Italy." In J-M. De Waele, R. Spaaij, S. Gibril, and E. Gloriozova (Eds.), *The Palgrave International Handbook of Football and Politics* (pp. 105–124). Palgrave Macmillan.

[2] Hildebrand, G. (1955). The Postwar Italian economy: Achievements, problems, and prospects. *World Politics, 8*(1), 46–70. doi:10.2307/2009097

[3] Giorcelli, M., & Bianchi, N. (2021, December). Reconstruction aid, public infrastructure, and economic development: The case of the Marshall Plan in Italy. *National Bureau of Economic Research*. Working paper 29537, 25, n. 37. https://www.nber.org/papers/w29537

[4] Ginsborg, P. (2003). *A history of contemporary Italy*. Palgrave Macmillan.

[5] Encyclopedia Britannica. (last updated 2020, January 1). Italy. In *Encyclopedia Britannica*. Retrieved December 4, 2020 from https://www.britannica.com/place/Italy/World-War-II

[6] Doidge, M. (2015). *Football Italia: Italian football in an age of globalization*. Bloomsbury. http://dx.doi.org/10.5040/9781472519221.0012

[7] La Galla, A. (2004). *Vomero. Storia e storie*. Guida.

[8] Archambault, F. (2006). Il calcio e l'oratorio: Football, Catholic movement and politics in Italian post-war society, 1944–1960. *Historical Social Research / Historische Sozialforschung, 31*(1), 134–150. http://www.jstor.org/stable/20762106

[9] Iaretti, M. (2014, December 27). Eraldo Monzeglio, calciatore in camicia nera. *Il Torinese*. https://iltorinese.it/2014/12/27/eraldo-monzeglio-calciatore-in-camicia-nera/

[10] Materazzo, G., & Sarnataro, D. (2014). *I campioni che hanno fatto grande il Napoli*. Newton Compton

[11] Carratelli, M. (2012). Una crisi a volo D'Angelo. *La Repubblica*. https://ricerca.repubblica.it/repubblica/archivio/repubblica/2012/10/08/una-crisi-volo-dangelo.html

[12] Tramontano.

[13] Acampora.

[14] Le scarpe spaiate di Achille Lauro. (2013, February 4). *Nuova Agenzia Radicale*. http://www.agenziaradicale.com/index.php/editoriale-e-commenti/1183-le-scarpe-spaiate-di-achille-lauro

[15] Matthews, J. (2004, October). Achille Lauro (1887–1982). *Naples Life Death & Miracles*. http://www.naplesldm.com/lauro.php

[16] Martin, S. (2012). Sport Italia: 150 years of disunited Italy? *Bulletin of Italian Politics, 4*(1), 49–62.

[17] Martin (2012), 52.

[18] Della Ragione, A. (2003). *Achille Lauro Superstar: La Vita, L'Impero, La Leggenda*. Mario Guida Editore.

[19] Archambault (2018).

[20] Signorelli, A. (2002). *Cultura popolare a Napoli e in Campania nel Novecento*. Guida Editori.

[21] Della Ragione.

[22] Acampora.

[23] Sengupta, S. (2017, December 2). The bank of Napoli: Hasse Jeppson. *The Gentleman Ultra*. https://gentlemanultra.com/2017/12/02/the-bank-of-napoli-hasse-jeppson/

[24] Id.

[25] Carratelli, M. (2009, January 18). Storia del centravanti che diventò un idolo e non se andò mai più. *La Repubblica*. https://ricerca.repubblica.it/repubblica/archivio/repubblica/2009/01/18/storia-del-centravanti-che-divento-un-idolo.html

[26] Materazzo & Sarnataro.

[27] Carratelli, M. (2017, February 2). Lo striscione d'amore al Vomero: la cessione indigesta di Vinicio; *Il Mattino*, https://www.ilmattino.it/sport/sscnapoli/napoli_bologna_lo_striscione_d_amore_al_vomero_la_cessione_indigesta_di_vinicio-2233690.html

[28] Notarnicola, V. (1958, January). Vinicio, il leone triste. *Corriere d'informazione*.

[29] Carratelli (2017, February 2).

[30] Id.

[31] Guarino, C. (1955, November 7). Tumultuoso epilogo dell'incontro Napoli–Bologna.*Stampa Sera*, 1.

[32] Foot, J. (2006). *Winning at all costs: A scandalous history of Italian soccer*. Nation Books, 65.

[33] Il miracolo di S. Gennaro si è rinnovato stamane. (1960, September 19–20). *Corriere d'informazione*, 14.

[34] E' morto Pontel Parò per l' Inter. (2003, January 7). *La Gazzetta dello Sport.* http://archiviostorico.
gazzetta.it/2003/gennaio/07/morto_Pontel_Paro_per_Inter_ga_0_0301071657.shtml
[35] Fiore, R. (2012). *Chi sono stato?* Roberto Fiore.

CHAPTER 5

Rollercoaster of Love

5

Italy's economic miracle began to flicker and fade in 1963, as the lira declined in value and foreign investment dropped dramatically.[1] Naples' postwar recovery also faltered. According to *Encyclopedia Britannica*:[2]

> That the city survived the postwar period without complete economic and social collapse can be attributed almost exclusively to the vitality and philosophy of its populace and to the Neapolitan ability to combine strong passions with a resilient endurance.

Napoli followed suit, dropping down into Serie B. While the club righted itself and spent the next dozen years comfortably ensconced in the mid-to-high table, the strikes, student protests, and social upheaval that rocked Italy in the late 1960s appeared to affect Napoli as well, as the club moved from president to president and coach to coach. On the other hand, many players elected to stay at Napoli season after season, absorbing the adoration of the fans and the belief that yes, *this* could really be the year they lifted the club's first *scudetto*. Further evidence of support was the formation of the first ultra groups, waving banners and singing in synchronicity. Despite the supporters' superstitions, they were right there behind the squad, trusting that soon it would be their year.

Moving on up

First, however, Napoli needed to climb their way back into the first division. Amidst the financial difficulties caused by the side's second relegation in three years, Napoli were unable to make any standout buys that might again boost them back to Serie A straight away, preferring to develop youth players such as Neapolitan Antonio Juliano, who had been given just three chances to show off his skills in Serie A, but had demonstrated his technical abilities and willingness to work hard on the training ground. Napoli did manage to sell off a few players: Luigi Bodi, who'd exerted little influence on the squad, primarily due to injuries, moved to Serie C side Novara; Giovanni Fanello, who had scored three goals in the European Cup Winners' Cup, and who would return the next season, went to Catania in Serie A; and Ugo Tomeazzi, who had been banned for a month the season before due to accusations of doping, also managed to join a Serie A club, Mantova, where he would become a legend.

Along with the shift toward youth, the Napoli presidency changed hands again, to Luigi Scuotto, one of Achille Lauro's most trusted advisors, and one who had held the position alongside Pasquale Russo and Vincenzo Savarese in 1945–1946. Bruno Pesaola left, replaced on the bench by Roberto Lerici, who was then sacked after 25 rounds as it became clear that Napoli had little ability to fight for a position toward the top of the table. His second-in-command, Giovanni Molino, took his place, but managed just 4 wins and 6 draws in 14 games, which resulted in an 8th-place finish.

In 1964, two factions fought for control of Napoli. One backed current president Luigi Scuotto, a businessman who attracted support from the Monarchist party, despite Lauro controlling it. Surprisingly, this was the group favored to take over the club. Meanwhile, Lauro wanted to hold the honorary title of Napoli president, but didn't want to perform the day-to-day work required. Yet despite his seeming inability to fulfill the role of president, he emerged victorious. The name of the historical Naples club was changed from Associazione Calcio Napoli to Società Sportiva Calcio Napoli (SSC Napoli), an S.p.A.[3] Now, interested parties could buy stock in the club. This club carried on the Napoli name. Lauro retained 40% ownership, while Roberto Fiore held 21%. The rest was shared between Achille's son Gioacchino and Lauro's friends Antonio Corcione and Mario Tardugno.[4] Meanwhile, Scuotto joined up with Giovanni Pronto, who created a new professional club from local side Cirio, calling it Internapoli, in homage to the "InterNaples" team that had existed before the war, as well as being a name they knew would annoy Lauro.[5]

– HEROES –
Antonio Juliano (1962–1978)

Napoli appearances: 505
Goals: 38

Antonio Juliano is Naples born and bred, and he's proud of it, and the fans are proud of it. Not only did the creative midfielder impress with his unsurpassable vision, superb long-range passes, and obvious technical skills, he became an unparalleled leader at Napoli during his 17 seasons at the club. If that weren't enough, he then returned as sporting director, and brought Diego Armando Maradona to Naples.

Juliano, from the coastal suburb of San Giovanni a Teduccio, made his senior Napoli debut in 1962, under Bruno Pesaola. Just 3 years later, at age 23, "Totonno" was awarded the captain's armband, a demonstrable token indicating the club's faith in his inherent leadership abilities. Juliano not only led his side on the pitch, but stood up for their rights in the presence of the board. He not only ensured his teammates were treated fairly, but also that the squad doctors, masseuses, and other workers were given bonuses alongside the players.[6]

Totonno's playing style was much like that of one of the most famous Italians ever to grace the world soccer scene, Andrea Pirlo. A fellow *regista*, the playmaker directed the flow of the game and carved open opposition defenses with his passes, providing other Napoli players with perfect goalscoring opportunities. His elegance drew the eye of even opposing fans. Juliano was the first Neapolitan player called up to the Italian men's national team, in 1966. But although he was present for three World Cups and part of the 1968 Euro-winning squad, he recorded just 18 caps for the national side—almost certainly because those selecting the starters often overlooked players from the South.

Inter made an offer for Juliano at one point, trying to convince him to move by noting he'd play far more for the national team. But the dedicated Neapolitan refused, unwilling to leave his boyhood club.[7] Conversely, his own club president, Corrado Ferlaino, tried to convince him to leave for Milan in 1969. Like most players, Juliano had his issues with the temperamental businessman, but no matter how much Ferlaino wanted the 800 million lire on offer from the *Rossoneri*, he knew that he could not sell Totonno without angering the fans, and so Juliano would have to come out and say he wanted the move— but he absolutely refused.[8]

Despite their tempestuous relationship, Ferlaino brought Juliano back to the club as sporting director, after he finished his playing career with a season at Bologna (saving them from relegation in the last game of the 1978–1979 season). Their relationship remained complicated at best, but there was no way Ferlaino would have let him get away. Juliano first brought in Ruud Krol, delighting the fans and taking some of the heat off Ferlaino. Then, in 1984, after several years of pursuing the Argentinian, Antonio Juliano convinced Maradona to come to Naples. For that alone, he will always be a Neapolitan hero, but given his hometown roots, his determination to stay with Napoli, and his graceful approach to the game, he should be placed on a pedestal almost as high as Diego's.

As SSC Napoli, the club entered their second straight season in Serie B, with Fiore taking over as the official president, and Lauro working behind the scenes. Fiore recalled Pesaola to the bench. Goalkeeper Walter Pontel, purchased for 75 million lire in 1961 after conceding eight goals in seven appearances for Catania, packed his bags for Palermo. Meanwhile, Glauco Gilardoni, an attacker who'd scored 12 goals in the 3 seasons he was at Napoli, left for Genoa in Serie A, where he was even less successful, while Rosario Rivellino, who'd only had 29 appearances in 3 seasons with the club, went to Parma in Serie B, where he got more playing time. Fiore brought in eight new players, including defenders Dino Panzanato and Mario Zurlini, who both ended up having long careers in blue; Pietro Adorni, a right-back who started nearly every Serie B game; Claudio Bandoni, who became the new keeper; and Gastone Bean, who made just 16 appearances but scored 7 goals, the 2nd most that season. The club had put together a strong enough squad to take on Serie B and very nearly win the league; they finished one point behind Brescia, but it was enough to send them to Serie A for the 1965–1966 season. Perhaps best of all, it seemed Napoli had finally found a strong backline, finishing with the fewest number of goals conceded in the league.

Heady heights

So back to Serie A the club went, with a much better squad than two seasons before. Napoli, still under Fiore and coached by Pesaola, made two extremely intelligent signings in the summer of 1965: two *oriundi*, José Altafini from Milan and Omar Sívori from Juventus. "The Fantastic Duo" quickly elevated their new side, and Napoli were in the title race from the beginning, unbeaten in their first five games—including a draw with Juventus. They beat Milan, Juventus, and Inter away, all with goals from their fantastic duo, most often edging their opponents by just one goal.

The *Nerazzurri* had won the European Cup the year before, and it was expected that *Grande Inter* would easily win the title again this season. Instead, that last game against Napoli very nearly was a title decider, as Inter had faltered at the end of April, and both Napoli and Bologna could have caught up. But Bologna could only draw with Juventus, and then lost control of the game against Lanerossi Vicenza, handing Vinício two penalties and letting him slide in another goal. The visitors left with a 3–1 win over Bologna, who finished four points out in 2nd. Meanwhile, Napoli stumbled a few times in the return fixtures. At the beginning of February, a goal from Altafini proved enough to allow the San Paolo fans to revel in a victory over Juventus, and they similarly triumphed over Milan at the end of the month, although it was Omar Sívori scoring a late goal that secured the points. However, they lost to Atalanta (12th) in March, to Lanerossi Vicenza (6th) thanks to their formerly adored Vinício in April, and to Sampdoria (16th, relegated) in May. That last loss, in the penultimate match, ensured Inter's trip to the San Paolo was rendered meaningless; they had secured the title. Napoli had a blast, though, in the second half. Altafini scored twice, bookending Juliano, and even Dino Panzanato joined in; unfortunately, the defender knocked his into the wrong net, bringing Inter's goal tally up to 70 for the season.

That 3rd-place finish, just five points behind that incredible Inter team, gave the squad confidence, especially as they'd just returned to Serie A.

As goalkeeper Claudio Bandoni said in 2013, *"A Napoli ci si divertiva così tanto che, al momento opportuno, ci è mancata la giusta cattiveria. Stavamo troppo bene. Quel Napoli divertiva e faceva divertire il pubblico"*[9] (In Naples we had so much fun that, at the right moment, we lacked the right wickedness. We were doing too well. That Naples amused and entertained the public). Such sentiments were echoed by many fans evaluating more recent Napoli squads.

That Napoli continued to be fun. In early summer 1966, they participated in the 6th edition of the *Coppa delle Alpi*, along with three other Italian teams and four Swiss teams. Though the tournament began as a friendly competition among Italian sides, they generously added clubs from Switzerland, seeing as the majority of the Alps are located in the country. The clubs from both countries were invited by the organizers without regard for geography or table position (for example, Catania had been relegated from Serie A just a few weeks before). In the five years previous, only Italian clubs had won. Napoli, the 1966 champions, playing under young manager Bruno Pesaola, beat each team they met, scoring 15 goals in 4 games. Juventus, meanwhile, could have done the same, had the match not been abandoned after their players walked off in protest when their opponents, Lausanne, were awarded a penalty kick; as a result, the Swiss side were gifted a 3–0 victory, while three Juventus players were temporarily banned from European competition.[10] Is it any wonder that the next year, the organizers introduced German teams into the competition (Schalke won) and, in 1972, abandoned all pretense of truly sharing the Alps and turned the competition into one exclusive to French and Swiss teams?

The thrill of coming in 3rd the previous season, the adrenaline produced from beating Juventus twice, and the excitement of lifting the Cup of the Alps gave Napoli a boost when entering the 1966–1967 season. They also had a new European competition to conquer, the *Coppa della Fiere* (Inter-Cities Fairs Cup). Originally, friendlies had been played by teams visiting other cities for international trade fairs, and the competition was open only to those hosting fairs, regardless of their position. But after 1968, it was referred to as the "Runners-Up" competition, as teams now qualified by virtue of their league position. Although not organized by UEFA originally, it is considered the predecessor to the UEFA Cup. Not yet designated the "Runners-Up" tournament, the 2nd-, 3rd-, and 5th-placed teams from the 1965–1966 Serie A season were still the ones invited to enter the Inter-Cities Fair Cup (Fiorentina, in 4th, were already in the Cup Winners' Cup); the tournament tended to select squads who'd placed high in their leagues. Napoli easily conquered their opponents in the first two rounds, but then fell to England's Burnley in February. While they lost to Milan ten days later, the European competition was not what threw a wrench in Napoli's title hopes but rather April (which would become a running theme in Napoli history): they fell 2–0 to Juventus, the eventual champions, managed to beat Mantova 1–0, and then closed the month drawing with Brescia, Cagliari, and Foggia and Incedit, scoring just three goals throughout. Except for that blip, Napoli remained entertaining, finishing 4th (and being selected for the Fairs Cup again) with 16 goals from Altafini.

The 1967–1968 season was likely the most exciting one most Napoli fans experienced—at least, those who were no longer around 20 years later. Pesaola, now their beloved *Petisso*, was still on the bench, and was credited with bringing in the players integral to creating a competitive Napoli side. Paolo Barison, a left winger from Roma; Ivano Bosdaves, an attacker from SPAL who demanded in return they send three players (including Gastone Bean) in a futile attempt to keep SPAL from relegation; Luigi Pogliana, a left-back from Novara who started the very first game for his new club and remained there for a decade, and—perhaps the most important acquisition for both that season and Napoli's future—goalkeeper Dino Zoff, who was good at Napoli for five seasons, but is most remembered for his job at Juventus, where he won six *scudetti* and the UEFA Cup.

As much as Napoli fans have claimed, and continue to claim, that the league is favored toward Juventus, Napoli converted four penalties that season, and a number of players they faced ended up scoring an *autogol*, their own goals likely coming from sustained pressure at the back. Napoli were 1st in the table after nine matches, beating Roma 2–0, Inter 2–1, and holding Juventus to a 1–1 draw, which could very well have been a win had Pogliana not scored an own goal. And Napoli could have turned a loss into a win had they prevented the Rossoneri attacker Pierino Prati from scoring a last-minute goal in the second half of the season. Yet Milan finished with 46 points to Napoli's 37, so the changes in those results would not have boosted the *Partenopei* to their first title win.

The unraveling

The previous season had been their best ever, and it's quite possible Napoli just did not know how to sustain such success, having never really been there before. In 1968–1969, things began to unravel, at Napoli, in Italy, across the world. Protests, led primarily by students, railed against the Vietnam War and the Cuban Missile Crisis, sprouted up all over the globe. In the United States, the assassination of Martin Luther King, Jr. kicked off the largest wave of violence experienced thus far during the civil rights movement. Back in Italy, the ideological differences shaped by World War II, the economic "miracle" that left many behind, and the migrations between the culturally different South and North brought to the surface issues that had been simmering for decades.

And in Naples, the diversion that was the city's only club had simply been spending too much money. Gioacchino Lauro, Achille's son, had inherited his shipping company, after the heir, Achille Jr., was imprisoned for insurance scams. Gioacchino's father then removed him from power after he lost 7 billion lire.[11] Much of the money was lavished on the players and club staff, whether gifts of fancy cars or upping their wages. Consequently, they adored him. He also invested in new players despite their signings resulting in bidding wars, such as when Gioacchino did battle with both Milan clubs for goalkeeper Dino Zoff.[12] After hemorrhaging money, Gioacchino decided to sell SSC Napoli, which set off a wave of chaos throughout the club. They were bought by Antonio Corcione, another owner the players loved, having been a member of the presidential team in 1967–1968, and prior to that, serving as a board member for at least five years—and clearly never intending to stop the money train Gioacchino was conducting. Corcione and Sívori became particularly close in his short time at the helm of the club.[13] However, Corcione soon died after an extended period of illness, paving the way for Corrado Ferlaino, Napoli's longest-running owner.

As would become clear later, it helps to be a bit of a character if one wants to own SSC Napoli, with all its own quirks, absurdities and—yes—complications. Ferlaino graduated from the University of Bari with a civil engineering degree, but before that he'd already become famous . . . for auto racing. His debut came in 1957, when he came first in his class at the *Sorrento-Sant'Agata dei Due Golfi* race. He also rode a motorcycle, competed in go-kart racing, took part in boat races, and won the *Targa Florio* (an endurance car race on Sicily), in 1964—three years after he'd responsibly formed his own construction company. In that same year, after playing amateur football for years, he was disqualified for life for "beating a referee."[14] By 1967, with his real estate development business flourishing, he became a partner in Napoli Calcio. Two years later, his friendship with Achille Lauro helped him take over as president; after buying Corcione's shares from his widow (just a month after Lauro's death) and then those of Roberto Fiore, with whom Lauro was on the outs, he became the majority shareholder on January 19, 1969.[15] Just 38 years old, he obviously had a passion for sport, a wide creative streak, and plenty of ideas for new business models.[16] Ferlaino served the club for 31 years, although he did step down as president more than once during that time.

Ferlaino's means of obtaining new players was also quite . . . distinct. Poetic, some might say, likening it to a bee drunk on nectar from a flower, who quickly sobers up, takes a swig of whisky, and leaves those he was trying to persuade in favor of discussing transfers with another club. He then jumps quickly back and forth among the teams whose players he is trying to woo, playing them off one another. Although this appears to be how he negotiated with Monza and Inter the summer before he had officially become president, the club brought in no one of real note at the time. Anyone in Naples with fortune-telling powers, however, knew to be patient—the best transfer in Napoli history would be conducted through Ferlaino.

In Ferlaino's first season as president, Bruno Pesaola left the club for personal reasons; specifically, the abuse he and his family had endured due to his role. He received phone calls filled with insults and demands for him to leave, and mail containing personal threats, causing him to flee Naples and declare that he'd never return[17] (which turned out to be untrue). His departure created a huge gap to fill on the Napoli bench. In Petisso's second stint at the club, he had stayed for four seasons, taking Napoli to heights they'd never experienced before. First, he helped them win promotion from Serie B to Serie A in 1964–1965. With Napoli now in the top division, he was able to woo José Altafini from Milan and Omar Sívori from Juventus. Those spectacular additions saw Napoli reach 3rd place in their first season back, and with the 4th-place finish in 1966–1967, they were back in European competition for two seasons. What's more, Petisso helped the team capture their first European trophy, the Coppa della Alpi, in 1966. The year before he left, he brought in the legendary goalkeeper Dino Zoff, and Napoli had their best season ever, finishing 2nd in the table in 1967–1968. Like others so beloved by the Napoli faithful, Pesaola adopted the city as his home, saying he considered himself *un napoletano nato all'estero* (a Neapolitan born abroad).[18] Still, this was likely little consolation for Napoli fans when he won the scudetto with Fiorentina the year after he left.

It was always going to be difficult to replace such a special *Mister*, and Ferlaino struggled. He first appointed Giuseppe Chiappella, who had done well at Fiorentina for the past few seasons.[19] Alongside him Ferlaino brought in Carlo Parola, who had been at Juventus with Sívori. The two had clashed from their very first training session at the Old Lady, back when they were both players.[20] Sívori's infamous hot-headed temperament had left him battling Pesaola constantly, but Ferlaino was hopeful he and the new assistant coach would be able to put the past behind them. Yet at Juventus, Parola had been unable to sufficiently motivate his star. The Old Lady finished 12th that season, 1961–1962, and Parola's reputation had taken such a hit that he moved on to Prato in Serie C, who he managed to bring up to Serie B. But rather than moving on up with Prato, Parola then went to Serie D side Chieri, who finished 12th in Group B in 1963–1964. Serie B side Livorno then took a chance on him, finishing 7th in 1964–1965. The next season, after a steep decline, Livorno sacked Parola, then took him back when his replacement failed to do any better. Ultimately, the team finished 16th, just above the drop. Yet Ferlaino appointed him to assist a side that had just finished 2nd in the top division, believing that Parola's calm demeanor would reduce his star's anxiety and help him perform better. It would prove to be a risky decision. The two continued their feud.

When Napoli beat Juventus 2–1 on December 1, 1968, the fallout was catastrophic. From the start, the home fans were up for a fight, throwing items such as a bottle and vegetables at the linesman, who managed to dodge the projectiles. Napoli were up 2–1 by the 37th minute thanks to a pair of goals by Vincenzo Montefusco. Then, just before halftime, Juventus midfielder Erminio Favalli pushed Omar Sívori, possibly the worst thing that could've happened in the tinderbox that was the San Paolo. The referee awarded Sívori the foul, but the Napoli man elected to deliver a sharp strike to Favalli's ankle instead. The kick earned Sívori a sending off, further igniting his temper. After snatching the ball and throwing it at Favalli, he then escaped the Napoli teammates holding him back and returned to kick Favalli once more. Chiappella ran to defend his player, screaming at and pushing the referee. The fight descended into a melee between both sides, with more players ejected. Chiappella himself was dismissed, but he pretended not to understand, remaining on the sidelines for a couple minutes, until the whistle blew for halftime. After the game ended with a Napoli win, Sívori stood in the tunnel, hurling insults at players from his former club. In the end, Sívori was suspended for six matches (though he was still allowed to train with the club) and Chiappella earned a two-month suspension—during which time Parola sat on the bench on his own, both during practice and at matches.[21]

The team then took a turn for the worse, clearly missing both their boss and Sívori, with just one win in six games. After scoring zero goals in four matches, a loss to Verona at the beginning of the *ritorno* proved to be too much, and Chiappella and Parola were sacked, just after the former had returned. Former Napoli player Egidio Di Costanzo took their place, but despite managing four wins and three draws in nine games, Ferlaino gave him the boot and reinstated Chiappella. It was likely the back-to-back losses against Juventus and Fiorentina sealed Di Costanzo's fate. Clearly, the team was missing Sívori, who'd traveled to Argentina at the start of his suspension and simply never returned. With all the turmoil they'd experienced in the 1968–1969 season, it's a wonder Napoli managed to finish 7th, but such a low position did disappoint their new owner. Like owners before and after, Ferlaino proved to be erratic, to say the least. Ferlaino floated the idea of changing the Napoli crest to that of the House of Bourbon. He often threatened to sell the club when things weren't going his way. He fought with a few players, including Antonio Juliano, and owed quite a few of them money. When Ferlaino informed the players they'd be going into *ritiro* before the Cagliari game in May 1969, the players—angry they were still missing wages from March and April—fought back, announcing they'd go on strike.[22] In the summer of 1969, the papers insisted he was shopping for a new coach and had met with Bruno Pesaola in February; however, he remained with Fiorentina and Chiappella stayed on the bench for Napoli.

And so it was time to approach the new decade, and Napoli fans were eager to see how the club would greet it. Unfortunately, it was with little change from the season before. Sívori was gone, of course, but the fans had had half a season to deal with the loss. Chiappella stayed on the bench, and Ferlaino retained ownership of the club. They sold Cané, the 30-year-old veteran attacker, to Bari, and bought 34-year-old Kurt Hamrin from Milan. The Swedish winger was the highest scorer in Serie A at the time, but hadn't played for the Rossoneri that season due to his being one of three foreign players in the squad, when only two were allowed. His age and Milan's willingness to sell made him a cheap December pickup; it helped that he was friends with Chiappella due to their time playing together at Fiorentina, and the head coach believed he could help the squad in a psychological sense, due to his experience in what was a relatively young side.[23] That said, he was relegated to the bench, where he scored just one goal in five appearances for Napoli that half-season. The younger Pierpaolo Manservisi, a winger, managed 3 goals in 22 appearances before being sold to Lazio, where he played as a defensive midfielder, to better success. Napoli also sold Claudio Sala to Torino for what seemed an outrageous price at the time, but the central midfielder became one of the biggest talents of the 1970s . . . for the *Granata*.

Given the financial troubles the club had faced over the past few years, it makes sense that Napoli's tactic seemed to be buy low and young, as Ferlaino had promised to keep the best players, and so he needed to reduce the wage bill in some way. Chiappella searched Italian stadiums for youngsters who could bring something to the club, or kept watch on loanees like Giovanni Improta, who had few appearances with SPAL, but impressed the coach enough to ask for him to be brought back to Napoli—in the end, Ferlaino still had final approval on all squad changes.[24] The Partenopei also tended to be as conservative on the pitch as possible, which was unfortunate, as the owner had also urged supporters to buy season tickets to support their club and the efforts to buy the best players. Unfortunately for the fans who showed up at the San Paolo that season, Napoli finished higher in the table than multiple teams that had scored more goals, and they recorded the fourth-most draws in the league—would it have killed them to be a bit more exciting on the pitch? Nine of their eleven draws were goalless, and Napoli failed to score in sixteen league games, making many of the matches utter slogs to sit through. They may have finished 6th, and they may have beaten Juventus at home 1–0, but it was clear they needed a striker who could consistently put the ball in the back of the net, and perhaps a coach that was willing to take a few more risks.

For today's Serie A fans, the season's winner likely comes as a surprise. Cagliari, from the island of Sardinia, had entered Serie A for the first time in 1964, and so their rise to their first (and only) scudetto in 1970 was something of a shock. But not only did they have the best defense in all of Europe, with just 11 goals conceded, they also could boast of the league's best goalscorer, with 21. The Sardinians had bought Luigi Riva in 1963, after their promotion to Serie B; Gigi, as he was nicknamed, would also go on to become the Italian men's national team's all-time leading goalscorer. A sense of stability may also have had something to do with Cagliari's ability to reach the top. The season before, when they'd finished 2nd, it had been a three-horse race with Fiorentina and Milan. This time around, Cagliari edged out Inter and Juventus. Protests in 1968 had left the North on edge; by fall 1969, the *Autunno caldo* (Hot Autumn) had settled over the North, where mass strikes by factory workers disrupted life in the industrial cities. Police officer Antonio Annarumma was fatally injured by an iron tube at the protests organized by the Communist Party in Milan in November 1969. Just a few weeks later, in the same city, far-right terrorists struck at the Piazza Fontana, killing 17 and wounding 88. This is generally considered the beginning of the *Anni di piombo* (Years of Lead), which would last almost 20 years and claim hundreds of lives. There is no denying that Cagliari's play demonstrated they deserved their title win, but Sardinia, isolated from the rest of the peninsula, was less affected by the turmoil in the North and how it leached into the stands and thus into the clubs.

Italy's social unrest led to the politicization of fan groups and the emergence of ultras as the "politics of the piazza" began to make their way into the stadiums.[25] The first ultra groups began to form in the late 1960s at Milan and Fiorentina. "Ultras" are fervent football fans who demonstrate their support through highly ritualistic means, characterized by an extensive display of flags and banners, the setting off of fireworks and flares, and the chanting of songs organized by a group leader or leaders.[26] While many other countries have adopted the "ultra" label for their fan groups, Di Biasi explains the unique displays seen in Italian stadiums. "Italians refer to the staging of a match as a *spettacolo* [spectacle]. No English word adequately conveys the spettacolo, but it involves creation of a special atmosphere characterized by a combination of color, vibrance and noise." These spettacolos often incorporate banners and flags that symbolize the ultras' political allegiances. A popular misconception is that Italian ultra groups are almost always politically aligned with far-right ideologies. While these ultras receive the most media attention, such groups can fall anywhere on the spectrum, including remaining politically neutral. At the time the ultra groups began to form, people in Italy were becoming much more vocal about their political beliefs, and so the ultras tended to adopt the ideology of their town or neighborhood.

Away from the upheaval in the North, Napoli pressed on, with Corrado Ferlaino continuing to oversee the club, and the pragmatic Giuseppe Chiappella on the bench. Somehow, in 1970–1971, they finished 3rd in Serie A, qualified for the UEFA Cup, and made it to the semifinals of the Coppa Italia. But in a year in which the *Capocannoniere*, Roberto Boninsegna of champions Inter, scored 24 goals, no Napoli player managed to make the top five in the list of top scorers. Once again, their top scorer was José Altafini, with 7 goals. What Napoli had was the best defense in the league. It may not have led to thrills and chills felt by the crowd in Naples, but with Dino Zoff in goal, right-back Carlo Ripari in the starting XI, and Luigi Pogliana bringing his fluid movements to the defense[27]—Pogliana even scored the winning goal when Napoli beat Juventus 1–0 in November—things were looking good at the back. Two seasons previously Mario Zurlini had been converted to a *libero*; there, he had the freedom to orchestrate the defense while providing extra cover in case his teammates allowed their opponents through. Both he and Dino Panzanato had been with the club since 1964, and the pair had gelled into an excellent unit. These defensive stalwarts managed to keep 18 clean sheets that season.

Then came the 1971–1972 season, and everything began to fall apart. Again. Oddly enough, the squad remained fairly steady, and Chiappella still sat in the dugout. However, Corrado Ferlaino wanted out; he'd been trying to exit since the early summer. It wasn't until October 1971, when the financial troubles weighing on the club, compounded by the bonuses the players insisted they were owed, that he finally stopped listening to Achille Lauro and quit.[28] Ettore Sacchi, Ferlaino's vice-president, promised that within weeks he would pay the owed wages, and the board voted him in as president on the 26th.[29] Amid the turmoil in the back room, Napoli's entire back line, which had performed so well in the previous season, remained in place. This is likely how they managed to secure 16 draws, compared to their 6 wins (5 of which came at home), and maneuver into 8th place. This time the Partenopei managed 13 clean sheets, but just 27 goals; this total was beaten by both Bologna in 11th and Vicenza in 12th.

Once more, in offense the team relied on José Altafini, who scored just eight goals in the league, the top-ranking goalscorer for Napoli yet again. Ferlaino sold attackers Ottavio Bianchi, Gian Piero Ghio, Kurt Hamrin, and Alessandro Abbondanza before the season began. In return, he bought 20-year-old Emiliano Macchi, who made just six appearances for Fiorentina in two seasons; brought back defensive midfielder Vincenzo Montefusco from Serie B side Foggia; found Andrea Esposito at Serie C side Policoro, only to see him play seven matches (although he scored two goals in those games); and recalled Gian Cesare Discepoli from a loan at Prato, a Serie D team, who encountered zero chances in the league over two seasons. Perhaps his smartest move was bringing Pierpaolo Manservisi back from Lazio, but the central midfielder made just 19 outings and scored 3 goals. In the last 11 games of the season, Napoli recorded only 1 win, scoring 4 goals. Unsurprisingly, the stadium remained around a third full for many of their matches. It was clear to everyone that what Napoli needed was a stronger attack; combined with the defense, they just might make it to the top three again.

The *Vesuviani* did have one significant triumph in 1971–1972: they made it to the final of the Coppa Italia for the second time in their history. Napoli, in Group 3, won three of their games, losing only to Sorrento in their first match of the tournament. They finished 1st in the group and moved on to the 2nd round, in Group B, where they won two, lost one, and drew three, which was enough for them to top the group again, barely edging out Fiorentina and Bologna. The final was next, quite a different format from the way Coppa Italia is played now. Napoli went head-to-head with Milan, the winners of Group A and the team that had secured 2nd place in Serie A, one point behind champions Juventus. Napoli did not have enough to break them down, or keep them from their goal, and wound up losing 2–0, with one of the goals an own goal from Dino Panzanato, who had become so solid in the backline since that fateful final game of 1966, when he'd scored one for title-winners Inter.

Things didn't get much better the next season. Giuseppe Chiappella remained on the bench, and Corrado Ferlaino returned as president during the summer. Unfortunately, the latter's goal was to keep the costs down as much as possible; he spoke of an "austerity plan" to lift Napoli out of their dire financial situation.[30] In 1972–1973, they sold off Dino Zoff to Juventus, who had been courting him since Napoli had begun their own relationship with the goalkeeper. Due to Napoli's enormous wage bill, Ferlaino finally had to accept Juventus' offer; Zoff himself was reluctant to leave, but wanted to help his club dig themselves out of trouble.[31] He became a major figure with the *Bianconeri*, with 330 appearances over 11 seasons, winning 6 Serie A titles and a UEFA Cup in his time at the Old Lady. He was also integral to the Italian national side, the oldest captain to manage Italy at 40 years old, in the 1982 World Cup, where he won the award for best goalkeeper of the tournament. The adored José Altafini also went to Juventus, which would end up coming around to injure Napoli at one of their strongest points. Carlo Ripari left the backline, to join Vicenza, as did Naples native Vincenzo Montefusco. Pierpaolo Manservisi, who'd been solid for Napoli the previous season, went back to Lazio. A reliable attacker, Angelo Sormani, headed to Fiorentina, as did Emiliano Macchi, a bench player for both sides. Giacomo Vianello, Mario Perego, and Gian Carlo Pulitelli all left as well, a total of ten players.

In return, Napoli went hunting for players to restock their side. They brought back 33-year-old striker Cané, as well as midfielder Alessandro Abbondanza, who'd been on loan for 5 seasons. Pietro Carmignani was signed from Juventus and although he was reliable between the sticks for five seasons, he was certainly no Dino Zoff, which is likely why Juventus threw in money and at least one youth player in the exchange for the extraordinary goalkeeper. However, the fans grew to love him, forgiving his mistakes and bestowing the nickname *Gedeone* (Gideon) upon him.[32] Carmignani says he's not quite clear why "Gideon" was chosen; Vinicio Viani began calling him Gedeone when he was at Como in the 1960s, and he believes because it rhymes with *Nasone* (big nose).[33] Via Angelo Rimbano, a defender picked up from the freshly-relegated Varese, Oscar Damiani, a midfielder from Vicenza, who had barely managed to escape relegation, and Giorgio Mariani, an attacker on loan from Verona—who'd finished *behind* Vicenza—were regulars in the 1972–1973 side, but were then sold the next season, Rimbano to Bruno Pesaola's Serie A side Bologna, while Damiani went back to Vicenza, who once again had barely escaped relegation. However, Juventus bought him in 1974, and he was a regular in the squad for two years. Damiani must have had some talent, as he scored 37 goals in 87 appearances in his 3 seasons at Genoa. Napoli brought him back in 1979, but his goal tally then dropped dramatically. Meanwhile, Napoli wanted to keep Mariani, but he was sold by Verona to Palermo, where he lasted four games before being bought by Inter, where he enjoyed a modicum of success over two seasons. Then there was Domenico Fontana, also from Vicenza, who made just nine appearances (but did score a goal), and was also sent straight back to where he came from. Finally, reserve goalkeeper Aldo Nardin was also picked up from Varese, the relegated club where he'd let in plenty of goals, and then quietly sent off to a starting position at Ternana, in Serie B.

In addition to Carmignani, Napoli bought two players that season who actually stuck around: Giovanni Vavassori and Giuseppe Bruscolotti. Vavassori came to Napoli via Atalanta, a team who'd been about on par with Napoli the season before. The defender spent five seasons with the club, used on and off as a starter, before returning to Atalanta and a spot in the starting XI. Bruscolotti, on the other hand, finished out his career in 1988 with Napoli after 16 seasons with the club. Corrado Ferlaino had his eye on Bruscolotti since he helped Sorrento move from Serie C to

Serie B, seeing how well he played under Giancarlo Vitali, whose system and reliance on defense was similar to that of Giuseppe Chiappella, which is why he went to great lengths to secure the defender.[34] Later, Ferlaino would call him one of the five best man-markers in Italy,[35] no small feat in a country known for its defending. Known to the fans as *palo 'e fierro* (the iron pole), he was not a huge man but one blessed with great physical strength, at the same time possessing such a strong sense of balance that he was almost always able to hold his ground while the attacking players fell on to it.[36] In essence, he was a classic central defender.

Despite his tough exterior, Bruscolotti cried when he was named captain of the club.[37] The defender captained the side for 5 years, and held the record for most appearances in a Napoli shirt—511—until Marek Hamšík surpassed him in 2018.[38] When he handed off the armband, it was to Diego Maradona. According to Bruscolotti, he told Maradona when he passed over the armband—for the defender was to remain at Napoli; it is rare for a player to step down as captain while still regularly playing for a side—that he was now the captain but he must keep his promise: Naples was waiting for the scudetto.[39] The two were good friends and in 1985 founded the Maradona-Bruscolotti Soccer School. Now called simply the Scuola Calcio Bruscolotti and presided over by its benefactor, the school aims to help keep the youth of the street engaged in not only football, but in learning the skills needed for a life outside of the career.[40] Bruscolotti was another who became an honorary Neapolitan, beloved by the fans for his long commitment to the club, his rock-solid dependability throughout those 16 years, and the enterprises he created in the city after he left the club.

The 1972–1973 season should be best remembered for bringing Giuseppe Bruscolotti to the club, for Napoli did little else that season worth celebrating—except, of course, defend. With Bruscolotti in front of new goalkeeper Carmignani, the squad allowed just 20 goals, the least in the league, 2 fewer than champions Juventus. But Napoli also scored even fewer goals than they conceded, landing just 18 in the net. Six of those came from top scorer Giuseppe Damiani. They played a dangerous game in the first half of the season, collecting just 12 points.

Naples is known for a football atmosphere that intimidates players and visiting fans alike; being a one-club town, the veneration of Napoli is evident before one even approaches the stadium. But that season, those in the stadium had little to celebrate, and thus the celebrated atmosphere faded away. Small groups began to form, organizing *scugnizzi* (young boys) in Curva B into clusters that sang and chanted—but different chants rang around the San Paolo at the same time, discordant and distracting. Led by Gennaro Montuori, the two largest groups, "Ultras" from Curva BI and "Commando" from Curva A joined together to create "Commando Ultra Curva B" (CUCB), Napoli's first official ultra group.[41] It is said that CUCB were the first ultra group to create the impressive "choreos" now seen at most Italian stadiums and elsewhere across the globe. Even if they did not lead the spettacolo charge, CUCB set the Napoli stadium on a path to greatness, a frightening cauldron that visitors would prefer to avoid. Other ultra groups helped ramp up the fear factor, dividing themselves between Curva A, regarded as the more violent of the two curvas, and Curva B, which, while not always successful, tends to continue to abide by a banner displayed there in the 1980s, reading, "Violence divides us, our passion unites us."[42]

Increased organized support notwithstanding, Chiappella found his job in danger due to his side's performance in the first half of the season. Things on the pitch improved a bit in the ritorno, but even in March, the rumors continued to swirl—this time, the papers reported that Chiappella was all but fired already, and that Heriberto Herrera would take over in the summer.[43] Not to be confused with the renowned Helenio Herrera, who brought the *Catenaccio* system to his Grande Inter, news that Heriberto had already committed to another year at Sampdoria was bound to have prompted supporters to thank Vesuvius for delivering them from suffering through tedious matches each round—and an even lower finish the next season. Napoli's 28 points saw them finish 9th, which may have been the halfway mark for Serie A, but was still only four points above the drop.

Hope, dangling on a string

Fans may not have been thrilled with the idea of Heriberto coming to Napoli, but were still relieved when Chiappella departed prior to the 1973–1974 Serie A season. The supporters' disdain was not entirely directed at the manager, however, although they weren't pleased that he couldn't form a decent squad with the players he'd been able to buy on a budget. They were also upset with Ferlaino because he had ordered that such a strict budget be adhered to and, most of all, that he had allowed José Altafini to move to Juventus.

Former Napoli forward Luís Vinício took Chiappella's place on the bench. Yet Ferlaino still seemed unable to open his wallet (he told the papers, on a number of occasions, that he wanted to make the club's finances healthier; one of the reasons fans were so dissatisfied is because he went about this by selling the biggest names, which by extension were the ones who made the most money)—except for two impressive players. The first was Sergio Clerici, a Brazilian with the nickname "*El gringo.*" A forward who at age 32 was getting on in years, Clerici had played for Lecco, Bologna, Atalanta, and Verona, averaging around nine goals over the past five seasons. In his first season under Vinício he scored fifteen, six of which came from the spot. It was the first season in years that Napoli had two players in the top ten Serie A scorers. The second smart acquisition, Giorgio Braglia, brought from Foggia, who'd just qualified for the top division (only to go right back down again), scored nine for Napoli. Braglia, too, played for a number of clubs before donning the blue shirt, but it was in Naples that he produced the best performances of his career, the fruits of which would be evident in the following seasons. The other five players Napoli added that season—attackers Rocco Fotia and Gaetano Troja, midfielder Riccardo Mascheroni, and defenders Spartaco Landini and even former Napoli player Carlo

Ripari, were all destined to primarily ride the bench during their tenure at the club.

As mentioned above, Damiani, Fontana, Mariani, and Ribano all left after one season. Central midfielder Giovanni Improta, Napoli born and bred, was sold to Sampdoria, which resulted in demonstrations against the management in his own neighborhood of Posillipo, and a banner that read "You can sell Vesuvius, but you can't sell Improta."[44] Turns out, they could. Dino Panzanato spent nine seasons as a (mostly) reliable defender for Napoli, but he seemed destined to be remembered for two incidents: being handed a nine-game suspension for his role in a brawl with Juventus players in the 1968 2–1 Napoli victory—the same December fisticuffs that saw Omar Sívori leave the club prematurely—and for being the player who scored an own goal in Napoli's loss to Milan in the 1972 Coppa Italia. Given that the team brought in five players who did next to nothing that season, it's surprising that the losses to the team were certainly not irreplaceable (despite what those in Posillipo believed), as the season went on to prove.

Napoli's success in the transfer market may have been down to bringing in Francesco Janich, a defender who'd spent most of his career at Bologna and helped the club secure the title in 1963–1964. Janich's official title was "general manager," yet given that he later became a sporting director and ran a talent-seeking agency, it is not a stretch to surmise that he fulfilled the role of sporting director at the time; in fact, in Italy the terms are often interchangeable. Vinício remained as coach, and the club ensured that the veteran goalscorer Clerici stayed on as well.

Napoli fans began the season with such hope, although as always it was tempered by superstition, as the supporters knew that everything that could go wrong in a season would somehow always manage to happen to their club. Given that a cholera epidemic raced through Naples in August–October 1973, claiming 12 lives, the pre-season willies were likely justified. Their second Coppa Italia group game, scheduled for September 2, was postponed due to Bologna's refusal to come to Naples, while the game at Genoa on September 16 was canceled, as the club did not want Naples players in the stadium (despite them all having been vaccinated[45]) but also refused to travel to the city. Because the authorities had deemed the match safe, Genoa's refusal earned them a one-point reduction, and Napoli were awarded a 2–0 win.[46] Although authorities were willing to let the *Rossoblù* play in Naples, they postponed the game at Avellino until the 27th, likely because the epidemic extended across Campania. They then required the match to be played on neutral ground in Bari, even though cholera raged through the Puglia region as well. Frustratingly, Napoli lost 3–2 to the side who struggled in Serie B, losing out on the top spot in the group because Bologna had scored one more goal.

But San Gennaro's blood did liquify in September 1973, signaling to the citizens of Naples that better days were coming; indeed, the city's health was restored the next month. (This was likely due to the vaccination campaign in which nearly 80% of the population was vaccinated within 5 days,[47] not the benevolence of a saint, but still.) Yet the first match of the Serie A season was a disappointment, failing to reflect the football that Vinício wanted his squad to play. Napoli were stifled by a goalless draw away at Cagliari. Then they came home, well aware that their first match at the San Paolo would be against Juventus, and would likely set the tone for the season.

Led by the ultras, last season's champions were almost certainly welcomed to the San Paolo with jeers, curses, derogatory songs, and banners praising the South and deriding the North. According to one fan, a child at the time, the unfriendly atmosphere boiled over when the visitors took to the pitch.[48] Men in suits "transformed," throwing out curses that no one would dare repeat outside the stadium. New players Clerici and Braglia, at "home" for the first time, were either going to be overwhelmed or bolstered by the noise of the stadium when they stepped onto the pitch. So many hopes were riding on this match, and the fans made that clear every time a Juventus player touched the ball. Their dreams depended on Napoli sending the ball into the back of the net. Then, at the stroke of halftime, the referee awarded a free kick. Juliano passed the ball to Clerici, who moved it on to Cané. The Brazilian shook off the Juventus players trailing him and sent a left-footed strike into the net to put the home side up 1–0.[49] The fans, jubilation barely hiding the anxiety still lurking in their bodies, were ecstatic when the players returned to the pitch. At 78 minutes in, Napoli earned a penalty, which Clerici neatly converted. 2–0. A little more than ten minutes to go, and Napoli looked like they should have been winning by an even bigger margin. This win was really happening. When the ref blew the whistle, the San Paolo roared. Napoli had played Juve off the pitch, while the visitors needed to rely on their opposition's former goalkeeper to keep them in the match.

Maybe this was finally Napoli's year.

Their ambitions began to take on solid form. The next week, they traveled to the San Siro to face Inter, who had finished 4th the season before. Helenio Herrera, who had returned to the club after stints with the Italian national side and Roma, was determined to bring the scudetto back to the black-and-blues, a trophy he had helped them capture three times in his previous time at the club, 1960–1968. This had been the time of Grande Inter, and the club had also hoisted two consecutive European Cups. He was a true superstar, and it was assumed his new Inter would at the very least be title contenders, if not a side that dominated the entire league. Napoli fans, still fueled by the adrenaline that flooded their bodies during the triumph over Juve, flooded into the city, looking as though they might cause trouble. It helped little that the citizens of Milan greeted the Southern interlopers as they would garbage in the street, decrying that they lived closer to Africa than their city and thus were barbarians, lacking true Italian character.[50] In other words, this was another match that could define Napoli's season.

In some ways, it perfectly encapsulates it: Inter's Adelio Moro kicked off the scoring before 15 minutes were up with a lovely solo effort. Just a few minutes later, Andrea Orlandini tackled Inter's star Sandro Mazzola, causing him to be subbed off.

Without Mazzola, the Nerazzurri became disjointed, allowing Clerici to pounce. The attacker had just been congratulating Inter's Lido Vieri on a great save when defender Adriano Fedele, still under pressure from Napoli, passed it back to the keeper. Clerici spotted his chance and slipped the ball past Vieri for the equalizer. Cané, too, capitalized on a mistake from the Inter defense, scoring after Vieri managed to deflect Giorgio Braglia's effort. Unfortunately for Napoli, the defense could not hold back Roberto Boninsegna. The Capocannoniere of 1970–1971 and 1971–1972 picked up a loose ball from a free kick to make it 2–2.[51] Still, Napoli fans were content having secured a point from such an esteemed side, especially given the hosts had surged forward again and again throughout the second half.

By December 16, Napoli sat top of the table, with five wins, two draws, and no losses. Was it time, yet, for Napoli fans to get their hopes up? Seven games seemed far too few, not with twenty-three to go, not for supporters hardwired with a sense of fatalistic dread, not with a crowd in which superstitious fans would chant *La Buonanima* (the Good Soul) when feeling the need to appease Vesuvius,[52] who feared it would be a poor season if San Gennaro's blood did not liquify. The miracle may have occurred, but they were right not to get their hopes up. That day Napoli came up against Lazio, who had lost to Juventus just after Napoli had beaten the Bianconeri, before going on to draw all of their games in November. However, they were on a high from beating city rivals Roma 2–1 in the *Derby della Capitale*. Napoli fought a hard battle in Rome, but Giorgio Chinaglia—who scored 34 goals that season—finally broke through with 15 minutes to go. Suddenly, Lazio were on the ascendency, and their win against Verona took them to the top of the table at Christmastime, while Napoli floundered and lost 2–1 when Milan came to town, both of the visitors' goals coming after a 50th-minute strike from Cané.

The teams clamoring for the top at the Christmas break were Lazio, Napoli, Milan, Juventus, and Fiorentina. That didn't last long. After a string of five winless games, Cesare Maldini joined technical director Nereo Rocco on the Milan bench, and despite being in 5th at the end of the *andata*, Maldini went at it alone in the ritorno. Then a loss to Juventus was followed by complete embarrassment in the *Derby della Madonnina*, which Inter won 5–1. The 3–2 loss to Fiorentina was the last straw, and Giovanni Trappatoni returned. But even he couldn't stop their bleeding at the end of the season, in which they won just 1 of their last 11 matches, finishing the year in 7th. Fiorentina, meanwhile, under new coach Luigi Radice, focused on youth, and had their fans ecstatic after losing just once in the first ten games. But the *Viola* couldn't maintain their consistency, and fell away from the pack. On the other side of Milan, Herrara had a heart attack in February, and was replaced by the second-team coach Enea Masiero. It took a few rounds for Inter to find their footing again, during which they lost to Napoli 2–1, but after winning at Lazio and dominating the Milan Derby, they still held out hope. The Nerazzurri then dropped important points against Fiorentina, Roma, and Juventus, and finished 4th.

Napoli were unable to gloat, however. The Partenopei found themselves touched by a scandal involving their star, Clerici. In the "*Scandalo della telefonata*" (Another imaginative name, the "Telephone Scandal"), Foggia offered to buy referee Gino Menicucci three Rolexes if he would assist them in their game against Milan, thus helping them avoid relegation. At the same time, Verona contacted their former player, Clerici, and offered to help him open a FIAT office in Brazil after his career ended, with the stipulation that he would have to do all he could to ensure Verona won against Napoli in April. Clerici informed the Napoli board and, although Verona did win the game 1–0, appeared to be the only *Azzurri* player actually trying to get something from the game. Verona and Foggia were eventually relegated to Serie B; Clerici was cleared, and he kept on scoring, three in the last four matches.[53] The team could not match his performance, struggling through a nine-game winless streak, their draws the one thing allowing them to keep pace with Inter. A 2–1 Napoli win at Genoa in the final game (both goals coming from Braglia in the first half) lifted the side to 3rd in the final table. Lazio had maintained a three-point lead over Juventus during the ritorno, and found themselves crowned champions of Serie A for the first time in the second-to-last round.

Yet in the end their enemies (primarily Roma) could celebrate one thing: Lazio got themselves kicked out of Europe. After losing to Ipswich in the 2nd round of the UEFA Cup, a game in which the Lazio players had already both fought among themselves and fought with the ref, their fans swarmed the pitch at the Olimpico, throwing beer cans and bottles as well as lighting rockets and burning an English flag.[54] The police were forced to use tear gas in an almost impossible attempt to contain the riot. This display saw the club banned from European play and served as an early warning of what *Biancocelesti* supporters would become.

Had Napoli fans ever dared to dream as they did in 1974–1975? In terms of personnel, the song remained much the same. Corrado Ferlaino at the helm, Franco Janich as technical director, and Luís Vinício in the dugout the entire season. Four men acquired the season before who had played very few matches left the club. The greatest loss was Napoli's sweeper, Mario Zurlini, who'd spent ten seasons as a staple in the squad. Once he'd moved to the libero position in the 1968–1969 season, he and Dino Panzanato had made life hell for opposing attackers, and Zurlini was crucial through much of the 1973–1974 season. However, after Napoli drew 1–1 at Cesena, Zurlini was driving home former Napoli board members Guido Guerra and Mario Russo, along with their friend Antonio Capobianco. In the early hours of April 1, a truck struck the car, killing Guerra and Capobianco and sending Russo and Zurlini to the hospital.[55] Zurlini was never able to play at such a high level again; after he recovered, he served as player-coach for Matera, in Serie D. In better news, Napoli were able to snap up Tarcisio Burgnich from Inter. Unsurprisingly, having been coached by Helenio Herrera, who had introduced catenaccio to Italy, Burgnich was an excellent sweeper, and he was also able to play every defensive position.

There seemed to be little in the way of excitement caused by other players the management brought in. Rosario Rampanti was snatched away from Torino, where he'd made the right wing his own, but despite being a regular on the pitch for Napoli, he was sold on to Bologna. He fared better than Ezio Vendrame, who played five games before Padova, in Serie C, bought him the next season. Nevio Favaro, from Fiorentina, played just one more game in a Napoli shirt, over four seasons, but given his role as a back-up keeper, six games over four seasons was certainly acceptable. Along with a new right winger, Napoli acquired Antonio La Palma from Brindisi in Serie B (Vinício had managed the club in 1970), who may have been the fastest player in Italy at that time, able to run 100 meters in around 11 seconds.[56] Although he functioned primarily as a left-back, he could also play in central defense, an indispensable trait after Giovanni Vavassori was injured (an injury to his right knee kept him out until the start of the 1975–1976 season, and he never reached his full potential again), and for briefer periods during his four seasons with Napoli. Finally, they picked up Giuseppe Massa, who was born in Naples and played his first season with Serie D club Internapoli, the "other" Neapolitan club, then helped Lazio win promotion to Serie A in 1971–1972 with 12 goals scored. It was not a number he would reach in his 4 years at Napoli, but he was a solid contributor, quick and with good technical skills, who scored 9 goals in his 26 appearances his first season at the club.

Napoli began the season on a high at home, with a nine-game unbeaten run that began with a 3–1 win over newly promoted Ascoli, the hosts' goals a hattrick from Giorgio Braglia, nicknamed *cavallo pazzo* (mad horse) by the fans for the thrilling runs he executed up and down the flanks—and the way his long hair flowed behind him as he galloped along. From the time he joined the club the season before, Braglia had struck up an excellent partnership with Clerici; not only were they spectacular to watch, but they had scored 23 goals between them. They'd only improve, scoring 26 in their second year as partners.[57] However, Braglia had a tendency to miss easy chances,[58] and he didn't manage to score again until the 8th round 5-1 thrashing of Cagliari at home. But given Napoli went unbeaten in the first nine rounds, Braglia's lack of production didn't feel like an immediate issue—particularly as he was absolutely adored by the Napoli faithful.

Then came Round 10 at the San Paolo on December 15, 1974. Against Juventus. The papers and assorted press outdid themselves in their search for superlatives to describe the upcoming match. In general, it was agreed that the week would pit the most interesting, most modern Serie A sides against one another.[59] Unfortunately for both Napoli fans and neutrals hoping for a hotly contested game, Juventus walked away victors in a 6–2 match that put a damper on Napoli's hopes for the season. The Old Lady proved they'd adapted to their new offensive style, modeled on Rinus Michels' Total Football in the Netherlands that Carlo Parola (in his third stint at Juve) had instilled at the club.[60] The first strike came from former Napoli hero José Altafini in the 28th minute. Giuseppe Damiani struck a brace before halftime, one of which was a converted penalty. Roberto Bettega added another five minutes into the second half. Napoli were down 4–0 (deservedly, given how tired they looked in the first half) when Clerici finally woke up, netting his first on the stroke of the 60th minute. He scored again 11 minutes later. However, Juventus' Franco Causio racked up their fifth goal in between. The death rattle sounded in the 84th minute when Fernando notched Juve's sixth. Quite honestly, it was a surprise the Southern side had conceded just six to the Northern giants. But should leeway exist in sport, Napoli deserved some here. While Juventus had faced Ajax midweek in the 3rd round of the UEFA Cup, the Partenopei were bested by Banik Ostrava 3–1 on aggregate. In addition to the loss, Napoli encountered difficulties returning from Czechoslovakia. They first needed to stop in Prague, where they could only get a flight to Rome, after which they took a bus to Naples, only arriving in their homes at around 6am the next day.[61]

The fans, however, accepted no excuses; after all, Juventus had played midweek and managed a 2–2 draw against Ajax, very recently thrice-European champions, which put them through to the next round thanks to the away goals rule. When Napoli lost so horrifically at the weekend, supporters vented their frustration by hurling objects at the pitch. One hit a linesman, resulting in the referee halting play a minute and a half before the end of the match. In return, Napoli were hit with a three-game penalty (later reduced to two) requiring them to play on neutral ground.[62]

Napoli struggled through the last five rounds of the andata, in three of which they failed to score a goal. In the last, Braglia and Clerici found their scoring boots again, while Salvatore Esposito added a third against Varese . . . the team that ultimately finished last in the league. However, their ability to secure draws—8 of their first 15 games ended as such—kept them in the race.

The ritorno opened in the same way: a nine-game unbeaten run, ended by Juventus, this time in Turin. However, the Partenopei collected five wins during that time, driving them up the table. And at least the Old Lady didn't humiliate them this time, winning just 2–1. That said, a defensive error by Luigi Pogliana allowed Franco Causio the first goal, putting the home side ahead. To be frank, Juventus could have scored one or two more in the first half, but Napoli were lucky. In the second half, Napoli turned the tables, threatening their former keeper Dino Zoff again and again. It didn't take long for Giuseppe Massa to dribble through the Juve defense and slide the ball to Juliano, who volleyed past Zoff for the equalizer. However, Altafini once more drove a stake into Napoli's heart, this time coming in off the bench to replace Oscar Damiani only to drive in the winning shot in the 88th minute. Had he not managed the goal against his former team, the season may well have ended with Juventus and Napoli tied on 42 points each. Instead, Altafini essentially clenched the title for Juventus against his former side. For 45 years, Napoli fans would feel no worse betrayal than the one that came at the feet of José Altafini.

Napoli took out their frustration on poor little Ternana, who in 1964 had been a team of semiprofessionals in Serie D, rising to Serie A for the 1971–1972 season and aiming for the new style of Total Football, filled with short passes and high pressing, promoted by coach Corrado Viciani. However, they had fallen down to Serie B before immediately coming back up to Serie A for 1974–1975, and the game at Napoli reminded everyone exactly why. Antonio La Palma opened the scoring before five minutes had passed. Giuseppe Massa needed less than 15 minutes to put in another. Salvatore Esposito contributed the third, and Sergio Clerici added a fourth before the halftime whistle blew. Ternana most likely got the hairdryer treatment in the dressing room, but it mattered little, as Giorgio Braglia struck within a minute into the second half. Things calmed for a while, but then Massa struck again in the 73th minute. Ferdinando Donati, unhappy at the humiliation his team was suffering, struck back two minutes later, a clear consolation prize—especially since Braglia scored yet another in the 87th, sealing a 7–1 win for Napoli. Napoli finished out the rest of the season with three wins and a draw. Juventus, meanwhile, lost 4–1 to Fiorentina in their penultimate game. A little extra pressure in Napoli's 1–1 draw with Torino, or a strike never taken by Altafini, and 1974–1975 could have been the year Napoli first raised the scudetto.

But then a young man named Diego Armando Maradona may have never come to Napoli.

[1] De Cecco, M. (1975, January). Italy's payments crisis: International responsibilities. *International Affairs*, 51(1), 3–22.

[2] Encyclopedia Britannica. (last updated 2020, January 1). Naples, Italy. In *Encyclopedia Britannica*. Retrieved January 6, 2022 from https://www.britannica.com/place/Italy/World-War-II

[3] Di Nanni, C. (1964, June 30–July 1). Tornano Lauro e Fiore per rilanciare in Napoli. *Corriere d'Informazione*, 10.

[4] Materazzo, G., & Sarnataro, D. (2013). *1001 storie e curiosità sul grande Napoli che dovresti conoscere*. Newton Compton.

[5] Di Nanni.

[6] Carratelli, M. (2007). *La grande storia del Napoli*. Gianni Marchesini Editore.

[7] Materazzo & Sarnataro, *1001 storie*.

[8] Id.

[9] Pelillo, M. (2013, October 4). Bandoni: "Il mio Napoli era fortissimo, vivevo a Forcella..." *TuttoNapoli*. https://www.tuttonapoli.net/le-interviste/bandoni-il-mio-napoli-era-fortissimo-vivevo-a-forcella-166934

[10] Gargiulo, A. (2015, May 30). Coppa delle Alpi 1966. *ForzAzzurri*. https://www.forzazzurri.net/coppa-delle-alpi-delle-alpi-1966/

[11] Bufi, F. (2008, June 29). Dal regno del Comandante al crac: Le sventure di una dinastia di potenti. *Corriere della Sera*. https://www.corriere.it/cronache/08_giugno_29/tragedia_lauro_bufi_c4081e0e-45ab-11dd-90eb-00144f02aabc.shtml

[12] Sacco, A. (2020, April 1). Gioacchino Lauro, quel figlio troppo generoso per i gusti del "Comandante." *Corriere del Mezzogiorno*. https://corrieredelmezzogiorno.corriere.it/napoli/cronaca/20_aprile_01/gioacchino-lauro-quel-figlio-troppo-generoso-gusti-comandante-a5001932-73f0-11ea-8c5c-17d34b758c9b.shtml

[13] Stamane ai funerali di Corcione; Sivori Piange. (1968, December 9–10). *Corriere D'informazione*, 8.

[14] Schiavon, D. (2019, December 19). Quando Ferlaino gettava scompiglio alla guida: il Presidente

pilota. *Napoli Today.* https://www.napolitoday.it/blog/ciuccio-stories/storia-ferlaino-pilota.html

[15] Bufi, F. (1995, June 25). Da Maradona alle bombe, la vita spericolata di Corrado. *Corriere della Sera*, 39; Caiazzo, M. (2016, July 31). Napoli 90 anni, le memorie di Ferlaino: "Io, gli scudetti vinti, da solo contro tutti..." *La Repubblica.* https://napoli.repubblica.it/sport/2016/07/31/news/napoli_90_anni_le_memorie_di_ferlaino_io_gli_scudetti_vinti_da_solo_contro_tutti_-145152822/

[16] Schiavon.

[17] Pesaola ribadisce: Ho chiuso per sempre, dopo Milan—Napoli me ne andro' a Sanremo. (1968, January 31). *Corriere della Sera*, 13.

[18] Carratelli, M. (2007, January 28). Bruno Pesaola: Il romanzo di Petisso, napoletano nato all'estero. *La Repubblica.* https://web.archive.org/web/20100903154607/http://static.repubblica.it/napoli/speciali/volti_archivio/precedenti/280107.html

[19] Grandini, C. (1968, September 14). Il momento di Chiappella e Pesaola dopo lo scambio delle panchine. *Corriere della Sera*, 21.

[20] Gandolfi, R. (2020, March 25). Omar "El Cabezon" Sivori: Un geniale farabutto! *Storie Maladette.* https://storiemaledette.com/2020/03/25/omar-el-cabezon-sivori-un-geniale-farabutto/

[21] E' un monito la sentenza del giudice, ma Sivori reagisce e attacca la Juve. (1968, December 6). *Corriere della Sera*, 21.

[22] Accatino, G. (1969, May 6). Finita la crisi del Napoli? *La Stampa*, 16.

[23] Hamrin al Napoli per un Miracolo. (1969, December 3–4). *Corriere d'informazione*, 6.

[24] Frossi, A. (1970, July 6). Chiappella—Sta sognando un Napoli molto diverso. *Corriere della Sera*, 13.

[25] Doidge, M. (2015). *Football Italia: Italian football in an age of globalization.* Bloomsbury Academic.

[26] Id.

[27] Sappino, M. (2000). *Dizionario del calcio italiano.* Baldini & Castoldi.

[28] Degni, F. (1971, June 19). 'Napoli-caos: convocato per martedì' il Consiglio di Amministrazione della societa'; Lauro a Ferlaino: "Resta!" *Corriere dello Sport*, 8.

[29] Sasso, E. (1971, October 27). L'eredità di Ferlaino. *Corriere della Sera*, 21.

[30] Ferlaino contestato dai tifosi napoletani (1972, June 23). *La Stampa*, 16.

[31] Zoff soluta Napoli rimpianto dai tifosi (1972, August 4). La Stampa, 15.

[32] Carratelli, *La grande storia del Napoli.*

[33] Carmignani, P. (2021, January 22). *Il Pallone Racconta.* https://ilpalloneracconta.blogspot.com/2008/01/pietro-carmignani.html?m=1

[34] Ciccarelli, L. (2012, October 24). La storia siete voi: "Pal 'e fierro" Bruscolotti. *TuttoNapoli.* https://www.tuttonapoli.net/rubriche/la-storia-siete-voi-pal-e-fierro-bruscolotti-120374

[35] Kühne, I. (1985). *Napoli passione mia.* Edizioni Lucarini.

[36] Ciccarelli.

[37] Ludden, J. (2018). *Once upon a time in Naples* (3rd ed.). CreateSpace Independent Publishing Platform.

[38] Burton, C. (2019, February 14). Hamsik brings record-breaking spell at Napoli to close as €20m China transfer is completed. *Goal.* https://www.goal.com/en/news/hamsik-brings-record-breaking-spell-at-napoli-to-close-as/193nwyv8h02e717w4n5zu9990e

[39] Ciccarelli.

[40] Mission Scuola Calcio G. Bruscolotti. (n.d.). Campo Sportivo Mariolina Stornaiuolo. Retrieved November 5, 2020 from http://www.mariolinastornaiuolo.com/cms/7-notizie/18-mission-scuola-calcio-g-bruscolotti

[41] Booze Brothers. (2020, February 29). Historija Ultrasa iz Napolija i zlatne godine Maradone. *Balkanskinavijaci.* https://balkanskinavijaci.com/historija-ultrasa-iz-napolija-i-zlatne-godine-maradone/

[42] Hall, R. (2014, October 16). Napoli: Serie A alternative club guide. *Guardian.* https://www.theguardian.com/football/the-gentleman-ultra/2014/oct/16/napoli-serie-a-alternative-club-guide

[43] Heriberto candidato a sostituire Chiappella al Napoli. (1973, March 31). *Corriere della Sera.*

[44] Materazzo, G., & Sarnataro, D. (2014). *I campioni che hanno fatto grande il Napoli.* Newton Compton.

[45] Napoli: vaccinati tutti i giocatori (1973, September 3). La Stampa, 13.

[46] Repetto, A. (1973, September 15). I giocatori del Genoa si rifiutano di andare a Napoli per la Coppa. *La Stampa*, 1.

[47] Roberts, H. (2021, March 1). Coronavirus brings back memories of Naples in the time of cholera. *Politico*. https://www.politico.eu/article/coronavirus-italy-vaccines-naples-cholera-outbreak-1973/

[48] Del Bono, S. (2018, April 3). Il mio giorno all'improvviso è Napoli-Juventus 2-0, nel 1973: gol di Cané e Clerici. *Il Napolista*. https://www.ilnapolista.it/2018/04/napoli-juventus-2-0-1973/

[49] De Felice, G. (1973, October 15). Napoli alla grande, Juve alla deriva. *Corriere della Sera*, 10.

[50] Ludden.

[51] Milazzo, R. (1973, October 29). L'inter combatte, il Napoli gioca. *Corriere della Sera*, 12.

[52] Ludden.

[53] Illecito nella gara Verona-Napoli? La smentita di Garonzi ma l'inchiesta procede (1974, May 25). *La Stampa*, 17.

[54] Gravi incidenti all'Olimpico per Lazio–Ipswich. Si scatena la folla: Bombe lacrimogene e feriti. (1973, November 8). *Corriere della sera*, 23.

[55] Zurlini gravemente ferito; trenta punti alla testa (1974, April 2). *Corriere della Sera*, 21.

[56] Di Nanni, C. (1974, September 3). Questa e' la sua prima scoperta. *Corriere d'informazione*, 10.

[57] Materazzo & Sarnataro, *I campioni*.

[58] Tina, P. (2017, February 29). Napoli, i 70 anni di Braglia: l'attaccante che fece impazzire il San Paolo per la sua anarchia tattica. *La Repubblica*. https://napoli.repubblica.it/sport/2017/02/20/news/napoli_i_70_anni_di_braglia_l_attaccante_che_fece_impazzire_il_san_paolo_per_la_sua_anarchia_tattica-158805534/

[59] De Felice, G. (1974, December 16). Dopo otto gol burrascoso epilogo al San Paolo; La Juve mette fuorigioco il Napoli. *Corriere della Sera*, 12.

[60] Chiesa, C. F. (2004, February). Il grande romanzo dello scudetto. Ventitreesima puntata: regno sabaudo tricolore. *Calcio 2000*.

[61] Calcagno, P. (1974, December 12). Il Napoli e' tornato all'alba; Siamo tutti stanchissimi. *Corriere dell'Informazione*, 11.

[62] Per una bottiglietta campo squalificato (1974, December 16). *Corriere della Sera*, 12; Napoli senza calcio per due mesi. (1974, December 20). *Corriere della Sera*, 12.

CHAPTER 6

Holding Out for a Hero

6

Both Napoli and their city were doomed to struggle before their savior arrived.

Several factors contributed to the next Italian economic crisis at the end of the 1970s and beginning of the 1980s, including frequent terrorist attacks and general political instability. The petroleum crisis of 1973 had subsided by 1979, but its repercussions were still felt. FIAT, for example, had to fire 20,000 workers. The South, in particular, struggled, and poor cities like Naples experienced a dramatic increase in youth unemployment.[1] Finally, the 1980 Irpinia earthquake left its marks on the Campania region, with around 3,000 dead and nearly 300,000 left homeless. Throughout those tumultuous years, Napoli lifted themselves up, only to plummet back down. Like the city, the supporters—who might be considered one and the same as the citizens of Naples—endured the ups and downs, remaining fevered fans, believing that what goes down must, eventually, go up. It was as though they knew that a saint greater than Gennaro himself would arrive in July 1984.

Shifting down a gear
In 1975–1976, things began to go downhill for Napoli, although the shift was smooth enough that most fans were able to cope. After all, their club won the 1976 Coppa Italia, holding the trophy high for the second time in their history. In the first group stage, they hit a slight bump in Group 3, with Cesena, Palermo, Foggia, and Reggiana. Only Cesena played in Serie A, and the goalless Cesena–Napoli game was the only time each dropped points. Fortunately for Napoli, they had a goal difference of +6, while Cesena's was +4. Napoli moved on to Group B, where three other Serie A teams—Fiorentina, Milan, and Sampdoria—awaited. The games were tougher, but Napoli, unbeaten in the six games (while Fiorentina floundered around with five draws and just one win) emerged victorious.

Then came the final at the Stadio Olimpico in Rome. Luís Vinício had left on June 19, due to a severely deteriorating relationship with Ferlaino. The coach alleged that Ferlaino was trying to take advantage of his depressed state and offered a contract that was unacceptable, so although he initially accepted the offer, later reflection prompted him to back out.[2] As such, the final would be coached by Alberto Delfrati, Vinício's assistant, and Rosario Rivellino, who oversaw the youth squad. At first, it seemed as though it would be impossible for Napoli to break down their opponents, Hellas Verona, coached by former Italian men's national team coach Ferruccio Valcareggi. In fact, much of the game was quite boring for spectators (even YouTube offers up other games and other finals; the famed commentator Sandro Ciotti's audio highlights has fewer than 3,000 views[3]). The minutes ticked on and Napoli barely got within sight of the goal. Neither side appeared to be playing with the urgency the crowd would have expected in a final; Napoli, in particular, seemed to be waiting for Verona to shoot themselves in the foot.[4] Fortunately for the *Vesuviani*, their strategy worked. In the 75th minute, on the back of a corner delivered by Salvatore Esposito, Verona goalkeeper Alberto Ginulfi put the ball in the back of his own net, and the floodgates opened. Giorgio Braglia scored two minutes later, Giuseppe Savoldi two minutes after that, and Savoldi closed the scoring with a second at the 86th.

Giuseppe Savoldi was likely the other reason Napoli fans didn't hassle their club or its owner—at least, not as much as they could have. Savoldi, then 28, came to Naples from Bologna for over a million pounds, a record at the time, while Sergio Clerici, 34, went the other way, along with a co-ownership agreement for Rosario Rampanti. Clerici failed to impress at Bologna and ended his career with 1 goal in 11 appearances for Lazio in 1977–1978. Rampanti, meanwhile, bounced from club to club, never really thriving anywhere. Then there was Savoldi, the top scorer in 1972–1973, one of the top scorers *ever* in Italian *calcio*, the second-top scorer from the previous Serie A season, with 15 goals. He could have scored 16 if it weren't for the infamous event at Ascoli, where he was denied a hattrick after a ball that had crossed the goal line was kicked back onto the pitch by a ball boy (very sneakily, one might say) before the referee had spotted the goal.

Much of what made Savoldi a great center-forward is that he capitalized on skills he'd developed playing basketball when he was younger. His skill in the air was legendary; he could time a header to perfection and was lethal in the area due to his ability to jump so high, and so quickly, from a still, standing position. In other words, there's no denying Napoli snagged an excellent player in exchange for one-and-a-half mediocre ones . . . and two billion lire.[5] Savoldi was bought with one purpose: to help Napoli win their first *scudetto*. He was known as "*Mister Due Miliardi*" due to the price the club paid for him, but Napoli sold 70,000 season tickets after his purchase—for comparison's sake, the club that sold the second-most tickets was Roma, with 19,000. The tickets sold paid for the arrival of Savoldi almost immediately.[6]

Savoldi was worth it to Napoli, but Napoli ended up not being as worth it to him as he'd hoped; he later said he'd come to Naples too late. While he led the team to their second Coppa Italia, scoring six goals in the tournament, he'd already had a hunch that this was a team on a downward trajectory, that this was not the Napoli that had just challenged Juventus for the title, that instead he'd landed at a club that was having a bit of an identity crisis. The 1974–1975 season had started well for Napoli, with Savoldi scoring on his Serie A debut against Como, and the club recording just one defeat in their first nine games. This likely helped the fans forget about their humiliation in the UEFA Cup, where they lost 4–1 at Torpedo Moscow away—Savoldi scored his first goal for the club there—and could only manage a 1–1 draw at home, dumping them out of the tournament in the 1st round. Then Savoldi was injured in the 1–0 victory at Lazio.[7]

From there, according to the forward, the season went downhill, the goalless draw in the next match against Ascoli more or less signaling the end of the season, just nine rounds in. Savoldi had a point—although he returned just three weeks later, Napoli won just once in their final eight games of the *andata*. Then, in the *ritorno*, the only sign of success was a 1–1 draw at home to Juventus. One major issue was Braglia, who had partnered so well with Sergio Clerici up top. Paired with Savoldi, Braglia's stats dropped dramatically. He was sold to Milan that summer, but he went out on a high, scoring his final Napoli goal in the 4–0 win over Verona that resulted in the *Azzurri* lifting the Coppa Italia for the second time.

Europe wasn't as much of a problem for Napoli the next year. They triumphed in the 1976 Anglo-Italian Cup over Southampton, winning 4–0 at home to wipe away Southampton's 1–0 victory in the first leg. In the European Cup Winners' Cup, Napoli defeated Norway's Bodø/Glimt, Cyprus' APOEL, and Poland's Śląsk Wrocław before being eliminated in the semifinal by Belgium's Anderlecht, who overcame a 1–0 deficit in the first leg to score two goals unanswered, ending Napoli's European campaign.

But things weren't so great at home. In the first half of the season, a fan hit a linesman with a bottle in Turin. The league punished Napoli by awarding a 2–0 win to Juventus (the score when the match was suspended) and announcing a three-game stadium suspension, which was later reduced to two matches.[8] Napoli subsequently drew 1–1 with Perugia at the Stadio Cibali, in Sicily. After coming so close to the Cup Winners' Cup, Napoli were actually in 3rd place, but after the loss to Anderlecht, they lost 3–0 to Inter at home, dropping to 5th. Napoli then drew 1–1 with Sampdoria, but lost their final four games to end the year in 7th. They would have finished level with Perugia in 6th, if not for another penalty, this time a 1-point reduction due to excessive fan disturbances at the San Paolo.

One such disturbance occurred in the final match, against Fiorentina. First, the fans threw things. Then, after a penalty wasn't given to Napoli and the visitors scored on the counter from what Napoli supporters were certain was an offside position, an angry Juliano kicked the ball off the pitch rather than take the kickoff after the Fiorentina goal, and was shown a red card. The furious fans not only pelted the field with objects, but started running toward the changing rooms. The referee suspended the match and Fiorentina were awarded a 2–0 victory.[9] Napoli ultimately finished three points behind Lazio in 5th; if they hadn't been penalized, a UEFA Cup place could've been theirs. Savoldi was practically the only player scoring goals; he finished the season with 17 in all tournaments; the next closest players were Luciano Chiarugi and Giuseppe Massa, with 7 each.

The problem may have simply been one of exhaustion, perhaps combined with a shallow bench. After all, Ferlaino was still in charge, and he'd brought the successful Bruno Pesaola back to the bench. But seven players had left that year, while five came in. One significant loss was that of Giorgio Braglia, who had scored 24 goals in 80 appearances with Napoli. Given that Napoli wanted Luciano Chiarugi, the "Mad Horse" of Milan, the two clubs swapped.[10] Aged just 29, it seemed as though Braglia likely had a few more years in him; instead he sat on the bench at Milan for the next two seasons, appearing only eight times for the *Rossoneri*. Perhaps he faced the same issues there as he had with Napoli: without Sergio Clerici by his side, he simply didn't play as well, and did not have enough tactical awareness to adapt his game and fit into the squad. With the exception of Luigi Boccolini, who'd been a regular in his one season at Napoli, the others who left were all bench players. Those that took their places played more regularly, suggesting Pesaola may have been rotating his players; however, the 9 he favored most played over 30 games between Serie A and the Cup Winners' Cup. Those numbers do not include the six games Napoli played in their Coppa Italia group—in which last year's champions finished second to Milan and equal to Bologna on points.

Google Translate's version of Napoli's 1977–1978 season is so poetic it cannot be improved upon: "The Neapolitan team in the league performs well, or rather without infamy and without praise." With a young new coach, Gianni Di Marzio, and many new faces on the pitch, Napoli finished 6th in Serie A. That was enough to get them into next season's UEFA Cup, but finishing 14 points behind champions Juventus did not exactly shower them with glory. Perhaps too many changes occurred at once: ten left the *Partenopei*, while eleven arrived. After five seasons at Napoli, goalkeeper Pietro Carmignani moved to Fiorentina. Carmignani had been the second-best keeper of the 1974–1975 season, in which Napoli came within spitting distance of the scudetto. Center-back Giovanni Vavassori went to Atalanta, versatile defender Tarcisio Burgnich retired, left-back Luigi Pogliana, who'd been with the club a decade, moved to Pistoiese, and defensive midfielder Andrea Orlandini transferred to Fiorentina. In other words, Napoli were essentially rebuilding their entire backline, as well as supplementing their midfield, where Salvatore Esposito, who had been with the team for five years, went to Hellas Verona, and ten-year veteran Vincenzo Montefusco—who may have spent a few seasons on loan, but who could be counted on as a solid defensive midfielder when in Naples—retired and moved into coaching.

Moreno Ferrario was brought in from Varese in Serie B to compensate for the loss of Vavassori, as he could function as a sweeper or a center-back. Francesco Stanzione, who'd spent the last season at Serie C side Paganese, primarily played as a sweeper for his only year at Napoli. Pogliana was replaced by 16-year-old Giuseppe Volpecina of Serie C's Casertana; that is, when he wasn't out on one of his many loan spells, or riding the bench at Napoli. The youngster actually spent his first three years at the Neapolitan club rather than being loaned out. His inexperience was supplemented by that of Pellegrino Valente, a 26-year-old bought from Sampdoria who could excel on the left, able both to contribute offensively but track back fast enough to help the defense. This new defense was overseen by Massimo Mattolini, a goalkeeper gained from Fiorentina in the exchange for Carmignani, who conceded 28 goals in 29 appearances in the Azzurri shirt.

A success? Not for a president who wanted more, more, more—6th place, even with European qualification, would not do. Savoldi, who'd scored five in a single game against Foggia, still led the team in goals. That was unsurprising, as all Napoli had done to add to their offensive power was bring in Luca Gabbriellini from Serie C side Pisa, a man so unremarkable that little can be found when researching him, and 24-year-old Antonio Capone from Avellino, in Serie B, who found himself regularly playing due to Luciano Chiarugi's injury issues and was essential in helping the side reach the Coppa Italia final.

The Coppa was Napoli's—and let's be honest, Di Marzio's—saving grace. The first elimination round saw Napoli easily triumph in the group, in which only one of the other four teams—Vicenza— were in Serie A. Although Vicenza *did* ultimately finish 2nd in the league, the games came before the season started, and Napoli had no trouble with them. Instead, they won all of their games, finishing with 8 points, 11 goals, and a +5 goal difference. The second elimination round was much more difficult. In May, Napoli found themselves in a group with Milan, Taranto, and their mortal rivals Juventus. Their round kicked off against the latter, and Savoldi led with a goal in the 1st minute, and had added three more by the 70th minute. Less than five minutes later, Livio Pin finished Juventus off with a fifth goal. Juventus answered not a one. Their dismal performance was explained by the fact that their Italian players—nine of them—had already been called home to prepare for the "Argentinian Tour," a series of friendlies.[11]

Next came Milan, against whom they had drawn 1–1 in the final game of the season days before. Savoldi began the festivities again. This time, his goal was answered by Milan's Giovanni Sartori in the second half, finishing the game with the same 1–1 scoreline. Taranto, from Serie B, were no trouble, with Capone scoring a hattrick. But then came Juventus, in Turin, and Napoli lost 1–0. In Naples, a late goal from Savoldi once more lifted Napoli to victory, 1–0 over Milan. The last round was a goalless draw at Taranto, putting Napoli on eight points, even with Milan. But Napoli had conceded just two goals, and Milan five, so the calcio world would not be treated to an intercity derby.

Instead, it was Napoli who would face Inter at the Olimpico in Rome. Antonio Juliano took a free kick in the fifth minute, utilizing a short pass to Luciano Chiarugi, but Renato Cipollini (who had not played for Inter in Serie A all season) blocked his sharp shot. Maurizio Restelli, a midfielder who hadn't scored all season, had no problems putting the deflection in the back of the net, giving Napoli a 1–0 lead. Soon after, Carlo Muraro sent a cross in from the right and—in a move hilarious to any non-Napoli fans—an onrushing Massimo Mattolini completely missed the ball, allowing Alessandro Altobelli to slide the ball into an empty net for the equalizer. Given the quick attacks by both sides, it looked like someone would break the deadlock in the first half; instead, the teams headed to the locker rooms with the score at 1–1. When they returned, Inter went all out, pressing a Napoli side that looked tireder every minute. In the 87th, Altobelli very nearly scored the exact same goal again, but somehow missed the open goal. Instead, it was Graziano Bini—another player with few goals to his name—who headed in the winner seconds later, the goal he was most famous for throughout his career, the goal that allowed Inter to lift the Coppa after 39 years.[12]

The 1978–1979 season saw many changes at Napoli, but alas they failed to translate those changes into the success that Corrado Ferlaino sought. Before the season started, the 1st round of the Coppa Italia began. The new format still began with a group stage, where Napoli were graced with four opponents from the lower leagues. Even so, they were only able to beat Rimini and Genoa, and were held to draws by Atalanta and Sampdoria. Still, they came out on top with six points, scoring just four goals throughout. Before the 1st round ended, the Vesuviani met Soviet club Dinamo Tbilisi in the UEFA Cup. The Georgians overcame the visitors 2–0 in Tbilisi. Then they came to the San Paolo, where Vitali Daraselia broke the deadlock in the 64th minute. Up 3–0 on aggregate, Dinamo Tbilisi finally slipped up, and Giuseppe Savoldi converted a penalty in the 79th minute. There was no amazing comeback for Napoli, though, and they fell out of Europe in the 1st round.

Longtime midfielder Antonio Juliano left for one final season with Bologna, while Antonio La Palma moved to Avellino for one game, before being sent to Lecce. Luciano Chiarugi headed to Sampdoria. And Francesco Stanzione, who'd been a steady presence at sweeper the year before, was bought by Serie B club Monza. To help ease pressure on Savoldi, Napoli brought in attacker Claudio Pellegrini from Udinese, who scored five goals that season, and Valerio Majo, from Palermo, who scored two. Clearly, Napoli were still short of the necessary firepower to move up in the league; they scored just 23 goals in the league.

It could've been all the changes at the club that kept Napoli from succeeding. While Gianni Di Marzio had guided Napoli through the first stage of the Coppa, he'd overseen yet another UEFA Cup failure. The club gave him two games in the league, and while they beat Ascoli 2–1 in their first match, they lost to Fiorentina by the same scoreline the next week. In what seemed like a surprise to all outside the boardroom, the club held a meeting in which they decided they didn't like the way the squad was performing—Ferlaino, in particular, wanted more spectacular football. Their belief was that the decrease in the number of season tickets sold was down to the manager's defensive style, and given he wasn't bringing in trophies, there was no reason to keep him around.[13] They fired Di Marzio and brought back Luis Vinício on October 9, 1978.

Little changed. This time around under Vinício, Napoli didn't even make it to the Coppa Italia final. The new format meant the next step after the group stage was a two-legged quarterfinal, and Napoli were paired with Perugia, who eventually finished 2nd in the league. But Napoli fought hard to protect their cup, and came back from 1–0 down to win 2–1 at the San Paolo, with goals from Attilio Tesser in the 81st minute and Roberto Filippi in the 88th. At Perugia, Napoli's strong defense kept the home team to a goalless draw, and thus advanced to the semifinals. But after beating the second-strongest team in Serie A, Napoli fell flat to Palermo, at the time in Serie B. Again, Napoli held the home side to a goalless draw, but in Naples, despite Beppe Savoldi leveling the score in the 43rd minute, Palermo's Filippo Citterio scored a second to give the *Rosanero* the win, and advanced them into the final. Then Napoli had to face the indignity of hosting the Coppa final, in which Juventus came to town. Vito Chimenti put Palermo up in the first minute, and though Juve equalized in the 83rd, the Serie B side hung on to a 1–1 draw almost to the end of extra time, when Juventus got their win through Franco Causio in the 117th minute.

It can always get worse

So Napoli entered the next decade in 6th place, the exact same position they had been in ten years before. But it was only going to get worse.

Beppe Savoldi went back to Bologna. Valerio Majo went to Catazaro, Claudio Pellegrini to Avellino, Livio Pin to Udinese. More or less all of Napoli's attacking power, which was weak already, left in the summer transfer window. The club brought back Walter Speggiorin, who'd scored just four goals in the last season he'd been there, but had been much more prolific at Perugia, and was there for their miracle season when they finished 2nd in Serie A in 1978–1979, with an unbeaten record to boot. Unfortunately, the center-forward managed just three goals this time around at Napoli. Similarly, Oscar Damiani performed exceedingly well at Genoa, although they'd been in Serie B the previous season, when he took the team's *Capocannoniere* title with 17 goals. As hoped, he was Capocannoniere for Napoli in 1979–1980 as well . . . with all of four goals. On the whole, the team scored 20. He performed much better for Milan two seasons later, so it felt as though Vinício, like Di Marzio, was simply too defensively oriented in his tactics. However, that was never the case; Napoli brought Vinício back because he played attacking football.[14]

Unfortunately, this was during the time of bland, boring calcio, in which most teams played defensive football, putting goals at a premium.

Napoli were excellent defensively, allowing just 20 goals, the third best in the league. It may have helped that Mauro Bellugi was the only addition to the back, the consistency possibly allowing for a more disciplined defense that communicated well. But while defense *can* win championships, the team still needs to be able to produce goals. It likely came as little surprise, then, when the team finished 11th—although they moved up a spot when Milan and Lazio were relegated to Serie B due to their involvement in the *Totonero* 1980 match-fixing scandal in which players took money to throw games.[15] "Totonero," a play on "*Totocalcio*," Italy's football lottery, involved 12 clubs and nearly two dozen individuals. On March 23, 1980, the police arrested 11 players and Felice Colombo, Milan's president, in the changing rooms after the day's matches. One of the best Italian forwards in history, Paolo Rossi, was picked up in the raid. He was originally suspended for three years, but the punishment was reduced to two to enable him to play in the 1982 World Cup. Italy wound up lifting the trophy that year, thanks in no small part to Rossi. Napoli were caught up in the net as well; although the club was acquitted, Giuseppe Damiani was suspended for three months, the shortest sentence imposed on any player. Meanwhile, Napoli's old friend Beppe Savoldi was suspended for three and a half years.

About all that Napoli could brag about when entering the 1980s is that they managed to beat Olympiacos 2–0 at the San Paolo, with a penalty converted by Damiani and a late goal by Andrea Agostinelli. That turned into the decisive goal, as in the return leg in Greece, Olympiacos won 1–0. Nevertheless, fans demanded that both Vinício and Ferlaino step down.[16] Napoli advanced to the second round of the 1979–1980 UEFA Cup, where Standard Liège beat them 3–2 on aggregate. That wasn't even close to enough for Ferlaino, and Vinício knew it. He asked to resign in February, but the owner convinced him to stay.[17] After Napoli beat Pescara 2–0 on April 5, Vinício offered his resignation again, saying he had a moral obligation to do so, after physically attacking a journalist from *Il Mattino* (to whom he has since apologized).[18] This time his resignation was accepted. Napoli had already been on a downward trajectory when Vinicio took over from Di Marzio, and he wasn't able to change that trend.[19] Nor could his replacement, another former Napoli player, Angelo Sormani, who oversaw three draws and a loss to end the season.

The fans despaired, wondering when the club (or Ferlaino specifically) would start shelling out the money for good players, so they could score a few goals, win a few more games, move back up the table, back into Europe, and most importantly, win their first scudetto.

It didn't seem too much to ask of the only club in town.

High, high, low

And the club, it seemed, had heard their frustrations, had felt them themselves. And Ferlaino was chasing such dreams as well. However, the club's finances had almost dried up. To avoid investing money they didn't have, Napoli promoted several youth players to the first-team side.[20] Given the fans' temperament, the move wouldn't be sufficient. So at the start of the 1980–1981 season, the suffering supporters were finally rewarded with something to get excited about—or someone, that is. Italy had decided to permit foreign players in the league once more, and each club was allowed one import. Napoli took advantage of this by bringing in Dutch sweeper Ruud Krol on loan from the Vancouver Whitecaps, where he'd spent one season after a long career with Ajax (Napoli then bought him outright in January).[21] Due to his versatility and his familiarity with Total Football, the club felt Krol would help the squad play the way they—and the fans—wanted Napoli to play.

Of course, Napoli also needed a coach able to get results while still entertaining the crowds. Rino Marchesi was not quite the man for such a job. A defensive-minded player who played in central midfield toward the end of his career, he had most recently led Avellino to 12th place in Serie A, a point behind Napoli in the 1979–1980 Serie A season. Marchesi played pragmatic football; like much of the rest of calcio, his Napoli did their best to never concede. What was better about the side he came to take charge of was that they brought in Krol. Not only was the Dutchman great defensively, but he also had a perfect long ball that opened up many opportunities on the counter and helped Napoli score a number of goals that season. In addition, it's thought that Marchesi was the first to use something that would later become known as the "false 9," although it was not identical to its use in modern football.[22]

It was a dramatic change, and no one knew for sure if it would work out. But Napoli were at least doing *something*, and that something seemed that it would help the club . . . even if it hadn't led to the increase in season ticket holders that the board had hoped for. The supporters held their breath in anticipation of the first game, played at the San Paolo. The opposing team were their Southern cousins from Catanzaro, who would've been relegated the season before had the match-fixing scandal not broken. Surely it would be an easy win, and the fans believed they'd finally get to see a more exciting brand of football.

Alas, not much seemed to go right for Napoli that day. First, the side was without Krol, making the club furious with the FIGC, as they felt they had filed his documents in time.[23] Things finally broke correctly when Napoli were awarded a penalty in the 59th minute, which Gaetano Musella converted. But the players didn't seem as though they'd quite mastered the artistry Marchesi was going for, and Catanzaro broke through their defense. Antonio Sabato scored in the 67th minute, and the game ended in a 1–1 draw.

It wasn't bad, but it wasn't great. Next they were on their way to Ascoli, a team who had surprised with a 4th-place finish the season before. Sure enough, they lost 3–2, and the first of the Napoli goals was an own goal by defender Donato Anzivino. Still, they had created a number of chances, and had bothered the Ascoli defense enough to confuse them into an error. Despite their own shaky defense, Napoli kept pressing forward, until at last a goal from Antonio Capone in the 88th minute lifted their spirits, even if it didn't lift them to a point. The game, at least, showed progress on Napoli's part. They knew Krol was still not in peak condition; in fact, Giovan Battista Fabbri, the Ascoli manager, remarked on Napoli's progress and stated that it was clear that once Krol settled in and Napoli became more familiar with their new style, they would be a very good side.[24] But the next game, a mere 1–0 win over newly promoted Pistoiese, secured in the 88th minute by Claudio Pellegrino, was rather lackluster. It didn't help that Napoli next traveled to Milan and were beaten 3–0 by Inter, the last of the home team's goals an *autogol* from the man who was supposed to be Napoli's new hero, Krol.

Then something clicked. After a week off, Napoli entered the San Paolo to face Roma, who had won the Coppa Italia the previous season, and were feeling confident of a title run this time around. Napoli broke them, their brand of fast, efficient football that often saw them inside the Roma area caused the visitors to falter, to the tune of two own goals. Pellegrini contributed another, and Enrico Nicolini finished the *Giallorossi* off in the 61st minute. Suddenly, the talk of a possible title challenge shifted away from Roma. Napoli supporters began to come around. Marchesi no longer tolerated being mocked for his methods such as having the players do yoga to prepare for games, saying it was clear it brought advantages to Napoli's play.[25]

Rampaging over Roma gave the players the confidence boost they sorely needed, but in the next round they were only able to come away from Cagliari with a drab, goalless draw. At home to Avellino in the next round, Pellegrini came alive once more, slotting in the team's only goal and subsequent winner. Two weeks later, the Partenopei headed to Bologna for Round 8. Again, Pellegrini gave his side the lead, but this time, the Napoli defense slipped up. *Rossoblù* forward Giuliano Fiorini, a man much better suited to the lower divisions, equalized less than ten minutes later. But it was after the game that disaster struck. Literally.

At around 7:30pm on November 23, 1980, as Napoli traveled back from Emilia-Romagna, the ground in Campania began to shake. The 6.9 magnitude Irpinia earthquake, lasting nearly a minute and a half, had three main epicenters and was followed by over 80 aftershocks. The province of Avellino, where Napoli had so recently emerged victorious in a *Derby della Campania*, was hit hardest. With nearly 3,000 killed, 8,000 wounded, and around 300,000 left homeless in the southern regions of Basilicata, Puglia, and of course Campania, it is considered Italy's greatest natural catastrophe.[26] Entire towns were left without means of communication, and vehicles, finding it impossible to move, filled the roadways, preventing anyone from leaving— or arriving. It took rescuers up to 48 hours to reach the more remote villages, where residents remained trapped under the rubble.

Naples suffered significant damage—buildings were leveled, the *L'autostrada del sole* (Highway of the Sun) was blocked, and neither it nor the train network were usable for days. The Napoli squad were traveling a relatively clear highway, and their bus driver broke the speed limit so they could all be reunited with their families as quickly as possible.[27] According to Sandro Ruotolo, a politician and former journalist for state-owned broadcaster RAI, the earthquake gave rise to further organized crime, and "entrepreneurial camorra," in which many of the funds meant to aid the city and region were instead funneled to corrupt politicians, business owners, and the Camorra itself.[28] While people from across the country aided in the rescue area and people across the world pitched in on reconstruction efforts, even 40 years on, rebuilding work in the regions remains unfinished, and Naples continues to pay off debts related to the disaster.

At least Naples survived. In Campania and Basilicata, entire villages were completely destroyed. Claudio Vinazzani and Giuseppe Bruscolotti, the latter of whom was a Campania native, and two members of the Napoli coaching staff, set off— with the club's blessing—on a drive as soon as the roadways were clear enough. Their mission was to visit the towns and villages hit hardest by the earthquake, distributing food, medicines, and other necessities.[29] Obviously football became an afterthought, and Napoli's next match, at home to Brescia, was postponed; the team did not need to play again until Torino visited on December 14. A hurting Napoli lost 3–1, drew 1–1 with Brescia in the rescheduled midweek match, and came away from Perugia with a goalless draw. Only after Christmas, away at Fiorentina, did Napoli manage another win, 1–0 with a goal from Gaetano Musella, a native of Naples.

Napoli came back to life but, even as title contenders who actually showed off some firepower this season, they were still in the habit of securing draws, which likely hurt them in the end. Even worse, most of these draws were only thanks to come-from-behind goals, such as the game at Juventus in Round 14, when Pellegrini put Napoli ahead inside two minutes, but Marco Tardelli made it 1–1 in the 58th. It was the fifth time in the season that Napoli failed to hold on to a lead. Was it that the players were simply still uncomfortable with the new tactics? Or was the defense so weak that simply bringing in Ruud Krol wasn't going to fix it?

During the ritorno, Napoli fared much better, going on a ten-game unbeaten run. This time, they beat the teams they were supposed to beat, like Ascoli, Pistoiese, Brescia, and Torino. In no game did a team come from behind to take points from Napoli. In fact, a Walter Speggiorin goal in the 78th minute in Rome canceled out Roberto Pruzzo's, and Roma and Napoli split the points in a 1–1 draw. Napoli also took revenge on Inter, with only one goal, from Mario Guidetti, necessary to best the *Nerazzurri* at the San Paolo. After beating Torino 1–0 with an early goal from Gaetano Musella, Juventus, Roma, and Napoli were all even on points at 35.

Then Perugia came to town. Just two seasons earlier, they'd been contenders for the title, finishing in 2nd place. This year, they were dead last, with 13 points before the Napoli match. However, if they hadn't had five points knocked off thanks to the previous year's scandal . . . they still would have been in the relegation zone. Perugia simply weren't good that season, and ultimately ended up relegated. Yet somehow, they beat Napoli. 1–0. At the San Paolo. Moreno Ferrario scored an own goal in the first minute, Perugia turning Napoli's strategy directly back at them. Napoli simply weren't able to recover.

It was the end of the line for the *Ciucciarelli*. Their title dreams crushed, they finished the season with one win in four games. It can almost be said that they handed Juventus the scudetto. When the *Bianconeri* came to town, they managed another 1–0 win from another own goal, this time netted by Guidetti. Juventus had just come off a draw with Roma, and if Napoli had come back to split the points with Juve, it's possible that Roma would have taken their final match against Avellino more seriously. Instead, they played out a 1–1 draw that allowed Avellino to stay up. A slight solace for supporters still repairing their province from the devastating earthquake.

It's probable. It's possible. It could have happened. All things Napoli fans felt throughout much of the season, but also arguments that leave calcio fans arguing in circles over what a team *might* have done should they *maybe* have had a chance of winning the title, even if said game occurred decades ago. It helps little that Italian teams have an informal policy of "You scratch my back, I'll scratch yours. Someday."

In other words, if a team that has already won the championship, been relegated, or are simply in a slot on the table where a win won't be of any help, often that team won't play their hardest if the other team needs a boost to the scudetto or a European spot (especially if it hurts Juventus), or if they require a helping hand to save them from relegation. Obviously teams refuse to take part in this gentleman's agreement if they are rivals, or if it would help their rivals. But when they sacrifice for such a noble cause, the result is often a *gemellaggio* (twinning)

Friendships with clubs are not limited to Italy, but usually those clubs are located in different countries, and there is certainly no word that describes the concept. The gemellaggio is almost as common in Italy as *gufare* (to root against)—something that most Italians do when Juventus or one of the Milan sides is playing. It also occurs when a local rival plays. "*Campanilismo*," literally the love of one's bell tower but meaning the love of one's hometown or neighborhood, created historic rivalries between villages and between regions, and football ultras tend to follow those same lines, making enemies of nearby clubs.[30] But at the start of the 1980s, Napoli had never been a truly successful club, having yet to capture a scudetto, and nearby clubs such as Avellino fell out of Serie A far more often than they rose up into it. Therefore, the only reason for calcio followers to wage a gufare against the side was due to the fact that they were a Southern team that Northern clubs despised simply for existing. Rather than despise Napoli fans for their politics, an identity that ultra groups had embraced in the previous decade, other supporters saw them as outsiders, not even part of the same country. Groups looked to their bell towers, embraced neolocalism, and as a result became more likely to adopt right-wing ideologies and, subsequently, their racist and xenophobic behavior.[31]

Gufare against Napoli was increasing, but at the start of the 1981–1982 season, the Partenopei had no friends to rely on to help boost their chances of making a successful run at the title (the gemellaggio between Napoli and Genoa, the longest-running in calcio, began when the whistle blew on the last game of that season). Ferlaino and Marchesi stayed on. Despite his introduction to Napoli fans being that own goal, Ruud Krol had had a good season in Naples overall, and he remained on as the club's foreign player. In fact, not much changed at the club in general. A few bench players, like Antonio Capone and Walter Speggiorin, were sold on. Surprisingly, Napoli made just one move to shore up the backline, perhaps hoping it would improve more quickly if few changes were made. They purchased full-back Filippo Citterio, too strong to remain with Lazio in Serie B. Napoli also bought Paolo Benedetti, from Pistoiese, a central midfielder who was able to play the way Marchesi desired, sometimes dribbling forward skillfully, sometimes dropping back to act as a *libero*. Antonino Criscimanni also stepped into a midfield role. A mistake was made in the purchase of Massimo Palanca, called "The Emperor" by adoring fans at Catanzaro. He was famous for his ability to score directly from a corner kick, performing the trick 13 times in his career, all with his left foot. Unfortunately, he was unable to replicate that famous shot at Napoli, and scored only one goal during the season.

– PIZZA –
The art of simplicity

Sometimes, you just need a break. Particularly when your team is performing poorly, and it seems the anguish will never end. And what better way to drown your sorrows, to escape from the world, than with pizza?

Well, limoncello, perhaps. But pizza never left anyone with a raging hangover the next day—and when your heart's already breaking, the last thing you need is for your skull to follow behind.

Pizza is a big deal throughout much of the world, but it was the Neapolitans who, if they didn't invent it, perfected it. The first record of something similar to pizza might be more akin to contemporary flatbread, meant more as an appetizer than a meal. The Greeks of Neapolis topped a circular round of dough with olive oil, herbs, and feta, establishing the foundations of pizza in the seventh century.[32]

So, in a way, pizza was invented in Naples. And long after the Greeks and their feta-olive-oil-covered dough departed, the people of Naples elevated the pizza to a fine art. As part of the Spanish Empire, the city was introduced to tomatoes in the 16th century. Naples never looked back. Tomatoes were not used as a pizza topping until 1738, but it's extremely rare to find a Neapolitan pizza without them today. And unlike other locales, Naples keeps it simple. The most common pizza is a Margherita, topped with tomato sauce, mozzarella, and fresh basil. Elsewhere, this functions as a base to hold multiple other toppings; in Naples, it remains beloved in and of itself.

Like in the United States, where New York, Chicago, and Detroit pizza styles dominate certain regions of the country, other Italian provinces began creating their own distinct pizza types in the 1940s, such as the square slices of fluffy dough served in Roman bakeries. The thin, yet chewy, crust with few toppings continues to dominate in Naples, home of the best pizza in the world. Pizza rose to prominence and remains immensely popular not only due to its tastiness, but its affordability and portability. Although the COVID-19 pandemic and its attendant supply chain shortages may have forced New York City to finally raise the prices on its famous dollar slices, this easy meal is still a bargain almost everywhere, and a portion can be grabbed and eaten on the go—especially handy for workers in the South simply doing their best to eke out a living.

Once again, Napoli started off the season poorly. The animosity between Ferlaino and Juliano, the latter of whom had been appointed sporting director, became obvious. Juliano even said "I hope Napoli lose every game so Ferlaino would finally leave." While lauded for having enticed Krol to come to the club, the supporters were certainly not okay with a club official wishing ill on the Partenopei. A hostile atmosphere pervaded both the stands and the locker rooms.[33]

On the pitch, Claudio Pellegrini scored just after the break to put his side ahead in the first game of the season, but once again Napoli dropped the ball, conceding a penalty to Catanzaro, converted by Edi Bivi. Away at Cagliari, Napoli picked up another 1–1 draw, once again Pellegrini putting his team ahead only to be let down by the defense with ten minutes left. Then Milan came to town, and Moreno Ferrario gifted them the only goal in the match. The 4th round was a goalless affair at Ascoli. It seemed as though this season would be a carbon copy of the last, although at least this time Napoli had made it out of the group stage of the Coppa Italia. However, they failed in Europe once again, falling to Yugoslav team Radnički Niš—Napoli actually came back from behind to secure a 2–2 draw at the San Paolo, but when they traveled to Yugoslavia, Radnički stuck fast, holding out for a goalless draw that saw them through on away goals. These matches may well have contributed to their poor performance at the beginning of the season, as could the 1st round of the Coppa Italia, given they beat Ascoli 2–0 on September 6, then traveled to Catanzaro on September 8. The Yugoslav games occurred shortly before the away draws at Cagliari and Ascoli.

And then, freed from UEFA Cup obligations and relieved to have a break from the Coppa until November, Napoli took off—in the 5th round, exactly as they had done the season before. Seven rounds unbeaten (although three of those were goalless draws) until they encountered Fiorentina. The *Viola* were making a serious title run, and they drew first blood with a goal from Francesco Graziani. However, Pellegrini was right there to answer the shot with a goal of his own. But once again Napoli failed to hold it together in the end, and a 76th-minute goal by Ricardo Bertoni led to Fiorentina's 2–1 win. Given that Fiorentina went on a 22-game unbeaten run that lasted until the end of the season, Napoli should have felt no shame at the loss. They finished out the first half of the season with a win over Roma, held Juve to a goalless draw, and then lost 2–0 to Genoa.

Napoli played well through the end of the season, continuing to earn points through hard-fought draws (15 in total) and scoring more goals than the fans had become accustomed to. But it was a difficult year to be a title contender, and Fiorentina found themselves heartbroken after losing out on the scudetto to Juventus by just one point. (Ever since, Fiorentina fans have been twinned with Catanzaro, with both teams believing Juventus was given an unfair penalty in their last round match with the latter, which naturally made them detest the Old Lady even more.) Napoli finished 4th, after Roma, but were granted a UEFA Cup spot because the Albanian team dropped out of the tournament.

Third place in 1980–1981, but knocked out of the Coppa in the 1st round. Fourth place in 1981–1982, but overcome after one two-legged tie in the UEFA Cup. As the 1982–1983 season began, the fans became restless once again. Their relationship with Ferlaino deteriorated further. Supporters knew exactly what they needed: someone who could score goals. Pellegrini's 11 goals in the Serie A season were nice, but Roma's Roberto Pruzzo had scored 15, and even Catanzaro's Edi Bivi had scored 12. Fiorentina, Inter, and even Cesena—who'd finished 10th—each had two players who had scored 9 goals. The closest Napoli could come was Mario Guidetti's four goals. Napoli needed a genuinely strong attack, with more than one finisher to take the pressure off the center-forward.

Instead, Ferlaino bought Ramon Díaz from River Plate, who had scored 57 goals in 3 seasons in Argentina. The move was about as helpful as the redesign of the crest, which now integrated the head of a donkey into the left side of the "N." Díaz came nowhere near his average of 19 goals per game in South America. He scored eight total, but only three in Serie A. His brace helped Napoli beat Lazio 2–1 to advance to the Round of 16 in the Coppa Italia, and his goal against Cesena helped them move to the quarterfinals. However, neither he nor anyone in the Napoli squad could get a goal past Torino while at the San Paolo, and they were dumped out of the domestic cup. Díaz also scored a vital goal at Dinamo Tbilisi, whom Napoli met once again in the UEFA Cup. That away goal carried them to the Round of 32, where Díaz scored against Kaiserslautern at the San Paolo, but the Germans won 2–0 at home, turfing Napoli out of another competition. In the league, he wasn't nearly as successful; the only one of his goals that could be said to be essential was the one he struck at Torino, but once again the Vesuviani let the home side come back, ending in a 1–1 draw.

Ferlaino also bungled the managerial appointment. Rino Marchesi, whose contract had run out at the end of the 1981–1982 season, left the club, although no one could quite figure out if he had refused Napoli's terms or if Napoli had not met his.[34] The club brought in 43-year-old Massimo Giacomini. As a young man, he'd coached the 1973–1974 Udinese squad, in Serie C, for seven matches. He moved to Treviso the next season and boosted them to a Serie D championship. He returned to Udinese in 1977–1978 to lead the *Friulani* to a Serie C championship, upon which they were promoted to Serie B. Astonishingly, Giacomini led Udinese to a double promotion, clinching the Serie B title in 1978–1979. This led Serie A champions Milan to poach him for the 1979–1980 season, in which he managed to guide the side to a UEFA Cup spot, but with three Rossoneri players caught up in the Totonero scandal, along with president Felice Colombo, the side were relegated to Serie B. Although Giacomini had brought in excellent players who'd helped Milan secure instant promotion, he was encouraged to resign after the penultimate game of the 1980–1981 season due to his issues with both Totonero and upper-level management. Because taking Treviso to Serie C, Udinese to Serie B, and Milan back up to Serie A labeled him a successful coach, Napoli took him on to be the one to finally lead them to the title. They perhaps should have looked at his performance in the 1981–1982 season, when Torino languished in 9th.

The fans *must* have been skeptical over the selection of Giacomini, especially when they brought back Antonio Capone, declared Díaz a young superstar who would follow in the footsteps of former Argentinians who had played for Napoli, and spent 600 million lire for co-ownership of Roberto Scarnecchia,[35] and 700 million for Paolo Dal Fiume,[36] both midfielders not known for their goalscoring prowess. But surely they did not expect a 10th-place finish.

Yet that's exactly what they got. Under Giacomini, and then Gennaro Rambone (whose greatest accomplishment was leading Catania to a Serie C1 championship twice), who even partnered with Bruce Pesaola in hopes of getting Napoli back to a decent position, it seemed like the club had reverted to their old, boring selves. They ground out draws. They failed to score, notching only 22, the same as relegated Cesena. It was enough to convince Ferlaino to finally give up his role as president in midseason. About the only positive note was that they were finally the team scratching out scrappy draws, coming from behind to make sure their team scored at least a point. And those 14 draws left them 2 points above the relegation zone.

Tenth. Place. For the fans who thought they were watching the ascendency of their great club, 10th place was not good enough. For a man who'd bought a football club for the greater prestige that it conferred, 10th place was not good enough. Yet despite the failures on the pitch, Corrado Ferlaino returned to head up the club. Of all people, it was Antonio Juliano who convinced him to come back.[37] Only Ferlaino and Juliano, working together, could salvage the next season. Napoli needed fixing, and it needed to be done over the summer—otherwise, who knew where they'd end up?

The answer was 11th. In other words, Napoli had actually managed to become worse in the 1983–1984 season, finishing only one point above the relegation zone. Dissatisfied with Giacomini (and no wonder), Ferlaino installed Pietro "Rino" Santin, a 50-year-old who'd spent 11 years as a manager bouncing, season to season, throughout the lower leagues, until he landed at Cavese for 3 years, where he very nearly got the side promoted to Serie A. It is possible there wasn't much choice in top-level managers at that time, and Santin's resume made it to the top of the pile. After one season, he'd head back to shuffling around the lower divisions, spending a year at each club once more. He didn't actually last the full season at Napoli, replaced on February 22 by Rino Marchesi once more.

Napoli shipped off Ramon Diaz to Avellino and tried again to find a strong goalscorer. This time the club landed on attacker Giovanni De Rosa, from Palermo. He'd scored 29 goals in 61 appearances for the Serie B side. But once again, buying from the lower divisions failed to pay off—De Rosa did score the club's most goals in Serie A, six, to add to the two he'd scored in the group round of the Coppa Italia, where Napoli were eliminated. Napoli also bought yet another foreign attacker, the Brazilian Dirceu. He scored five goals in the Azzurri shirt—about average for him over the course of his career—despite playing every match that season. Bringing back Luigi Caffarelli from his loan at Cavese proved to be a good choice, as he showed himself to be a solid midfielder who remained at the club until 1987. Simone Boldini, from Ascoli, was meant to help strengthen the backline for two seasons, but defense was never Napoli's real problem. Yet still they brought back defensive midfielder Pasquale Casale, who bailed at the end of the season, asking to be sold as he was tired of coming into almost every match as a substitute.[38]

Napoli supporters were warned from the start how the year would go. The squad failed at the first stage of the 1983–1984 Coppa Italia, beaten by Udinese and Varese, the latter of whom were in Serie B. Then came Serie A's opening day, and Napoli were away at Florence. They returned with heads hanging in shame, having lost 5–1 to Fiorentina. Even when Massimo Palanca finally scored in the 79th minute, the Viola compounded the humiliation by answering with one last goal of their own. Napoli also lost 5–1 to Roma in the 7th round, and suffered back-to-back 4–1 losses to Sampdoria and Udinese, before closing out the andata with a 0–0 draw with Torino. Napoli collected three wins in the first half of the season. From there, Napoli drew the first three matches of the ritorno, including a 1–1 with Juventus in which the Partenopei came from behind to equalize through De Rosa, and then beat Avelino. It looked like Santin might keep his job, despite sitting 15th, just one point above the relegation zone. Then came the humiliating 2–0 loss to Inter at home, which was finished off with a Moreno Ferrario own goal. Santin got the sack, Marchesi returned, and Napoli just barely managed to stay in Serie A, thanks to a few good wins and, of course, a number of draws.

The early 1980s were not a good time to be a Napoli fan. Or Rino Marchesi. Or—perhaps especially—Corrado Ferlaino. Imagine a city of three million people, almost all of whom utterly despise you. The club Ferlaino had bought to enhance his prestige was now dragging his name through the mud. It was crucial he do *something*. Backroom machinations began.

In fact, they had begun much earlier than the summer of 1984.

[1] Encyclopedia Britannica. (last updated 2020, January 1). Italy. In *Encyclopedia Britannica*. Retrieved February 1, 2022 from https://www.britannica.com/place/Italy/The-economy-in-the-1980s.

[2] Corbo, A. (1976, May 20). Vinicia fa a braccio di ferro con Ferlaino ma i napoletani non si scaldano piu tanto. *Corriere dell'Informazione*, 12.

[3] NapoliTube. (Accessed 2022, January 6). *Napoli - Verona 4-0 | Coppa Italia 1975-76, finale | commento di Sandro Ciotti e 2 gol di Savoldi* [Video]. YouTube. https://www.youtube.com/watch?v=c1J6ujG0qoE&ab_channel=NapoliTube

[4] DeFelice, G. (1976, June 30). Finalmente il Napoli ha finto qualcosa. *Corriere della Sera*, 20.

[5] Winter, S. (2016, February 23). The Story of Giuseppe Savoldi, football's first million pound player. *These Football Times*. https://thesefootballtimes.co/2016/02/23/the-story-of-giuseppe-savoldi-footballs-first-million-pound-player/

[6] Olivari, S. (2019, September 13). Da Savoldi ad Ancelotti, gli abbonamenti del Napoli. *Guerin Sportivo*. https://www.guerinsportivo.it/news/calcio/2019/09/13-2373426/da_savoldi_ad_ancelotti_gli_abbonamenti_del_napoli

[7] Sasso, E. (1975, December 8). Savoldi e' rotto a il Napoli trema. *Corriere dell'informazione*, 7.

[8] I provedimenti del giudice per il guardaline ferio al San Paolo: Tre giornate di squalifica al campo del Napoli. (1977, January 13). *Corriere della Sera*, 16.

[9] Incidenti al San Paolo per la sconfitta del Napoli. (1977, May 23). *Corriere della Sera*, 20.

[10] Lajolo, G. (1976, July 13). Chiarugi al Napoli, Braglia e G. Morini al Milan. *Corriere della Sera*, 15.

[11] Juventus a Napoli con tanti giovani. (1978, May 13). *La Stampa*, 17.

[12] Bianchini, M. (1978, June 9). Grande delusione dei tifosi napoletani all'Olimpico Bini all' 88' e la Coppa all'Inter. *La Stampa*, 17.

[13] Visioli, E. (1978, October 11). Vinicio per il Napoli e' come San Gennaro; Gli hanno gia' chiesto di fare due miracoli. *Corriere della Sera*, 22.

[14] Petrone, N. (1979, December 4). Il Napoli guida la crisi del calcio italiano; Si gioca sempre peggio, si segna poco, il pubblico giustamente protesta. *Corriere della Sera*, 22.

[15] Quando il pallone si sgonfiò: 40 anni fa lo scandalo-Totonero che travolse la Serie A (2020, March 23). *Sport Mediaset*. https://www.sportmediaset.mediaset.it/calcio/quando-il-pallone-si-sgonfi-40-anni-fa-lo-scandalo-totonero-che-travolse-la-serie-a_16452898-202002a.shtml

[16] Corbo, A. (1979, November 5). Il Napoli sempre piu' giu', ma Vinicio non si tocca. *Corriere dell'Informazione*, 17.

[17] Vinicio si dimette, Ferlaino lo trattiene. (1980, February 12). *La Stampa*, 18; Luise, A. (1980, April 6). Il Napoli vince, Vinicio si dimette. *La Stampa*, 21.

[18] Luise.

[19] Il Napoli di Vinicio. (2020, April 7) *Guerin Sportivo*. https://www.guerinsportivo.it/news/calcio/2020/04/07-2903476/il_napoli_di_vinicio/#:~:text=Ritornato%20nell'ottobre%201978%20a,il%201973%20e%20il%201976

[20] Bianchini, M. (1980, August 4). Marchesi punta sui giovani per fare un nuovo Napoli. *La Stampa*, 14.

[21] Krol resterà al Napoli Accordo col Vancouver. (1981, January 31). *La Stampa*, 21.

[22] Schioppa, R. (2019, October 12). L'eleganza di Krol che faceva sognare tutta Napoli. *La Repubblica*. https://www.repubblica.it/dossier/sport/la-partita-della-vita/2019/10/12/news/l_eleganza_di_krol_che_faceva_sognare_tutta_napoli-238316081/

[23] Cisternino, A. (1980, December 14). Il Napoli senza Krol per un Bisticcio di telex. *Corriere della Sera*, 25.

[24] Bertolani, L. (1980, September 22). Krol ancora a mezzo servizio, Bertoni a tempo pieno. *Corriere della Sera*, 21.

[25] Cisternino, A. (1980, October 20). Marchesi, e' merito della yoga il grande successo? "Troppa ironia su questo argomento, ma io ci credo." *Corriere della Sera*, 18.

[26] 40 years on, why the Irpinia earthquake is remembered as Italy's "worst catastrophe." (2020, November 23). *The Local Italy*. https://www.thelocal.it/20201123/40-years-on-why-the-irpinia-earthquake-is-remembered-as-italys-worst-catastrophe/

[27] De Falco, D. (2020, November 3). Quarant'anni dal terremoto in Irpinia, l'azzurro Pellegrini ricorda la trasferta a Bologna: "Vivevamo nella paura." *Spazio Napoli*. https://www.spazionapoli.it/2020/11/23/claudio-pellegrini-terremoto-irpinia/

[28] 40 anni dal terremoto dell'Irpinia, de Magistris: "Ricordo quel cielo viola e la parete di casa che si aprì." (2020, November 23). *Napoli Today*. https://www.napolitoday.it/cronaca/terremoto-irpinia-cosa-successe-napoli.html

[29] Vinazzani, Bruscolotti e i massaggiatori: Quattro del Napoli nei paesi disastrati (1980, November 29). *La Stampa*, 20.

[30] Doidge, 145–50.

[31] Id., 155.

[32] Choi, D. (2020, October 20). How pizza was invented. *History of Yesterday*. https://historyofyesterday.

com/how-pizza-was-invented-1e1311e66388

[33] Raio, V. (1981, October 13). I tifosi del Napoli amano ma pretendono. *La Stampa*, 25.

[34] Raio, V. (1982, April 16). Marchesi lascerà il Napoli o viceversa? *La Stampa*, 23.

[35] Scarnecchia—La Roma inaugura il Mercato; Scarnecchia ceduto al Napoli. (1982, September 29). *Corriere della Sera*, 26.

[36] Gandolfi, G. (1982, July 14). Novellino all'Ascoli, Maltiera alla Roma? *La Stampa*, 17.

[37] Cisternino, A. (1983, June 8). Ferlaino di nuovo presidente del Napoli, Juliano puo' cominciare la ricostruzione. Corriere della Sera, 26.

[38] Raio, V. (1985, July 24). Ora Maradona fa il dribbling a Napoli. *La Stampa*, 20.

CHAPTER 7

A Dios le Pido

7

On July 5, 1984, at least 75,000 people packed the stands of the San Paolo, all having paid a thousand lire for the privilege of seeing a man who'd yet to kick a ball for their team, Napoli.[1] The supporters had been buzzing about the player throughout the month of June. When the helicopter arrived, "like an angel descending from heaven,"[2] a short, squat, dark-skinned man with a pile of wild curls emerged, looking nothing like a traditional deity. But a god is exactly what Diego Armando Maradona became to Naples.

Tempting a deity

At the time, Serie A was the strongest, the richest, the most powerful league in the world. Yet the FIGC decided to impose a temporary block of three years on Italian clubs bringing in players from outside Italy. This was done in order to "protect" *calcio*, because if there were no limit on foreign players, the smaller clubs would lose the revenue stream funneled to them by bigger teams buying their players. Clubs were permitted to buy players from non-Italian clubs until June 30, 1984, after which they could only buy from teams within the country.[3] Almost every team, even the weaker ones, paid to bring in the best from around the world. From Juventus and Michel Platini, Michael Laudrup, and Zbigniew Boniek to Udinese and Zico, other European teams were left far behind.[4]

Even with official permission, it was beyond difficult to bring Maradona to Napoli. Barcelona wanted him gone, for a number of reasons, yet they also didn't want to lose much of what they'd invested in him when he transferred from Boca Juniors, an amount worth roughly 7.3 million euros. Then they recalled the cocaine, the orgies, the gambling, the way he'd hurt Barça's image.* Still, this would mean letting go of the best football player to have emerged over the last generation. However, they soon had little choice: Maradona heard about their lack of commitment and publicly declared he would never play for Barcelona again.[5] Word went out throughout Italy that the Catalan club would sell Maradona. And luckily for Napoli, there was not much competition. Juventus declared he would never succeed in Italy, citing his short stature.[6] The nearly 9 billion lire in debt Napoli were said to be carrying would almost certainly be an obstacle.[7] But President Corrado Ferlaino had a dream, and he wasn't willing to let it go yet. Besides, Maradona *wanted* to come to Napoli.

Knowing just how much they coveted the Argentinian, Barcelona squeezed Napoli for everything they had, to the point where citizens on the street were giving

*See the following chapter for more information on Maradona, the man.

their change to those collecting for the cause. It didn't matter that Maradona hadn't performed all that well at Barcelona; they had seen him in the 1982 World Cup and knew of his talent. They also discussed his performances in Argentina, where he won the title with Boca Juniors. It wasn't just the club who believed Maradona was the missing piece in their *scudetto* puzzle. The fans were convinced from the start that he would be the one to bring the trophy south.

And so the negotiations began. Perhaps surprisingly, Napoli were led to Maradona by Pierpaolo Marino, the sporting director of Avellino. Much like Napoli, they had just barely escaped relegation to Serie B on the final day of the 1983–1984 season. The ambitious club had arranged a friendly with Barcelona during the summer of 1984, but without Diego Maradona, whom Avellino were told would not remain in Spain, the fans would not be enthusiastic about attending; they'd lose the large gate funds Avellino had counted on, the funds they had wanted to advance their club and develop it into one that regularly remained in the top half of the Serie A table. Thanks to his Argentina colleague Ricardo Fujca, Marino was the first one in Italy to know Diego was available. Marino cracked a joke about Avellino buying Maradona, but obviously they did not have the means, so he picked up the phone and called Juventus' sporting director. Giampiero Boniperti said that, due to his reputation, Maradona was not a good fit for the Serie A champions.[8] Paolo Mantovani of Sampdoria, mired in midtable in their first year back in the top division, said that Maradona was far too expensive.

Enter Antonio Juliano. The club knew he'd be able to work with the huge egos on the Catalan side, clearing obstacles with ease, and his native Neapolitan streetwise knowledge helped him know who he would need to bribe. The man's finely honed diplomacy helped him soothe tempers and cast aside threats. Ignoring Napoli's debts, or perhaps believing that bringing in Maradona would help erase them, he began negotiating for the purchase of Maradona on May 11, 1984. Juliano is one of those people who can immediately put almost anyone at ease, even the best players in the world.[9] While some suggest this trait ensured the negotiations went smoothly from the start,[10] that only applies to the talks with Maradona, his manager Jorge Cyterszpiler, and Juliano. It is said they agreed on 1 billion lire plus 600 million each year for 6 years; in addition, Diego would have the opportunity to earn bonuses. From there, Juliano took on the bulk of the negotiations with Barcelona, while Corrado Ferlaino worked with Cyterszpiler.[11] The latter was the easy part, as Napoli were willing to give Maradona almost everything he asked for. Barcelona, not so much.

Barcelona President Josep Núñez knew they needed a high return on their payment for Maradona, particularly as the fans were likely to be upset by the transfer. He demanded 13 million, in US dollars, as the currency was more stable than the Italian lira. Juliano, the most loyal of Neapolitans, was convinced his club needed something incredible to quiet the disparaging voices of the North, and knew Napoli absolutely could not pass on a player that was more or less priceless. So Napoli agreed, to the great surprise of Barcelona, who wondered how a team so in debt, in an impoverished city still attempting to rebuild after the 1980 earthquake, could be counted on to come through with the cash. Yet if there was one thing Ferlaino was good at, it was establishing and maintaining relationships with the political and economic elite of the city. Vincenzo Scotti, the mayor of Naples, intervened on the club's behalf to convince a group of banks to guarantee the loans that would secure the purchase of Maradona.[12]

It should have ended there, but President Núñez's hatred of Maradona forced him to take an extra step. Barcelona demanded Napoli pony up 600,000 more US dollars. At this point the fans, fed up with the drama and desperate to have the man they already believed was a god join their club, started giving of their own money. Yes, in this city that so many associate with complete poverty, where the early 1980s truly had not been kind—in addition to the Irpinia earthquake, many large factories closed or scaled down production, and where organized crime is often believed to take whatever money there is out of citizens' pockets anyway, Napoli supporters parted with whatever spare change they had. (It was also rumored that the Camorra helped in securing the funds for Maradona; if so, many believed they had bestowed further benevolence on the people of Naples, although, as time would tell, they would also extract their pound of flesh in exchange for bringing in the famous footballer.)

Then, on June 29, the day before the transfer window closed, Barcelona had the temerity to ask for even more.[13] It seemed like the transfer was about to break down. Enter Maradona, almost as excited to go to Naples as the people of Naples were excited to have him there. On the 30th, Maradona went to Núñez's house, stood beneath his window, and yelled about how he had to leave Barça. Diego then contacted Ferlaino. He'd put up the 15% required to be given to the player at the time of the transfer. Núñez, however, couldn't be found, and Joan Gaspart, his right-hand man, told the press that there was no deal. Ferlaino, furious, flew to Milan the next day and handed the Lega Calcio office an empty envelope, pretending it contained Maradona's contract. Maradona again spoke up, demanding to be let go, pointing out that Barcelona didn't even want him anymore. Finally, shortly after 5pm, Núñez counted the money. It was all there. By 10pm the press finally confirmed the transfer, and Naples burst into celebration. Ferlaino, and the signed contract, hopped the plane back to Milan, where the Lega office was closed. However, he caught the eye of a security guard, begged, pleaded, and was able to replace the old, empty envelope with the new one containing the contract bearing Diego Armando Maradona's signature.[14]

#SquadGoals

Yet one player, no matter how deified he is, does not a squad make. Especially as Napoli had let go of ten players that summer, including Claudio Pellegrini, Dirceu, and—most surprisingly—Ruud Krol, a true superstar in his own right. They would rebuild with players from their youth squad, such as defender Ciro Ferrara, and reliable players like Ascoli midfielder Walter De Vecchi. Maradona's presence attracted bigger talents as well, including attacker Domenico Penzo from Juventus, midfielder Salvatore Bagni from Inter, and Daniel Bertoni from Fiorentina, who had won the World Cup with Argentina in 1978.

– HEROES –

Ciro Ferrara (1984–1994)

Napoli appearances: 247 (league) • 322 (total)
Goals: 12

Depending on which Napoli fan is asked to name the club's heroes, Ciro Ferrara could well be the first homegrown player that lived up to the title. He was born in 1967, in the neighborhood of Posillipo, and joined the Napoli *primavera* squad in 1980. He began playing senior games in 1984, when he was just 18, and became part of the starting XI in 1985–1986. All this despite the fact that he was diagnosed with Osgood–Schlatter disease and confined to a wheelchair when he was 14. Such a disease often affects those who play sports involving quite a bit of running, producing a painful bump under the knee, with flares that can last months. Fortunately, most adolescent males grow out of the pain by age 16. Unlike many current players, he seemed uninterested in the bling playing football could bring him—when he signed his first professional contract, he asked his father for the loan of his Fiat 126, only because the Napoli fans mocked him for turning up to practice on a moped.[15] He also slipped away from journalists after playing his first Serie A match to go see an aircraft carrier that had arrived in Naples. Charismatic and intriguing, his off-the-field antics still couldn't compare to his talent on it.

Considered one of the best center-backs of his generation by age 20, most articles about Ferrara mention his fairness and decency—even when paired with adjectives like "gritty."[16] But Ferrara, a fair man on the field and off, was also celebrated for those talents that saw him win two *scudetti* and a UEFA Cup, as well as receive a call-up to the 1990 Italian national World Cup squad, by the time he was 24. His ability to perform in different roles—full-back, stopper, even sweeper, along with his enviable performances at center-back— and to adapt to many formations and tactics, despite particularly excelling as a man-marker, is praised in a number of articles written during his time at Napoli.[17,18] In fact, many reporters and pundits enthused about the Napoli and Italy defender, praising his technical skills and ball-playing ability;[19] labeled him a "complete" defender due to his speed, athleticism, strength in the air, and excellent ability to read the game; discussed his ability to make last-minute, clean tackles in both a man-marking and zonal-marking system;[20] and noted that his strong personality combined with his professionalism made him a trusted leader, both on the pitch and in the dressing room.[21]

Much of the credit for Napoli's scudetti wins in 1987 and 1990, and their UEFA Cup victory in 1989, is—naturally—assigned to Diego Maradona. But no matter how stellar one player is, he's not enough to lift a club to their greatest heights on his own. A modern example would be Cristiano Ronaldo, purchased to ensure Juventus could finally hoist the UEFA Champions League trophy; the fact that they did not do so before they sold him on to Manchester United illustrates the importance of a squad's talent and depth. Without Ferrara, who was there for the entire reign of Maradona, Napoli might have shipped far more goals than they did as they climbed their way to the top. The *Partenopei* conceded the third-fewest goals in both 1986–1987 and 1989–1990. The necessity of a strong back line was emphasized by Milan in 1988–1989, when Napoli scored 12 more goals, but the former let in just 17 goals. Milan's title showed that cautious play (the *Rossoneri*, under the famed Arrigo Sacchi, recorded 11 draws and just 2 losses) could lead to the scudetto.

Ferrara's aforementioned leadership and flexibility enabled him to lead a strong defense, but he was also considered a quality offensive defender, an opinion made fact during the 1989 UEFA Cup final, in which Ferrara scored the second goal in the second leg of the final, keeping Napoli in the match. They ultimately drew with Stuttgart 3–3, but won 5–4 on aggregate. This was Napoli's first significant European honor.

Finally, in May 2005, Ciro Ferrara proved his value as a human being, not just a footballer. Together with Fabio Cannavaro, also from Naples, he created the Cannavaro Ferrara Foundation, which cares for disadvantaged Neapolitan children.[22]

At first, the team didn't seem to gel. It didn't help that the first game of the 1984–1985 was to be held up north, at the Bentegodi in Verona. As life-long Napoli fan Randy Grice says, the Neapolitans and the Veronese "sure seem to hate each other." This statement has been confirmed many times and in many ways, and it is based on the division between North and South in Italy. As both internal migration and external immigration increased, racist chants and banners became more widespread, reflecting the change in politics and revealing the racism that pervades much of Italian life, even to this day. In the stadium, ultras hurled racist abuse at rivals, battering them psychologically to give their own team the advantage.[23] Maradona, the dark-skinned boy who grew up impoverished in another southern country, was a perfect emblem for the North-South divide. Insulting the color of a player's skin was not simply an overarching insult; it was a means of deriding their fitness levels or their technique.

Not-so-fair Verona

The city of Verona was home to a great number of Lega Nord members, who believed that the richer North should no longer go on paying for the poorer South. Then and now, a number of people took such ideas to the extreme, believing the richer regions of the country should split off from the others. According to sport historians Papa and Panico, matches between Northern and Southern teams "took on all the significance of a cultural and ethnic challenge."[24]

It's essential to remember that not all hurling racists insults in the stadium are ultras, and not all ultras are racists. According to Italian supporter expert Mark Doidge, "The emotional energy created by participating in an ultra group connects participants in a way that transcends their everyday life and creates a bond among members." Many participated in ultra groups simply to find friendship and share a common passion. However, certain clubs have earned a reputation for racist behavior; the ultras looked for any and all ways they could "conceptually violate" the competition, including insults against skin color and against place.[25] While it would only grow worse in the 1990s, Doidge believes that Verona engages in "instrumental" racism, which is ideologically motivated; far-right groups in the mid-'80s were already using the stadium to promote their political beliefs. At the same time, the Lega Nord, strongly rooted in the region, utilized and encouraged the fans' anti-Southern biases.[26] Many Hellas Verona fans fall on the far right end of the political spectrum, and like to express their leanings with "clever" banners. The one unfurled when Napoli visited on that first day of the 1984–1985 season read "Welcome to Italy"—the unspoken meaning referring to their belief that Naples was not part of their country, but instead an African nation.

Unfortunately, while the traveling Napoli *tifosi* managed to strike back with a banner of their own—"Juliet was a whore," in reference to William Shakespeare's play set in Verona—they were unable to counter the home supporters' jabs with a devastating win. As the teams dueled on the pitch, the Verona crowd continued to hurl insults from the stands. In their minds, Napoli was from the "black South," and so they started monkey chants and filled the ultras *curva* with chants of "*Seig Heil*," which matched the swastikas on their banners.[27] The season's eventual champions beat them 3–1 that day, with Napoli's only goal coming from winger Bertoni. This, despite the fact that the Verona players isolated both Argentinians, Bertoni and Maradona, and subjected them to brutal attacks.[28] Given that Diego had been brought in thanks to the collection of cash from the fans, possible involvement from the Camorra, and certainly with the help of the financial institutions of the city, whenever Maradona went down, the visitors would shout that the Banco di Napoli had also fallen.[29] This became a tradition whenever the star was fouled. But during that first game, it did not help that, in their clear-out of players that summer, Napoli had remained loyal to their 39-year-old goalkeeper Luciano Castellini. While still capable of pulling out an amazing save, he'd become sloppy, and he was a major part of the reason that Napoli lost to the Northerners that day.[30]

Disappointed, but not down

Napoli had sold 67,000 season tickets thanks to Diego Maradona (comparable to the number sold in 1975, when Giuseppe Savoldi debuted to much excitement; the total number of tickets sold during Diego's first year exceeded that sold in 1975–1976[31]), and the first home game, against Sampdoria, was nearly impossible to get into unless a fan was in possession of one of them, with the few thousand that were left sending the supporters trying to get into the match into a frenzy.[32] Unfortunately for the thousands of Napoli supporters packed into the San Paolo, the game ended 1–1. But those same supporters could dine out forever on the story they told, that they had witnessed Maradona's first Serie A goal in a Napoli shirt, a penalty he converted with a chip shot that dinked over Ivano Bordon. According to Ludden, the San Paolo shook as thousands danced on the terraces, while fireworks and rockets were let loose from the *curvas*. The home side was unlucky to be left with the draw; a shot hit the post, the Sampdoria goalkeeper made a few terrific saves, and the referee denied Napoli a second penalty.[33] Still, Maradona was in Naples, and all would soon be right in the world.

But not that season. Maradona's first goal from open play came in the 4th round, when Napoli hosted newly promoted Como. Their new star flew up the left flank, latched on to a pass placed perfectly in his path, and let fly a shot that landed in the upper right corner of the Como goal. He celebrated by leaping the advertising boards and running across the track to the home fans, a celebration that would occur again and again at the San Paolo throughout Partenopei history. Napoli's only other win before the Christmas break also came at home, to another newly promoted side, Cremonese. Daniel Bertoni scored the only goal in the 1–0 victory. At this point the Napoli tifosi were starting to wonder—was Diego Maradona really their savior? Or had they ponied up their own cash for a player who was turning out to be a dud? After all, they were fighting not to go down, not rocketing toward a scudetto.

Prior to signing, when journalists mentioned that Napoli had finished just a point above the drop the season before, Maradona stated the club seemed closer to the second division than to a championship. Half a season later, that's where they remained. Maradona should have kept his own words in mind: "I knew I was going to suffer—a lot—but I also knew that the harder something is, the more I like it. The less faith they had in me or us, the angrier I was and the harder I played."[34] This was in direct contrast to Giuseppe Savoldi who, after signing on in 1975, essentially gave up on the club, declaring they were on the precipice of falling apart—after which Napoli won the Coppa Italia and qualified for the UEFA Cup twice before he left in 1979.

But Maradona, too, was having regrets. He went back to Buenos Aires during the break, and spoke of the shame he felt for the way he was playing.[35] The people of Napoli had welcomed him unconditionally, treated him as one of their own. They made him a messiah from the start, the man that would lead their team out of the desert and cause them to hold their heads high when encountering the vicious fans of the North. He was failing. Maradona was a player who assumed responsibility for both his own personal playing problems and that of the team as a whole, making Naples love him even more. He needed to get his team back on track.

And that's exactly what he did when he returned to Naples. The first game after the Christmas break was at home, against Udinese, who were also fighting the drop. Twelve minutes in, Edinho put the visitors in front through a well-taken spot kick. Napoli fans crumpled. How long could this continue? Ten minutes later, Maradona went down inside the area, and converted the penalty awarded by the referee. 1–1. His compatriot Bertoni then gave Napoli the lead, and supporters were feeling confident until Udinese brought the two sides level again through Paolo Miano, a mere three minutes after the Partenopei had managed to get out in front.

This time, Maradona was determined to live up to the image the tifosi had of him. He threaded through Udinese's tackles and challenges, beating the defense again and again. With 15 minutes to go, Maradona put his side up yet again, through another penalty, and again Bertoni shortly followed. It was a good thing, too, because the visitors had one more left in them, a screamer from Marco Billia with just five minutes to go. It wasn't always safe and it wasn't always pretty, but in the end Napoli came away with the 4–3 win. And this time, they would go on not to win only two of their games, but to lose just two through the end of the season. Diego played his best game against Lazio, in Round 20. It took until the 58th minute mark for Maradona to get on the board, but he scored a fantastic hattrick in less than 20 minutes, including a delicious chip and a goal straight from a corner; most believe it should've been four goals, but the 78th minute goal was credited as an own goal by Daniele Filisetti.[36] Overall, Maradona scored a respectable 14 goals in his first season at the club. They finished 8th, but Napoli fans had seen what their Argentine god could do, and they were pleased.

Maradona's maneuvers

That summer, Diego became part of the transfer team. He was a man who more or less scouted as he played.[37] He, Ferlaino, and new manager Ottavio Bianchi (who had spent five seasons as a solid Napoli midfielder in the late 1960s) would gather for meetings, and Maradona would point to specific traits that would make a player an excellent addition to the Partenopei squad. The three put together a list that would help advance Napoli by recruiting players that excelled creatively and could be counted on to provide the final touch often needed in the area.[38] Ferlaino was wise enough to bring Pierpaolo Marino into the fold as general manager as well, considering he was the one whose connections allowed him to be the first in Italy to find out that Maradona was available.

Maradona helped in the selection of Bruno Giordano, a Lazio player with excellent dribbling abilities and a superb shooting strength. He was the type of player Napoli desperately needed to take some of the pressure off Diego, one who could combine the traits of a typical center-forward with those of a playmaker. He was also a street kid, having grown up playing football in the alleys of Trastevere, a neighborhood in Rome. Giordano would go on to say that his parish church saved him—the streets in the mid-'70s were heaving with drugs, but his priest would gather up the boys and make them play calcio all day, leaving them too worn out for drugs or petty theft. Maradona later called him the best South American born in Italy; as it turned out, he'd been begging Giordano to join him at Napoli before he'd even left Barcelona, sending him telegrams telling him, after he picked up a devastating injury at Lazio in 1983, to heal quickly because he wanted Giordano to move to Naples as well.[39]

Napoli let go of six players that summer, including Simone Boldini, Paolo Dal Fiume, and Walter De Vecchi. Their performances for the team had been solid, but not up to the standards Napoli were now measuring its players against—players that would push the squad into a run for the title. Luciano Castellini, the goalkeeper who had been an obvious weak spot the season before, retired. In came Claudio Garella, the keeper who had just led Hellas Verona to their first title. Garella had an unorthodox style, seemingly wanting to use every part of his body *except* his hands to block the ball. His theory of goalkeeping was to do whatever he could to keep the ball from going into the back of the net, and if that meant his gloves never got dirty, oh well.[40] As Garella tells it, it was Maradona who ensured his transfer to Naples, telling Ferlaino to take a chance on him, not just for his abilities but for his manner as well.[41]

This would also be the season that Diego would take on the captain's armband.[42] Giuseppe Bruscolotti passed it on while telling him that he would need to win the scudetto in exchange. When questioned about the change in captaincy, Maradona responded that he had answered "yes" to the question of whether he'd like to be the Napoli captain one day, but had certainly not asked for it then. He told reporters that it was disrespectful of them to suggest he'd asked Bruscolotti, a veteran Campanian who had been with Napoli for more than a decade, if he could be the one to steer the ship.

In the end, though, it was the shift from Rino Marchesi to Ottavio Bianchi that really turned Napoli around. The attacker had more than 100 appearances in the Napoli shirt, and from that time he had ideas of how the club should be managed. He felt Marchesi had been using Maradona incorrectly, and set about making changes. According to Bianchi, "It would be a mistake to deploy Maradona in a merely tactical position." He continued, "He must be left to express himself as best he can. He must be allowed to show his class, his characteristics. It's up to us to create a concrete block around him, to highlight these qualities."[43] With Giordano, Maradona was relieved of having to fight through defenders for possession in the middle of the area. Instead, he could concentrate on the attack—really, he could become the focal point of the attack altogether. The style was similar to the way Argentina played in the World Cup in 1982.[44]

Maradona ended up scoring 13 goals, 1 fewer than the previous season, while Napoli scored 35 altogether, 1 more than the year before. It would take time for Bianchi's plan to truly work, but the fans were certainly satisfied with the 3rd-place finish, made all the sweeter by last year's champions, Verona, falling to 10th. And they were ecstatic that they'd beaten Juventus 1–0 in November, Maradona's cheeky free kick flying directly into the top corner of the net. Five fans were so excited about the win—their first in twelve years—that they fainted.[45] The win was for all Neapolitans, challenging the premise that the South was inferior. After all, the man who'd made the difference hailed from the Southern hemisphere, and he'd decided to make Southern Italy his home. Although the *Bianconeri* would go on to win the title, victory over Juve was a triumph for the little guys against the establishment.

After the season ended, it was time for the 1986 World Cup in Mexico. Maradona may have made more than a few enemies throughout that tournament, but it was when he transformed from a star into a legend.[46] In the quarterfinal against the Three Lions, set against the backdrop of the Argentina–United Kingdom war over the Falklands, Diego showed how quickly he could turn from a conniving devil to an actual angel. Argentina's first goal came via Maradona's hand, although for decades he maintained it was scored through his head and God's hand. While England were infuriated, and remain so to this day, much of the world simply shrugged off his "Hand of God" moment, believing that any way you can get a goal past the referee, plus the linesman, was perfectly okay. It helped that the people of Mexico were willing to throw their support behind Argentina, one downtrodden Latin nation to another—especially as Mexico had just been kicked out by West Germany. It also helped that Maradona scored not just a perfectly legitimate goal four minutes later, but the "Goal of the Century." That goal cemented the public's view of Maradona as a player of incredible talent. One of the best descriptions of Diego's goal was given by Bryon Butler of BBC Radio:

> Maradona . . . turns, like a little eel and comes away from trouble. Little squat man, comes inside [defender Terry] Butcher, leaves him for dead, outside [the other center-back Terry] Fenwick, leaves him for dead—and puts that ball away. And that's why Maradona is the greatest player in the world. He buried the England defence![47]

Argentina went on to beat Belgium handily in the semifinal, with Maradona scoring both goals. For the final, Mexico's Azteca Stadium was filled with over 100,000 fans, almost all supporting the blue-and-whites against the team that had defeated Mexico. It was the most difficult game that Argentina played. They went up 2–0 before the hour mark, but then West Germany came roaring back. Two goals in seven minutes brought them level. It would have been a fantasy cup had Maradona been the one to score the winner, but instead he provided the pass that allowed Jorge Luis Burruchaga to score the winning goal. Maradona had captained his side to another World Cup trophy, and captured the world's attention while doing it. Napoli fans, already aware of his talents, celebrated Argentina's win late into the night.

The world sings a song that never dies

Now Napoli were *the* team to watch. Considering that most matches were not broadcast on TV in the 1980s, Diego fans had to make pilgrimages to the San Paolo. It's safe to say that in the 1986–1987 season, few were disappointed. Neither were the club, whose gate receipts continued to grow—receipts that had already swelled to the highest in Italy, even when the team were performing poorly, simply because Maradona was the one to watch.[48]

The season would become, and at time of writing remains, Napoli's best, most exciting ever. Prior to its start, Corrado Ferlaino, Italo Allodi, who had been promoted to general manager, Pierpaolo Marino, who had shifted to director of sport, Ottavio Bianchi, who remained on the Napoli bench, and of course Diego Maradona gathered to determine what pieces still needed to be filled, and who could fill them. Daniel Bertoni was the only significant player who left that summer, sold to Udinese. Andrea Carnevale came the other way to strengthen the attack, and Napoli bought the appropriately named Fernando De Napoli from Avellino, a defensive midfielder who enabled the creativity of Maradona and his teammates through his ability to break down the opposition, win possession, and quickly return the ball to the attackers. "Rambo" was nicknamed not only for his stamina, but for the way he took on the other team, being unafraid to get into their space and win the ball—while still remaining a player who others considered both fair and professional.[49] Finally, Francesco Romano from Triestina came in during the winter transfer window.

Unlike the modern-day Coppa Italia, which begins in August with the two lowest divisions of Italian calcio, in 1986–1987 even the 3rd-placed team from the season before were required to begin the tournament before the Serie A season began. Napoli were placed in a group with Lazio, Cesena, Vicenza, Taranto, and SPAL. The former three were competing in Serie B, the latter two in Serie C1. Napoli won each match, and only conceded goals against Cesena and Vicenza. Maradona scored in three of the matches, and Napoli easily topped the group. The perfectly put together Napoli side had won five games before Serie A even started. Unfortunately, their first matchup in Europe was intertwined with the start of the league season.

The 1986–1987 Serie A season began for Napoli on September 14, when Diego Maradona's goal lifted them to a 1–0 win to newly promoted Brescia, who were doomed to go right back down. Three days later, French side Toulouse came to town. Napoli came out the winners in a close fight, with Andrea Carnevale scoring the single goal. Then came their first home game, against Udinese. De Napoli struck first, but his shot was answered by Francesco Graziani early in the second half. The *Vesuviani* couldn't get themselves back in the match, and it finished 1–1. A team had once again come from behind to snatch points from Napoli. The fans were dismayed. Was this really a new squad, one that could conquer all of Serie A, or were they watching a replay of the last two seasons? Would Napoli ever hold a lead? Would European competition once again prevent them from reaching their domestic dreams?

The next two games were even more disappointing. Napoli traveled to Avellino and came home with a goalless draw. Then it was time for the second leg of the 1st round of the UEFA Cup, in France. The 1–0 victory again went to the home side, with Yannick Stopyra striking what would be the winning goal in the 15th minute. However, the teams were level on aggregate, 1–1, and after they played through extra time, penalties were on the agenda. Stopyra stepped up first and sent the ball over the bar. Napoli were elated. Bruno Giordano, Moreno Ferrario, and Alessandro Renica all beat the Toulouse keeper, Philippe Bergeroo. So did the next three from Toulouse. Napoli fans didn't worry; Salvatore Bagni was up next. But the midfielder saw his shot saved. Alberto Tarantini, part of the 1978 World Cup winning Argentine team, was up next. He converted. Toulouse were up 5–4. No matter. Maradona was up next. But the God of Naples hit the post, and Napoli were kicked out of the UEFA Cup in the 1st round.

Failing to advance in European football may have been the best thing to happen to Napoli that season. Their first game back, they beat Torino 3–1 at home, coming from behind with goals from Bagni, Ciro Ferarra, and Giordano. That set off a ten-game unbeaten run, the best of which was the win over Juventus, the previous season's champions. In the 9th round, Juve began the scoring with a cross from Antonio Cabrini, which Claudio Garella managed to punch out of the area, but the ball fell straight to Michael Laudrup's foot, and he easily slotted it home. Prior to the Laudrup goal, Juventus had already got the better of Napoli for nearly an hour. The visitors began the game with only one forward, Giordano, and a strong midfield, as though Bianchi would have been happy with a goalless draw. After Juventus drew first blood, Napoli fought back with a tenacity not often seen in Turin. They sent in shot after shot, but Juve keeper Stefano Tacconi kept denying the *Azzurri*. Then, in the 73rd minute, Ferrario scored after a corner taken by Francesco Romano fell at his feet. His left-footed strike first collided with the post, then hit the back of the net. The very next corner, this time taken by Maradona, found Alessandro Renica at the front post. The midfielder headed the ball to the far post, where Giordano volleyed it in from a few yards away. 2–1 to Napoli. Finally, Juventus went all out in attack, but Napoli caught them on the counter. Andrea Carnevale took the ball in the host's half, positioning him and Giuseppe Volpecina against the entire Old Lady defense. Volpecina curled his left-footed shot into the right corner of Tacconi's goal to deliver a 3–1 victory, delighting both the visiting players and the fans who had followed them to Turin.

The win propelled Napoli to the top of the table. Although closely tailed by Inter, with whom Napoli had played out a goalless draw just before the Juventus win, it was the Partenopei who claimed the midseason winter championship. Such an accomplishment does not come with a trophy—it is a phrase that sells more newspapers, and makes the fans excited—but the *campione d'inverno* has, more often than not, lifted the scudetto.

In the middle of a five-game winning streak from Napoli, the next Coppa Italia round began. On February 25, Napoli beat Brescia 3–0 with goals from Maradona, Giordano, and Ciro Muro. The weeks passed. Napoli scored all of four goals in the six Serie A rounds before the second leg of the cup tie at Brescia, on April 8. They lost to Inter, who had stumbled in their chase for the title, thanks to a late goal by Giuseppe Bergomi. However, to much excitement in Naples, they did the double over Juventus on March 29. Renica scored first, but Juventus came back early in the second half with a goal by Aldo Serena. Napoli did well to push back, though, and Francesco Romano scored eight minutes later. The game ended 2–1 for Napoli.

The next week they didn't fare so well. A goalless draw at Empoli, a relegation candidate, did not leave the impression that Napoli would be continuing their run for the scudetto. Midweek, another 3–0 win over Brescia, with a goal from Carnevale and a brace from Giordano, showed that, at the least, Napoli were after the Coppa Italia. Unfortunately, it led them straight to an ugly defeat at Verona, featuring a 21st minute goal from Marco Pacione, then an own goal by Renica, and finally a penalty, converted by Preben Elkjær. Was this how it would end for Napoli supporters, in a season they had had so much hope for? With a 3–0 loss in unfair Verona, a city where fans' uncivil blood seemed to make all unclean?

A 2–1 league victory over Milan, who were desperately competing for—at the least—a trip to Europe, raised hopes yet again. Next came yet another Coppa Italia win, a 3–0 victory over Serie B side Bologna, with goals by the usual suspects: Carnevale, Giordano, and Maradona. Napoli fans' dreams soared again, until they remembered a Coppa Italia match always seemed to affect their play in the league. Sure enough, the next round brought a 1–1 draw with Como, a perpetually midtable team that scored only 16 goals that season. One of those came from Salvatore Giunta, who began the scoring. Fortunately, Carnevale was there to equalize less than 15 minutes later.

Then came the second leg at Bologna, in which *Rossoblù* midfielder Giancarlo Marocchi opened the scoring in the 40th minute for the home side. In the second half, in the 55th minute, Luigi Caffarelli (a player so unremarkable as to not even have his own Wikipedia page, despite playing for a scudetto-winning team), equalized. Five minutes later Giordano put Napoli ahead, and a converted penalty from Maradona seemed to wrap up the game in the 70th minute. Then Lorenzo Marronaro scored a second for Bologna deep in the second half. But Napoli tuned back in, and Giordano scored a second in the 88th, for a 4–2 Napoli win (7–2 on aggregate) that sent them into the semifinal of the cup.

It was the first week of May, and Napoli were still top of the table, but it seemed like they were destined to falter, to once again lose out on the title. After all, before their Coppa meeting with Bologna, they'd managed just a 1–1 draw with Como, who were closer to the bottom than the top. With two more rounds to play, Juventus and Inter were too close for Napoli fans' comfort, and it appeared as though the Partenopei would once again lose out to the North after coming so close to the trophy. For decades—for centuries, if the years after the *Risorgimento* are factored in, as they should be, given the South was closed off from many of the benefits the North received—the people of Napoli, who had often felt rescued by whichever noble had taken control of the kingdom, were again waiting on the one person who could elevate their city above those throughout the peninsula. They wanted to feel the respect of the entire country of Italy, and they believed it would be Maradona who would bring that respect. But that rough patch had made the tifosi a bit nervous.

Some, however, had little doubt that Diego would bring the scudetto to Naples. Gennaro Montuori, a fan who operates his own Napoli supporters' channel and appears on Calcio24 Napoli, says, "We didn't have any doubts. We knew Maradona, who would be able to dribble a small orange up three flights of stairs for fun, would be able to win [the title]. I used to play some football, and when I saw him for the first time I knew that we would get the *scudetto*."[50]

Napoli held on. On May 10, 1987, Fiorentina came to the San Paolo. A solidly midtable team, the *Viola* had one (huge) thing going for them—Roberto Baggio, their No. 10, who would later be lured to Juventus, then Milan, then Inter. Napoli were desperate to seal the scudetto at home, in front of their own supporters. They wanted the party. When Carnevale scored in the 29th minute, the stadium exploded. Unfortunately, the miraculous Baggio scored just ten minutes later. The crowd was then reduced to gnawing on its nails, willing Baggio—or any other Fiorentina player—not to score again. Because of Inter's failings against Ascoli the previous week, all Napoli needed was a draw and the title was theirs.

Books and articles focusing on the Maradona period all describe the explosion of noise when Napoli won their first scudetto, the excitement, the flares and fires and the dancing in the stands, but some must wonder if the ecstatic Napoli supporters began the celebrations before the final whistle. Having witnessed the spectacle of the San Paolo even without a scudetto at stake, it feels unlikely that the fans tempered their enthusiasm; surely, the curvas continued to chant and cheer, the crowd continued to roar. Yet given how superstitious the people of Naples are, particularly the fans, they almost certainly did not sing about a scudetto or send off any flares before the title was officially theirs. The party *did* begin before the Napoli match ended, as news spread around the stadium that Inter had lost to Atalanta, confirming that the title belonged to the Partenopei.

Despite previous attempts, it is perhaps impossible to explain Napoli supporters' joy at winning the scudetto. The noise from the stands was tremendous, feet stomping, hands clapping, and loudest of all, the screams, yells, shouts, chants, and sobs of happiness. They set off flares and fireworks, filling the curvas with smoke. After the whistle, Maradona said:

> For me, this title means a lot more than winning the World Cup. I won a youth World Cup in Tokyo and I won the World Cup in Mexico last year but on both occasions I was alone, I had no friends with me. Here, all my family, the city of Naples are with me because I consider myself a son of Naples.[51]

Diego took a lap of honor around the stadium to show his appreciation to the fans who had loved him and supported him beyond all reason, in every season, before he had even arrived in Naples. Seemingly everyone in the city took to the streets. The 70,000 fortunate souls who had been among the crowd at the San Paolo joined the party, car horns honking, Vespa riders wearing dark curly wigs and trailing banners darting around pedestrians, fans wrapping themselves in Napoli flags, climbing atop buses and dancing. Traffic was a snarl as everyone tried to find a way to celebrate, with some abandoning their vehicles outright to join in the party growing around them.

Napoli are unique in that they are a one-city club from the South, with no nearby rivals. In celebration, the entire city rejoices, and the ironic comments and clever insults Neapolitans dish out so well is all directed at the North. Napoli *is* the city and the city is their country, one that does not extend beyond the *Mezzogiorno*. In the stadium, the songs they sing, the banners they wave, are all related to the social institutions tied to the Naples working class. Fandom is inextricably connected to Neapolitan culture and collective identity, which is necessarily set in opposition to the rich North.[52] It is fitting, then, that supporters held a mock funeral for Juventus that night, complete with a fake coffin and death notices that announced, "May 1987, the other Italy has been defeated. A new empire is born."[53]

The party didn't end that Sunday evening. The city continued to revel in its glory with a huge feast to which everyone was invited. Even those who had passed on were there in spirit, with *Napulitano* graffiti appearing in cemeteries reading, "*Guagliu! E che ve sit pers!*" (Y'all don't know what you're missing!). In some neighborhoods the parties continued for a week—at minimum. Superstitious persons asserted that the title win had been preordained through the lottery. First, Maradona's current kit number, 43, came up the week before. Then the week the team clinched the scudetto, 43 came up once more, followed by 61, the number of years Napoli Calcio had been waiting for a title. One freshly painted mural showed Diego Maradona in the arms of the city's saint, San Gennaro (whose blood had, in fact, liquified the week before[54]).[55] Others traveled back to 1926, the year the current iteration of the club was born, and adopted the donkey as their costumes, the donkey being the animal other clubs had used to mock the squad that had earned just one point during their first league season, 60 long years ago.

The celebrations still hadn't ceased. Napoli officially won the title on May 10.

After they drew with Ascoli 1–1 on May 17, and the Serie A season officially ended, the parties began again. But the players couldn't let down their guard completely; ten days later, Napoli had to travel to Cagliari for the first in a two-legged tie in the semifinal of the Coppa Italia. A late goal from Maradona gave Napoli the lead, with an away goal for extra insurance. Fortunately, such insurance was unnecessary. When Cagliari came to town, Carnevale scored first, followed by Giordano two minutes later. Shortly before the break, Giordano scored another. Finally, Ciro Muro scored the fourth in the 68th minute. The game finished 4–1 thanks to an own goal by Romano.

The miraculous year was not yet over. With the win over Cagliari, Napoli qualified once again for the Coppa Italia final, against Atalanta. In 1987, it was a two-legged final with a match played in both cities, rather than a winner-takes-all at Rome's Olimpico. The first leg took place in Naples on the evening of June 7. Although ticket prices were high, 60,000 still attended the match at the San Paolo. Napoli were the better team throughout the match, but Atalanta held fast in the first half, defending well. Napoli began to increase the pressure in the second half, and in the 67th minute Atalanta finally cracked. Renica, Muro, and Bagni all scored within ten minutes, setting up a second leg in Bergamo in which *La Dea* would need to score at least three goals. Renica took a speculative shot from distance that somehow ended in the back of the net. Three minutes later, Muro and Maradona performed a one-two to put Muro in front of Ottorino Piotti, Atalanta's keeper, who was unable to save his diagonal shot. And in the 77th, Bagni's header ended up behind Piotti's back and in the net. In the second leg, Napoli didn't need to do much of anything, but they still went full out in attack.[56] Giordano scored in the 86th off a Maradona free kick to make it 4–0 on aggregate. A month after officially securing the scudetto, Napoli had wrapped up the Italian double. It didn't seem as though they could reach higher heights.

La magica

Shortly after Napoli hoisted the scudetto in 1987, Pierpaolo Marino resigned, citing irreconcilable differences between himself and Corrado Ferlaino.[57] Yet the Neapolitans remained convinced the 1987–1988 season would be MaGiCa. On either flank were Bruno Giordano and Diego Armando Maradona, and in the center was Napoli's newest purchase: the Brazilian Careca. MaGiCa.[58] The newcomer created a perfect balance in the front line. First, he made friends with Maradona, knowing that it was essential for a good working relationship. He was admired for the fact that he was an "atypical Brazilian," in that he was excellent without the ball and always willing to be a team player, while at the same time, whenever he saw an opening, he would shoot.[59] The balancing act could be credited to new sporting director Luciano Moggi, just arrived from Torino, bringing Careca with him.[60] (Moggi was later banned from Italian football for life due to playing a central role at Juventus during the *Calciopoli* scandal, and should never be excused for the homophobic attitudes he held, but he did attract many talents to Napoli.)

Prior to the season's start, last year's Coppa Italia winners triumphed in their group, beating Modena, Livorno, Udinese, Padova, and Fiorentina to move to the next round. After winning 14 games on the hop last season, the extra 5 took them up to a new level, setting a record for the number of wins in a row in the Coppa with 19.

Napoli's first Serie A game was an easy one, at newly promoted Cesena. A goal from Salvatore Bagni was sufficient, though perhaps not as decisive as Napoli had hoped. After all, in just a few days they would be playing their first-ever match in the European Cup, at Real Madrid—the first team to ever hoist that particular trophy. The Spaniards easily dispatched the Neapolitans 2–0, and although Napoli managed a 1–1 draw at the San Paolo with a goal from Giovanni Francini, the 3–1 aggregate scoreline dumped Napoli out of the cup in their 1st round.

Napoli may have lost their first match in the European Cup but any sadness they felt was quickly erased by their hot streak in the league. The Partenopei secured 21 of 24 points in their first 12 rounds. But in 1987–1988, it was a different Serie A Napoli were playing in, compared to just the year before. With the league shaken up after the Southerners secured last season's championship, the big clubs made big changes. Milan brought in Arrigo Sacchi, who knew how to harness the power of Ruud Gullit and Marco van Basten.[61] Their rivals kept Giovanni Trapattoni, who couldn't quite manage to instill enough discipline in his players to hold such a strong line, when it seemed that's what all of Italy was doing. Napoli, still coached by Ottavio Bianchi, scored 55 goals, 12 more than league winners Milan. But the Rossoneri conceded only 14, and that season was about, for most teams, the defense. Sacchi, on the other hand, had a team of creative players whom he had to manage carefully in order to ensure the team slowed down and fit within the rigors of his system.[62]

In the capital, Roma snatched up Milan's old coach, Nils Liedholm, who was a proponent of the zonal marking system, and relied on Rudi Völler for their attack. Fiorentina brought in Sven-Göran Eriksson, whose system was over-reliant on Glenn Hysén and thus didn't improve much. Meanwhile, Sampdoria made few changes, relying on Gianluca Vialli and Roberto Mancini. Vujadin Boškov remained at the helm, a coach that held on to a man-marking system, and won the Coppa Italia with it.[63] Then there was Juventus, who had finished within three points of Napoli the season before. They retained Rino Marchesi, believing that his techniques would help them continue to threaten for the title. But the Old Lady made a huge mistake—the superstar signing of the season, Ian Rush, from Liverpool. He was meant to replace the great Michel Platini. Rush had devastated teams in England, but couldn't adjust to Serie A. He scored just 7 goals in 29 appearances. Instead of fighting for the scudetto, they needed to fight Torino for 6th place; the playoff game ended in a goalless draw, with Juventus winning on penalties, 4–2, and securing a place in the UEFA Cup.

Napoli played it their way, and the reigning champions were top of the table from Round 3 all the way to Round 27. Maradona was crowned Serie A *Capocannoniere* with 15 goals, while Careca and his 13 took second place. The low goalscoring titles revealed just how much the league had shut down teams' offenses. So did Milan's eventual win. Napoli's fall began when Juventus managed to beat them 3–1, with one of Rush's rare goals for the Bianconeri. Next, Verona managed to fight off the Vesuviani with an equalizer by Roberto Galia, answering Diego Maradona's earlier goal. Then came the May 1 match against Milan and the real demise of Napoli. Sacchi's team had finally fused, and the result was a tough game against a team with incredible offensive power, who followed a zonal marking system, utilized a four-man defense, and applied continuous pressure all over the pitch. Napoli's plan appeared to be to defend their one-point advantage over Milan; to do so, Bianchi moved right-back Francini further up the pitch to contain Roberto Donadoni. Napoli looked as though they were in control and could continue to hold Milan off, but then Pietro Virdis began the scoring, somewhat luckily. In the 36th minute, a ball from a free kick bounced around inside the area. Virdis was quicker than Napoli's defense and somehow, clumsily, put the ball behind Claudio Garella. Maradona answered with basically the last shot of the first half—a fantastic free kick from about 18 yards out that sailed over the wall and right under where the post and the bar meet.

At halftime, Sacchi made a sub that essentially won the game for Milan. He pulled Donadoni and brought in another center-forward, van Basten. Gullit moved back a little, behind Virdis and van Basten. The Rossoneri manager's tinkering completely changed the game. All of a sudden, Milan was pressing like crazy. No longer did Napoli's over-defensive squad seem a problem for the visitors. Bianchi didn't have the tools to respond to Sacchi's changes. He did bring Giordano on, but that didn't help much. In the 68th minute, following a perfect cross by Gullit, Virdis scored and Milan took the lead again. Bianchi then brought Carnevale on, but again it did little to help Napoli. Milan were simply a better team. A few minutes later, Gullit sent a low cross from the right side for van Basten, who put Milan up 3–1. At that point, it was clear to almost everyone that Milan would win the title. Careca's header 13 minutes before the final whistle didn't change much. The game finished 3–2, and Milan moved to the top spot. It seemed as though Napoli had given up. Their final two games were a 3–2 loss at Fiorentina and a home loss, 2–1, to Sampdoria. They had enough of a lead, however, to retain 2nd place, four points ahead of Roma and with a spot secured in next season's UEFA Cup.

However, many still believe that the Serie A season was rigged, particularly Napoli supporters. The Camorra could have placed illegal bets on Milan to win the season, and used their influence within Napoli to convince them to throw the season. The breakdown seemed inexplicable—how could a team that had performed nearly perfectly from the start of the year simply fold?[64] The only theory that made sense seemed to be the one in which illegal meddling occurred. However, according to Salvatore Bagni, there were no threats, no payments, and no one asked them to lose the title.[65] Napoli were simply tired and fell out of form toward the end of the season, and it didn't help that the relationship between certain players and coach Bianchi had soured.

We're at the top of the world

In 1988–1989, the Serie A format changed from 16 to 18 teams, with four teams coming up from Serie B, and four going back down at the end of the season. Napoli were once again 2nd, but came nowhere near threatening Inter, instead finishing 11 points behind the champions. Still, it wasn't the worst season for Napoli. In the summer of 1988, they shed a few veteran players, such as Salvatore Bagni, whose weakened flair as a winger had shifted him to the center of the field; Moreno Ferrario, who'd been with Napoli 11 seasons; and Bruno Giordano, who'd scored 23 goals in 78 appearances for the club. Napoli bought Massimo Crippa for 7.6 billion lire from Torino in order to strengthen the midfield, playing alongside Fernando De Napoli and Francesco Romano. Defensive midfielder Alemão, deemed the best foreign player in Spain's La Liga the previous year, joined from Atlético Madrid for 4.6 billion, Napoli apparently snatching him right out from under Juventus' noses. While he suffered from injury earlier in the season, he remained essential to the squad.[66] Giancarlo Corradini, a central midfielder, finished out his career at Napoli and would add crucial goals in Napoli's next season. Giuliano Giuliani replaced Claudio Garella in goal, just as he had done three years earlier at Verona. Finally, Luca Fusi, another central midfielder, was purchased for 5.8 billion lire, and helped Napoli with their amazing feats in 1989 and 1990 before moving on to Torino, then Juventus.

That year, Napoli recorded a few amazing wins, such as a 4–1 victory over champions Milan, which followed a 5–3 victory over Juventus; in the two matches Careca scored five goals. He also scored two against Pescara, as did Maradona. Carnevale scored a hattrick. Alemão chipped in with one of his own. Napoli gave away two penalties that Pescara converted, but given the 8–2 victory it mattered little. Napoli may have outshone most of the teams in Serie A, and finished awfully high in the table, but the 1988–1989 season was more about other tournaments than where they finished in the league. Specifically, it was about the UEFA Cup.

The UEFA Cup is now known as the Europa League, and is no longer taken seriously by major clubs—three or four teams from each of the big leagues qualify for the Champions League (previously the European Cup), meaning these strong leagues send clubs to Europa who may have finished 7th or 8th. Or, in some instances, a relegated team heads to the Europa League because they won their domestic cup. But back in the day, when only the top team went to the European Cup, the UEFA Cup was a big deal. And Napoli were finally ready to take on the challenge.

The furthest Napoli had made it in the UEFA Cup was the 3rd round, back in 1974–1975. But Ottavio Bianchi and Corrado Ferlaino had built a team they were confident could conquer Europe this season. PAOK of Greece traveled to the San Paolo for the first elimination round, where one converted penalty from Maradona was enough for Napoli to emerge victorious. In the second leg, Careca started things off with a goal in the 17th minute. In the 65th minute, Georgios Skartados equalized, but it wasn't enough for the Greeks. Napoli won 2–1 on aggregate and moved on to the

Round of 16. Given other aggregate scores in the 1st round—Aarau 0–7 Lokomotive Leipzig, Sliema Wanderers 1–8 Victoria București, Bayern Munich 10–4 Legia Warsaw, Partizan 10–0 Slavia Sofia—it didn't seem as though Napoli would make it much further, as was usual in their European adventures.

The goals slowed down in the second round. Nails were chewed to the quick in Naples, as Lokomotive scored first in Leipzig. Fortunately for Napoli fans' cuticles, Giovanni Francini equalized less than five minutes later. Yet, despite the away goal, the 1–1 result left the Partenopei faithful with nibbled fingers for the next two weeks. Things were going well in the league, but supporters were anxious for their team to lift a European trophy. Fortunately, Francini scored in the second minute of the second leg. Yet, one goal from Leipzig could've sent them to extra time and then penalties, so the Napoli fans' fingernails weren't safe yet. Fortunately, an own goal by Heiko Scholz at the hour mark gave everyone a little breathing space, and 30 minutes later, Napoli were on their way to the next round.

Napoli were paired with Bordeaux in the third elimination round. The game in France was a tough one. Carnevale scored in the 5th minute, but then the visitors needed to hold back *Les Girondins* until the 90 minutes ran out. That was made easier when defender Alain Roche was sent off in the 56th minute, but then again, Fernando De Napoli was sent off at the same time, reducing Napoli's defense as well, as he was tasked with guarding the backline. Bordeaux came to the San Paolo, where Napoli knew they were clinging to the slimmest of 1–0 leads. Once again, a Bordeaux defender was sent off, but this time it was in the third minute when Jean-Christophe Thouvenel was shown a straight red. The hosts hunkered down and Napoli didn't manage to score, but Bordeaux weren't able to break through, either. The goalless draw meant Napoli's 1–0 win in the first leg sent them through to the quarterfinals—a first for the Vesuviani.

Yet another two-legged encounter was needed to make it to the semifinal. And this time it was against Juventus, who had also had a difficult time getting to the quarters, facing Belgium's RFC Liège and coming out 1–0 in each game. Though Juventus seemed to have a slightly easier time in the cup, Napoli had been beating them regularly in the league. That's why it was rather surprising to see Juve win in Turin, with a goal from Pasquale Bruno and an own goal from Napoli defender Giancarlo Corradini, ending 2–0 for the Old Lady. Despite all the recent times Napoli had upended Juventus, fans still remained nervous as they filtered into the San Paolo, knowing Napoli needed at least three goals to move on, and that if Juve scored one, the Partenopei would need to score an extra goal for every visitor ball put in the net. The fans were barely breathing easier at the end of 90 minutes; a Maradona penalty and a 45th minute goal from Andrea Carnevale left the two teams in a 2–2 deadlock. The sides headed into extra time. It looked like the tie would be broken through a penalty shootout, but Napoli kept pressuring Juventus, dancing around the penalty area for minutes before, in the 120th minute, Careca won the ball on the right. He crossed it in the middle of the box, where Renica was waiting to head the ball in. And to give Napoli the win.

On to the semifinal, where Napoli hosted Bayern Munich in the first leg. The Bavarians had lifted the UEFA Cup trophy way back in 1966–1967, and the European Cup in 1975–1976. Yet Napoli defeated Bayern, the eventual Bundesliga champions, easily at the San Paolo, 2–0 with goals by Careca and Carnevale. The second leg produced a few more nerves, as nothing much happened in the first hour. Then Napoli fans paying attention back in the south of Italy could once again take a breath when Careca scored, but two minutes later, Roland Wohlfarth equalized. Over two legs, the score was 3–1, and while it was unlikely Bayern would manage to catch Napoli, the born-and-bred fans knew that leads could evaporate quickly, even leads in which only 30 minutes remained. Careca scored again in the 76th minute and Napoli fans were finally dreaming of a trip to the final when Stefan Reuter equalized in the 81st. It was highly unlikely Bayern would find three goals in the little over ten minutes that were left, but again, Napoli fans tend to be on guard. Nevertheless, the match in Bayern ended 2–2, which, added to Napoli's 2–0 win two weeks before, meant the Azzurri were finally headed to the main stage, the two-legged UEFA Cup final.

The first leg would be played on May 3 in Naples. The trouble was that Napoli still had eight more rounds of Serie A to go, and although Inter had held the top spot since Round 5, the Partenopei were still hoping to close the gap and somehow overcome the *Nerazzurri*, as well as conquer the UEFA Cup. As the first game took place at home, the squad had a bit more confidence in the buildup. The week before, Napoli had emerged the 1–0 victors from a dreadful game against Verona in which Napoli had looked to be focused on the UEFA Cup first leg throughout. A further boost came from the knowledge that their next domestic match was against Bologna, who were struggling to survive. The city and the fans were euphoric.

As Bianchi said himself, it was clear that their heads were already in the Stuttgart game.[67] But it was Napoli's worst performance in the UEFA Cup, at least for the first 70 minutes. In the 17th minute Stuttgart took the lead through Maurizio Gaudino, and then parked the bus. Napoli weren't able to create many chances, much less an equalizer. However, after Maradona converted a penalty (Stuttgart complained that before the ball hit their defender on his hand, it first hit Maradona on his; watching the match reveals this certainly did happen), the Partenopei were revitalized. Prolonged pressure by Napoli led to an 87th minute goal, in which Careca got on the end of a Maradona cross to give Napoli the 2–1 win.

How serendipitous that turned out to be. Napoli were coming off a 1–1 draw in which Roma sent in a rather late equalizer, to the disappointment of the over 80,000 watching at the San Paolo, and it seemed that match cast a cloud of pessimism over Naples. Neither the players nor the fans truly believed that the UEFA Cup was in reach. But then came the second leg in Germany. The fans were right to be a bit skeptical, as the match turned out to be rather difficult for the visitors. Stuttgart started off a bit more brightly, but Alemão managed to score first. Maradona received the ball in Napoli's half and passed it off to the Brazilian to spring the counterattack. Alemão performed a quick one-two with his compatriot Careca before finding himself one-on-one with Stuttgart goalkeeper Eike Immel and scored a very lucky goal to put Napoli up 1–0 in the 18th minute. The midfielder had been injured a few minutes before scoring and had continued playing through the pain, but had to be subbed off in the 31st minute.

Just eight minutes later, an error from Napoli's keeper allowed Jürgen Klinsmann to equalize from a corner. Giuliano Giuliani came out to grab the ball but misjudged its trajectory. Klinsmann was waiting at the far post to tap in. The Neckarstadion held its breath—both sides continued attacking, and one would surely score soon. It was the Napoli fans who got to celebrate in the 39th minute. Their deity, Maradona, sent in a corner that was cleared by Stuttgart, only to fly straight back to Maradona. Diego headed the ball into the penalty area. Ciro Ferrara volleyed it past Immel to put Napoli back in front.

In the second half, Stuttgart pulled out everything they had, pinning Napoli back in their own half. But after one of the home side's attacks, a clearance from inside the penalty area flew straight to Careca, who was almost at the halfway line. He headed it to Maradona, who had a clear run on goal. Yet the Argentinian waited for his fellow forward to catch up before sliding the ball over. Careca executed a perfect chip over Immel to make it 3–1 to Napoli (5–2 on aggregate) in the 62th minute. As Napoli had more away goals, Stuttgart would need at least four goals to triumph. Superstitious Napoli fans still refused to celebrate, which was likely wise, as Fernando De Napoli made a mistake and scored an own goal in the 70th. Still, Napoli were up 3–2 with just 20 minutes left. Was it time for even the most skeptical fans to start getting excited?

No, because this is Napoli. Supporters remained nervous. Napoli needed to make sure it would be impossible for Stuttgart to score, because they needed just three goals now, and the Partenopei just had that kind of (bad) luck. While some of the 20,000 Napoli fans in the stands began to celebrate, others attempted to shush them. Sure enough, in the 89th minute, Olaf Schmäler capitalized on another mistake by De Napoli, who crossed the ball into his own penalty area, where the Stuttgart forward headed it in to make it 3–3. Napoli still led on aggregate, however, and just had to play out a few minutes without Stuttgart scoring twice. They closed ranks and battered down the hatches. In a short time that felt like hours for the Napoli supporters, Maradona was lifting Napoli's first significant European trophy.

The UEFA Cup caused huge celebrations in Naples. Before the whistle blew, fans back at home tossed aside their worries and a concert of car horns started up, accompanied by firecrackers. Chants rose up in every neighborhood. The noise only grew after Napoli had officially won the cup. Not only had an Italian side battled a number of difficult teams to get to this stage, but it was another triumph of South over North. Four Italian teams were part of the 1988–1989 UEFA Cup, and the other sides knew Napoli were a force to be reckoned with. With or without the strong English sides (without, as the English were still barred from international competition), the

Italians were still considered the most fearsome; Serie A was deemed the best in the world at the time. In the quarterfinals, Dynamo Dresden eliminated Roma, while Bayern Munich needed to rely on the away goals rule to knock Inter out. And Napoli needed extra time in the semifinal to dig in their heels against the Juventus offense while still creating as many chances as possible. Then, it was just Napoli left against the East and West Germans. Napoli had likely found themselves relieved to draw Bayern Munich, as Stuttgart had been a thorn in each club's side, while Dresden were racking up goals.

This was worth a party to rival that of the one that followed Napoli's first scudetto, exactly two years before. Almost all of Naples poured into the streets, where they reveled in their joy until dawn. Long lines of honking cars drove up and down the streets, avoiding the drunks dancing in the middle of the road. The next day, the papers compared the scenes in Naples to Carnival in Rio, combined with New Year's Eve antics.[68]

If you only knew what you missed

Few would have predicted it, particularly given the final standings, but the 1989–1990 season was the beginning of the end for this magical Napoli side. Ottavio Bianchi left for Roma, where he led the team to a Coppa Italia win in 1991 and to a UEFA Cup final, where they lost to Inter. Ferlaino enticed Alberto Bigon, an up-and-coming, tactically focused Italian coach who had managed Cesena for the past two seasons, to move to Naples (Maradona disagreed with the decision, telling Ferlaino and Moggi they'd finish 4th or 5th at best[69]).

The season began with the first Coppa Italia match against Serie B club Monza. Monza, shockingly, opened the scoring with a goal by Massimiliano Cappellini. It took until the 84th minute for Andrea Carnevale to equalize. The goal saved Napoli, but the game dragged into penalties. It took the Partenopei ten penalty goals to finally knock out a team that would be relegated to Serie C that season.

When examining the list of penalty takers for Napoli, one name is notably absent—Diego Maradona. The other South American players who played at the 1989 Copa América were also late in returning, but Maradona decided to stay a week longer in Buenos Aires, threatening not to return at all. It wasn't an empty threat; Diego refused to play the first month of the season because he was upset over the club rejecting a proposed transfer to Marseille, which would double his salary.[70] Moggi said the club had never received an official offer from Marseille, but if they had, they wouldn't have considered it.[71] Maradona responded by telling Napoli he was prepared to involve lawyers in an attempt to cancel his contract. He also stated he was prepared to sue anyone who talked about him in the context of drug use or the Camorra. But Moggi told him he had to return to play for Napoli, or he would not be allowed to play for anyone. This was around the time that the club started to wonder just how far Maradona could go, and still be worth keeping. As for Maradona, he said in 2009,

"After a four-hour meeting, Ferlaino said that if we won the UEFA Cup I could leave, but we won it, and he blocked the move anyway."[72]

And then Maradona returned, and Napoli won their second scudetto. Napoli reinforced their side with strong purchases, such as midfielder Massimo Mauro from Juventus, who had excellent technical skills. They also strengthened the defense with the purchase of Marco Baroni from Lecce. Most importantly, Moggi saw the talent in Gianfranco Zola, playing at Torres in Serie C1. Maradona would become Zola's mentor, grooming him to take over his position. They would spend hours after training practicing one-on-one, with Zola absorbing just how Maradona sent in his free kicks.

Napoli made a strong start to the season, an unbeatable run of 16 games, managing to stay top from Round 7 to Round 24, even through their UEFA Cup matches (although they usually drew a game on one side of those games, often against much weaker competition). In the semifinal, they got kicked out by Werder Bremen, with a rather embarrassing 5–1 loss to seal their fate. They then drew 1–1 with Bari in the next round of Serie A. Fortunately, that was just before the winter break, and Napoli came back strong until Round 24. That week, Milan beat Napoli 3–1 in the second leg of the Coppa Italia semifinal, knocking them out of the tournament. Then Napoli traveled to Milan for their Serie A match. They came back having lost 3–0 and knocked from the top of the table by the Rossoneri. Napoli held on tight over the roller-coaster ride that comprised the next six weeks. They lost 3–1 at Inter and 2–1 at Sampdoria, drawing 1–1 at Lecce in between. However, they also beat Roma 3–1 and Genoa 2–1 at the San Paolo, Zola scoring a late winner in the latter. Fortunately, Milan lost back-to-back games to Juventus and Inter.

Napoli saved their season on March 25 with a 3–1 victory over Juventus. The Old Lady had lost to Hamburg in the UEFA Cup midweek, and were feeling the sting, regardless of having advanced to the next round. The European tie left Juventus tired, and it showed, as they were unable to stop the Napoli attack. Maradona played absolutely brilliantly. scoring the first two goals before the half hour was up. Napoli gave up a penalty at the hour mark, and Luigi De Agostini converted. But just a few minutes later, defender Giovanni Francini ensured the win. After beating the Bianconeri, Napoli won their final four games, keeping a clean sheet in three. One was a result awarded by the FIGC; Napoli were playing out a goalless draw with Atalanta when the Brazilian Alemão was hit by a 100 lire coin thrown from the crowd, resulting in the visitors being awarded a 2–0 win. The Partenopei finished a mere two points over Milan, whose loss to Verona in the penultimate game caused them to just miss out on another scudetto. Controversy still surrounds this match, with many certain that the referee was paid off, given the number of poor decisions made. Marco van Basten still thinks the FIGC did everything they could to make Napoli win the scudetto.[73] The former Milan star argues that in many games Napoli were favored, supposedly because the FIGC wanted to have two teams in the European Cup next season: Milan, who had won it the season before, and Napoli, entered by virtue of winning the league. Napoli needed just one point against Lazio to capture the title, and when Marco Baroni scored in the seventh minute, they essentially won the scudetto, despite Milan's 4–0 win over Bari.

The papers describe the last round of the season against Lazio as a "carnival." The fans were jubilant, even though there was, of course, fear that Lazio could win, particularly as the *Biancocelesti* arrived in Naples ready to spoil their party, and played very well against the hosts. Lazio spent the majority of the game attacking, because Napoli scored in the seventh minute, off a header from Baroni, who got on the end of a Maradona free kick. The God of Naples did his best to provide the supporters (and his mother, who was in attendance) with a good show, dribbling, adding little flairs, entertaining the masses. In the end, though, the only goal was from Baroni, but it was enough for a 1–0 Napoli win—and their second scudetto. Diego would later say he considered the championship to be the South's revenge on the way the North treated him, Napoli, and the Mezzogiorno in general. As soon as the whistle blew, the fans invaded the pitch.[74] The city was colored in blue minutes after the game ended. Car after car waving blue flags made it impossible to actually drive through the city, while many neighborhoods were so flooded with fans that even attempting to walk through them proved pointless.

The fans hosted a fake funeral for Milan; even before the game, people were selling vials filled with Berlusconi's tears. They also erected a makeshift chapel to mourn Berlusconi's demise. Someone hung a banner that read, "Better to lose as *terroni* than to win as Berlusconi."[75] People wore masks and danced in the streets, once again creating the impression of a Brazilian carnival. The city celebrated the whole night. A boat floated through the Bay of Naples carrying Napoli players, who celebrated with the fans on land. Tragically, one man died during the celebrations, in a motorcycle accident.[76]

Napoli were again at the top of the world. At least the fans, and likely the players, believed they were. But in truth, Napoli were struggling to stay afloat. The expense of creating and maintaining a team that could consistently compete for the title was beyond difficult for a Southern club, especially one without a powerful investor backed by extensive holdings. Adjustments would need to be made.

But first, goalkeeper Giuliano Giuliani needed to be replaced, and they chose Milan's keeper Giovanni Galli. In addition, they sold Andrea Carnevale, who had paired so well with Maradona and Careca, but was persuaded by Ottavio Bianchi to move to Rome. Five games into his career at Roma, he was suspended for a year due to illegal drug use. This would soon become painfully ironic.

Crimes and punishment

As Napoli made their moves, there was a World Cup being played in Italy. Maradona's Argentina, holders of the cup, faced off against Cameroon in the first match, played at Milan's San Siro. Maradona was despised in the North, a symbol of the dirty South and all that was wrong in the region. He had also brought Napoli two scudetti, even daring to challenge Milan the previous season. When the crowd of 80,000 saw Maradona's face on the big screen during the Argentina national anthem, they booed and whistled. Surprisingly (given the long history of racism in Italy, particularly in the North), they cheered Cameroon throughout the match, believing that the "blacks" from the South were a more suitable target.

"I feel like I represented a part of Italy that didn't count for anything," he said in *Diego Maradona*, the 2019 documentary by Asif Kapadia. This belief should have spurred him on while playing in front of hostile Italian crowds, but by that time his drug use and partying ways had caught up to him, and he seemed listless, even washed-up, during the tournament.

Then came the semifinal in his home, in Naples, against Italy of all opponents. Prior to the match, Italian officials begged the Neapolitan crowd to put country before club and cheer on their nation. But during his press conference, Maradona appealed to his adopted city, reminding them that most of the time, Northerners called Southerners "Africans." While many insist that nearly half the crowd at the San Paolo that day sided with Argentina, or rather Maradona, others argue his every touch was booed by the entire stadium. Watching the game today, both cheers and whistles can be heard when Maradona is in possession. What is clear is that the game continued through extra time until penalties, tied at 1–1. Both teams scored their first three penalties. After Roberto Donadini saw his attempt pushed away, Maradona stepped to the spot, and of course his low shot fooled the keeper and put Argentina in the lead. Aldo Serena, the next Italian spot-kicker, was denied by reserve goalkeeper Sergio Goycochea. and Argentina were through to the final. There they met West Germany and lost 1–0, passing the cup on to the new holders.

Diego Maradona waved goodbye to Naples in the spring of the 1990–1991 season. Or, more accurately, Napoli actively devised ways to get him away from the city.

They pushed him away when the team was actively floundering, not helped by Maradona's own play. The season started well enough—Napoli beat Juventus 5–1 in the Supercoppa Italia, the third year of the tournament that pitted the winners of the Serie A title against the winners of the Coppa Italia (if they were one and the same, the Coppa Italia runners-up would challenge the scudetto winners). In the league, Napoli won just two of their first ten matches. Maradona, now 30 years old, could no longer force his body to withstand the pressures of his lifestyle, and he bore little resemblance to the player who helped guide his team to a second scudetto the previous season.

Diego also began flaunting and utterly disregarding the rules set out by the club. For example, after Napoli beat Hungarian side Újpesti 5–0 over two legs in the European Cup (Maradona provided two goals in the first leg, played at the San Paolo), they played out a dismal goalless draw at home to Spartak Moscow. That match came after they had drawn with Milan at the weekend, Napoli's only goal coming from a Maradona penalty in the 83rd, after which the defense made way for Milan's Ruud

Gullit to equalize five minutes later. This time, the team would be flying to Moscow after a 1–0 win against Fiorentina, their spirits bolstered and the squad coming together as a team rather than a number of individuals.

Except for Maradona. His partying had made him both foolish and reckless; fed up after his shamelessly late arrival in Moscow, Bigon had no choice but to finally bench his star. The rest of the team performed poorly against the Russians once again, and did not improve by much when Diego came on in the second half. The game dragged into extra time and then penalties. Spartak made all their shots, but Napoli defender Marco Baroni failed to covert his. Losing 5–3 on penalties during the Round of 16 is not how Napoli had planned to bow out of their second-ever European Cup, the continent's top championship, but unlike last time they at least made it past the 1st round.

The league's defending champions scored just 15 points in 17 matches in the *andata*, losing to Inter and Juventus, drawing with Milan and Roma, and also losing to Parma, drawing with Genoa, and beating Torino, all three of whom topped both Juventus and Napoli in the table. Their two meetings with that season's eventual champions, Sampdoria, resulted in a combined scoreline of 8–2. The club was in a bit of turmoil as the tabloids splashed Maradona on cover after cover, bringing up his use of sex workers and his cocaine habit, as well as his alleged drug trafficking for the Camorra.

Meanwhile, the FIGC was determined to get Diego out of Italy, given the shadow he was now casting over calcio. With no love left for Maradona, the star was repeatedly drug tested. It was after the 1–0 win over Bari in March 1991 that they finally caught him. Napoli's star was slapped with a 15-month ban. Rather than stick it out in Italy, he packed his bags and headed home to Argentina, knowing the club would never allow him to play for Napoli again.

Yet the fans still adored him. He was a god who had brought two scudetti and a UEFA Cup to Naples. More than that, he had shown Italy that the South were able to succeed in calcio, using their own talents like hard work, grit, intelligence, and unique ways for finding the money to pay for such stars. The complexities of Maradona the man were understandable to the people of Naples, and his big heart and love for the city would always be more important to them than his flaws.

[1] Copa90. (2017, May 11). "Maradona is immortal here"—Napoli fans 30 years after winning their first Scudetto." *Guardian*. https://www.theguardian.com/football/copa90/2017/may/11/diego-maradona-naples-napoli-fans-30-years-scudetto

[2] Ludden, J. (2018). Once upon a time in Naples (3rd ed.). CreateSpace Independent Publishing Platform.

[3] Perucca, B. (1984, January 29). Stranieri, un'estate di trattative poi il blocco Federcalcio. *La Stampa*, 25.

[4] Garganese, C. (n.d.). Diego Maradona: The god of Naples. *Goal*. https://www.goal.com/story/diegomaradonathegodofnaples/index.html

[5] Ludden.

[6] Foot, J.(2006). *Winning at all costs: A scandalous history of Italian soccer*. Nation Books.

7 Pastore, G. (2016, July 5). Operazione San Gennaro. *Sky Sport*. http://www.ultimouomo.com/operazione-san-gennaro/

8 Bifulco, L. (1984, June 30). Il Napoli acquista Maradona. *Il Mulino*. https://www.rivistailmulino.it/news/newsitem/index/Item/News:NEWS_ITEM:5273

9 Pastore.

10 Id.

11 Bifulco.

12 Id.

13 Storie di Matteo Marani: "1984, ho visto Maradona." (2019, July 2). *Sky Sport Italia*. https://sport.sky.it/calcio/approfondimenti/maradona-storie-marani

14 Pastore.

15 Cito, V. (2005, 4 June). L' ultimo ciro. *Gazzetta dello Sport*. http://archiviostorico.gazzetta.it/2005/giugno/04/ultimo_ciro_sw_0_050604407.shtml

16 Zenga. (1987, September 12). Fratelli d'Italia. *La Repubblica*. https://ricerca.repubblica.it/repubblica/archivio/repubblica/1987/09/12/fratelli-italia.html

17 Di Fabrizio, B. (1988, September 2). L'Italia d'autunno. *La Repubblica*. https://ricerca.repubblica.it/repubblica/archivio/repubblica/1988/09/02/italia-autunno.html

18 Cerboncini, R. (1995). Due novità, Ferrara e Petruzzi. *La Stampa*, 30.

19 Bedeschi, S. (2014, February 11). Gli eroi in bianconeri. *TuttoJuve*. https://www.tuttojuve.com/gli-eroi-bianconeri/gli-eroi-in-bianconero-ciro-ferrara-81418

20 La mia sfida è anche ai pregiudizi. (2009, November 11). *La Repubblica*. https://ricerca.repubblica.it/repubblica/archivio/repubblica/2009/11/11/la-mia-sfida-anche-ai-pregiudizi.html

21 Da Maradona al terzo Millennio la carriera di un campione infinto. (2004, January 5). *La Stampa*, 29.

22 Obiettivi. (n.d.). *Fondazione Cannavaro Ferrara*. Retrieved November 8, 2020 from https://www.fondazionecannavaroferrara.it/chi-siamo/

23 Doidge, M. (2015). *Football Italia: Italian football in an age of globalization*. Bloomsbury Academic. <http://dx.doi.org/10.5040/9781472519221.0012>

24 Foot.

25 Das, V. (2007). *Life and words: Violence and descent into the ordinary*. University of California Press.

26 Doidge, 157.

27 Ludden.

28 Id.

29 Looch, C. (2019, June 10). How Maradona changed Naples forever. *The Culture Trip*. https://theculturetrip.com/europe/italy/articles/how-maradona-changed-naples-forever/

30 Rispo, R., & Bell, H. (2020, May 1). Maradona era episode 1: 1984/85 season. Far from Vesuvius. [Audio podcast episode]. In *Far From Vesuvius*. https://farfromvesuvius.podbean.com/e/maradona-era-chronicle-episode-1-198485-season/

31 Maresca, P. (2021). *Almanacco storico fotografico del calcio Italiano 1898–2022: La storia del calcio in Italia: I campioni, i club e la nazionale*. Paolo Maresca.

32 Ludden.

33 Rispo and Bell.

34 Garganese.

35 Ludden.

36 Rispo and Bell.

37 Id.

38 Carter.

39 Veltroni, W. (2016, December 6). Veltroni intervista Giordano: "I trionfi con Diego. Ora sogno la Lazio." *Corriere dello Sport*. https://www.corrieredellosport.it/news/calcio/2015/12/05-6469468/veltroni_intervista_giordano_i_trionfi_con_diego_ora_sogno_la_lazio

40 Ormezzano, T. (2010, July 28). Fare miracoli con i Piedi. *La Repubblica*. https://ricerca.repubblica.it/repubblica/archivio/repubblica/2010/07/28/fare-miracoli-con-piedi.html

41 Id.

[42] Gibelli, M. (1985, August 3). Maradona sara' il nuovo capitano del Napoli. *La Stampa*, 18.

[43] Carter.

[44] Id.

[45] Garganese.

[46] Balague, G. (2020, November 26). Diego Maradona dies: Guillem Balague on "the magician, the cheat, the god, the flawed genius." *BBC*. https://www.bbc.com/sport/football/55084504

[47] Vickery, T. (2020, November 25). Diego Maradona: How tormenting England made him an Argentine deity. *BBC*. https://www.bbc.com/sport/football/55074235

[48] Looch.

[49] Ciccarelli, L. (2013, July 16). La storia siete voi: Nando De Napoli, Rambo. *TuttoNapoli*. https://www.tuttonapoli.net/rubriche/la-storia-siete-voi-nando-de-napoli-rambo-156803

[50] Looch.

[51] Garganese.

[52] Bromberger, C. (1993). Fireworks and the ass. In Steve Redford (Ed.), *The passion and the fashion: Football fandom in the new Europe* (pp.89–102). Avesbury.

[53] Carter.

[54] Miracolo di San Gennaro in un minuto. (1987, May 3). *La Stampa*, 6.

[55] Carter.

[56] Sommani, N. (1987, June 14). Il "grande slam" del Napoli, la Coppa dopo lo scudetto. *La Stampa*, 29.

[57] Mercato. (1987, June 25). *La Repubblica*. https://ricerca.repubblica.it/repubblica/archivio/repubblica/1987/06/25/mercato.html)

[58] Mura, G. (1987, December 9). Ancora lezione straniera? La Repubblica. https://ricerca.repubblica.it/repubblica/archivio/repubblica/1987/09/12/ancora-lezione-straniera.html

[59] Id.

[60] Careca Racconta: "Moggi mi voleva a Toro, poi andammo insieme a Napoli: che gioia! Amo passeggiare per la città, ma ancora mi nascondo. (2017, September 12). *Calcio Napoli 24*. https://www.calcionapoli24.it/le_interviste/careca-racconta-moggi-mi-voleva-al-toro-poi-andammo-insieme-a-napoli-n322488.html

[61] Mura.

[62] Id.

[63] Chicken, S. (2020, December 18). Fallen giants: A tribute to the great Samp of Vialli, Mancini, Lombardo & co. *Planet Football*. https://www.planetfootball.com/nostalgia/fallen-giants-tribute-great-samp-vialli-mancini-lombardo-co/

[64] Garganese.

[65] Rasullo, F. (1994, March 12). Perdemmo per stanchezza. *La Repubblica*. https://ricerca.repubblica.it/repubblica/archivio/repubblica/1994/03/12/perdemmo-per-stanchezza.html

[66] Gandolfi, G. (1988, July 9). Napoli brinda con Alemao e Crippa. *La Stampa*, 19.

[67] Ralo, V. (1989, April 30). Bianchi: "Eravamo con la testa già a mercoledì, è chiaro." *La Stampa*, 17.

[68] D'Errico, E. (1989, May 18). Un grido su tutti nella notte della grande festa: "Evviva il 17 maggio." *Corriere della Sera*, 37.

[69] Raio, V. (1989, July 11). Bigon a Napoli spinto dalla moglie. *La Stampa*, 18.

[70] Id.

[71] Fabbricini, M. (1989, August 31). Maradona licenzia il Napoli. *Corriere della Sera*, 23.

[72] Diego Maradona: Italy's Naples remembers its "barrio boy." (2020, November 26). *Al Jazeera*. https://www.aljazeera.com/news/2020/11/26/i-represent-the-nobodies-maradona-the-barrio-boy-in-naples

[73] Scudetto 1990, Van Basten accusa: "Hanno fatto di tutto per darlo al Napoli." (2021, March 18). *La Repubblica*. https://www.repubblica.it/sport/calcio/2021/03/18/news/scudetto_1990_van_basten_accusa_hanno_fatto_di_tutto_per_darlo_al_napoli_-292818969/?ref=search

[74] Maradona in campo: "Ho vinto lo scudetto contro il razzismo." (1990, May 3). *La Repubblica*. https://ricerca.repubblica.it/repubblica/archivio/repubblica/1990/05/03/maradona-in-campo-ho-vinto-lo-scudetto.html

[75] D'Errico, E. (1990, April 30). E poi e' scesa in Campo Napoli. *Corriere della Sera*, 16.

[76] Gramellini, M. (1980, April 30). Napoli folle e scatenata per lo scudetto bis. *La Stampa*, 29.

CHAPTER 8

An Imperfect God

8

Diego Maradona's years at Napoli saw him surrounded by other extremely talented players, but there is no doubt that he was the one who led Napoli to two *scudetti* and a UEFA Cup trophy. His personality also made him the face of Napoli, and to the fans he could do no wrong, no matter where his excesses led.

"You have to understand Naples to understand how the arrival of Maradona changed the city," says Fiammetta Luino, an archive producer and assistant producer for *Diego Maradona*, the 2019 documentary on the Napoli star.[1] The history of the city—set out in the prologue to this book—and its cultural quirks and legacies, scattered throughout this work, are crucial to understanding how Naples embraced their superstar. Why did they pay a thousand lire each to see a man who'd yet to don a Napoli jersey?[2] Why did they scream "*Ho visto Maradona, ho visto Maradona*" ("I saw Maradona, I saw Maradona")? Why did they chant "Ole, Ole, Ole, Ole, Diego, Diego" before they'd seen him touch a ball? And why, in turn, did Maradona embrace the city immediately, throwing kisses in response to the crowd's adulation?[3] The newest, most famous addition to the Napoli squad finished his display in front of the enormous crowd at the San Paolo by kicking the ball high into the stands and yelling "*Forza Napoli!*"[4]

The great shining beacon of hope

The entire city seemed downtrodden: impoverished, still recovering from the 1980 earthquake, bitter about the inequity of wealth distribution in Italy, tired of being put down by unrelenting Northerners who had the cash, the industry, the influence. Neapolitans chipped in to bring Maradona to Napoli not only to earn sporting success by clinching their first *scudetto*, but also because they knew *El Pibe de Oro* (The golden boy) would help the economy. Or at least, the businesses that invested knew what gold the star would put in their pockets. As did the Camorra, who almost certainly helped fund the transfer.[5] The question of whether they did provide assistance, put forward at the press conference at which Maradona was presented, was not permitted by the powers that be at Napoli; French journalist Alain Chaillou, who raised it at the start, was thrown out by President Ferlaino and put on the next plane home.[6]

What's essential to remember is that Maradona *wanted* to move from Barcelona to Napoli, and that in itself endeared him to the supporters instantly—this player whose talents were said to be beyond those of anyone in the world was choosing their club. Maradona had a tough time at Barcelona. There, he was constantly called a

Sudaca, a term that, at the time, was deployed pejoratively toward people from South America with dark skin like his.[7] Meanwhile, Maradona said of Naples:

> Naples was a crazy city—they were as crazy as me—soccer was life itself. A lot of things reminded me of my origins. There had been hunger strikes and people had chained themselves to the fence at San Paolo stadium, begging me to come. How could I let them down?[8]

He couldn't. And the feeling was mutual—the people of Naples would not back down from their mission to bring Maradona to the city. They saw him as a new means by which to defend their honor. While most fanbases turn to regional rivalries to express their beliefs and demarcate their territory, Napoli fans had no consistently strong rivals in the South. As the ultra groups expanded, so did the idea that their curvas—and in Naples, the curvas were an extension of the city itself—were sacred spaces.[9] These spaces had to be protected at all costs, particularly against successful Northern clubs. Buying Diego would elevate Napoli, strengthening their defenses and bringing honor to the entire city.

Fortunately, Barcelona wanted to get rid of Maradona. By the end of the 1983–1984 season, they knew full well about the cocaine habit and the orgies hosted at his home. They also knew about his gambling addiction which, combined with his drug use and fixation on prostitutes, was eating up his 3 million US dollar salary. Soon Diego found himself in debt to "the type of people in Barcelona who didn't accept IOUs."[10] But what was worse was his behavior at the Copa del Rey final, which pitted Barcelona against Athletic Bilbao in a match played before the King of Spain. Maradona once again faced off against Andoni Goikoetxea, who had performed such a vicious tackle, catching Diego's ankle, that he was suspended for 18 rounds, and Maradona was out for a little over 3 months.[11] When Athletic won 1–0, Maradona's desire for revenge grew stronger. His target was José Nunez, whom he headbutted. After the on-field brawl, Maradona was suspended for three months. Now that neither his on- or off-field antics could be ignored, Barcelona decided this display at the Copa del Rey was the final straw.[12] Maradona may have been on his way to becoming the greatest player in the world, but Barcelona are "*més que un club*" (more than a club) and they had a reputation and a specific philosophy to uphold.

Antonio Juliano, the club's sporting director, was truly the one who managed to bring Maradona to Naples. "Operation San Gennaro" began on May 11, 1984. The ex-captain was a great fan favorite; however, he and Corrado Ferlaino didn't always get along. But Ferlaino was savvy enough to pick up on Juliano's best traits, like his ability to get along with almost anyone. In addition, Juliano was from the slums of Naples, so he was able to relate to Maradona immediately. Diego was from a similar background, the *barrios* of Buenos Aires, and, after two years of feeling excluded at Barcelona, he was eager to return to a place where he felt at home, where many of his most ardent supporters were from poor and working-class neighborhoods. He also empathized with the people of Naples constantly feeling discriminated against by the richer, more powerful North.[13]

Més que un kid

Maradona had grown up in Villa Fiorito, a neighborhood of extreme poverty. There were no paved roads and no clean running water. His father was a bricklayer who went to work at 4am each morning and, according to Maradona, "came home dead."[14] Maradona was the fifth of eight children and, despite the fact that the family was poor, he received his first football when he was just three years old.[15] Like many boys who grow up in poverty around the world, Maradona dreamed of a success so great that he could support himself and the family who'd given so much to him. Unlike most of those other boys, Diego was selected, at age ten, for Los Cebollitas, one of many youth teams under the prestigious Argentina Juniors umbrella. He was truly a prodigy, leading the squad on a 136-game unbeaten streak, an amazing run. During halftime, he'd perform tricks for the crowd. He was then selected for the senior team and made his debut just before he turned 16.[16] In 1976, he was the youngest player to take to the field in Argentina.[17] Both sets of fans applauded the No. 16 that day; *El Gráfico* suggests they knew they had seen something magical, that the nutmeg he fooled Juan Domingo Cabrera with shortly after coming on to the pitch was extraordinary, that his dribbling would one day melt the blood in the veins of all who watched.

In 1981, it was time to move on. Buenos Aires' club of the upper crust, River Plate, offered to make him the best-paid player on their squad. But for Diego—who would certainly come to love money and the lifestyle it afforded him later in life—it was more important to go to the working-class club, his boyhood club, the club he'd always dreamed of playing for. Boca Juniors made an offer of 4 million US dollars, which was accepted, with a bit of string pulling by Maradona, who made it known where he'd most like to go.[18] Somewhat ironically, Maradona's first game for Boca also came against Talleres de Córdoba, the team he'd first played against at the professional level. This time, he scored twice in his side's 4–1 win. He also scored a dribbly, mazy goal in his first *Superclásico*, in which Boca beat River 3–0. In the end, Maradona and his beloved club won the *Metropolitano*, the first national championship of the season, in the Argentina 1981 Primera División season.

Maradona's brilliance could not stay hidden in South America. In 1978, while still playing at Argentina Juniors, he was watched by fellow Argentine Nicolau Casaus, the newly appointed vice-president of FC Barcelona. Although the Catalan club made several offers, the Argentina FA did not want their star player to leave the country before the 1982 World Cup. In the end he technically did: Maradona put pen to paper on June 4, 1982, for a then world record fee of around 7.5 million US dollars. Less than two weeks later, he took to the field in the Argentina stripes, playing his first World Cup tournament in his new country, Spain. While he may have made history as the world's most expensive player, Diego stayed at Casaus' home for a time while he settled into *Blaugrana* life. He later said of the Barcelona vice-president, "For me, Mr. Casaus was my sporting second father, after my manager, he is the person I love most and trust in most."[19]

Maradona was excited by the news, as he believed that at Barcelona he'd earn proper fame and recognition. Instead, he became sick with hepatitis, suffered through the broken ankle caused by Goikoetxea, which kept him out for three months, and he (re?)discovered easy access to cocaine and women. Even if he had not been taking illegal drugs before moving to Spain, clubs had been injecting him with a number of substances (many unknown to Maradona; however, many argue he never took performance-enhancing drugs), even when he was in Argentina. Diego's issue was that his skills acted like a red flag to a bull, and defenders knew there was little they could do to stop him except execute crushing tackles.[20] To keep him playing, he needed consistent pain medication to stop his ankles and legs from crumpling.

Barcelona's desire to get him off the books spread to Italy, and Juliano almost immediately picked up the phone. It took nearly two months of negotiation. After all, the city had just come through the 1980 earthquake, gang activity had accelerated after the disaster, and the people remained poor and uncertain of the city's future.[21] Yet, amazingly enough, Napoli got their man. And it seemed that almost the entire city had brought him to the club, considering that Barcelona's demand for an extra 600,000 US dollars led to a collection among the citizens to find the extra cash (and likely the help of the Camorra, whose district of Forcella was transversed for coins and who knew the benefits of having an association with the world's best player).[22] When the deal was announced in Naples on June 30, the last day of the transfer window, the city exploded, fans celebrating by buying t-shirts that had already been printed with Diego's face, and by spontaneously conducting a torch-lit procession.[23] Diego knew he had found his home, a dysfunctional city that suffered mass unemployment and far-reaching poverty, under the thumb of organized crime, and considered a major blemish on Italy's reputation by the aristocratic, industrial North.[24]

Idolize and scandalize

Maradona had exactly what he wanted—a place where he could feel at home, where he could become a "living god" whose people would lay down their lives for him, and where he could serve up a dream to the poor street kids who reminded him so much of himself. On his first day at the club, Diego said, "I want to become the idol of the poor children of Naples because they are like I was when I lived in Buenos Aires."[25] Of course, it also helped that Maradona was in such a great amount of debt from supporting his entourage and family, indulging in gambling and drugs, and supporting his overall lifestyle, that he desperately needed the transfer fee that came along with his move to Naples.[26]

It is almost impossible to believe that Maradona did not know the role played by the Camorra in his transfer. From the start, he became friends with the Giulianos, considered the most powerful clan within the Naples city limits. They practically owned the neighborhood of Forcella; although photos of Diego in a hot tub at the family compound emerged, he denied that he knew the Giulianos had ties to organized crime. Later, he was purported to say, "They were like something out of *The Untouchables* movie about Al Capone. 'Any problem you have is my problem,' they would say. They said they would protect us. It was like something from the movies."[27]

Maradona's first season at Napoli began with a less-than-cinematic start. He scored just three goals; the team had won just two games. When Christmas came around, he was ashamed, yet he also spoke out (from the safety of Buenos Aires) against his new club, saying it felt like he was playing for a Serie B side. He wondered what the hell he'd gotten himself into, agreeing to sign for this club of little prestige. Could something be done? At that time, Maradona didn't make a fuss; there was no way to bring in new players in the winter and there did not seem to be much else that could be done.[28] Instead, Maradona went back and worked on fixing his own approach. The *tifosi* began to see the deity they had expected, and Maradona lifted them to a fairly respectable 8th in the table, with only two games lost after the break. This may have been due to another deity, as it was Maradona's ritual to pray to the Madonna di Pompei, one of the series of cards that lined the tunnel the players used to enter the field.[29]

But that wasn't enough for Maradona. He wanted to play a much bigger role in deciding who'd be playing around him, to ensure the quality of the squad Napoli would field would be up to his standards. He began to function as a director of football for the team, discussing his choices with the owner and manager and using his connections and influence to convince players to come to Napoli.[30] He was personally invested in seeing Bruno Giordano transfer to the club, sending him telegrams saying he felt they were very alike. And it was true—the two came from similar neighborhoods, had experienced similar social conditions. They fit in well at Naples. They became good friends, and while Giordano acknowledged his flaws, he called Diego a generous, selfless person.[31]

That is probably not how many of the players at the 1986 World Cup saw Maradona. Back in Naples he may have anonymously donated toys to children in hospitals at Christmas, but on the field he was a threat, a menace, and—as many would soon believe—a cheat.

Demon to some, a ghost to no one

The 25-year-old Maradona captained the Argentina side. In Group A, Maradona already dominated the 1st round. He dominated the game against South Korea, which Argentina won 3–1. He scored in the next match, a 1–1 draw with Italy. Finally, Argentina dominated Bulgaria 2–0 and finished top of the group, progressing to the Round of 16. There they met their neighbors and rivals, Uruguay. The *Albiceleste* won 1–0, advancing them to the quarterfinal match with England.

During the game, Diego displayed his willingness to take chances, his desire to rebel against those in power, and the incredible talent he had with the ball at his feet. Others, especially those from England, might object to this characterization of "taking chances" or "rebelling" against the system. But in the famous "Hand of God" goal, Maradona went up against England goalkeeper Peter Shilton, who had eight inches on him. So Maradona took a risk—he used his hand to parry the ball into the goal. The worst that could have happened is audacity being rewarded with a yellow card. Instead, Tunisian referee Ali Bennaceur awarded the goal, and the second linesman backed him up amidst the fury of the England players. Four minutes later, Maradona put Argentina up 2–0 with the "Goal of the Century." Almost every football fan has seen the goal, and no words can truly do it justice. Maradona grabbed the ball in midfield and, in his typical fashion, dipped and dodged as he dribbled, winding his way to goal. His stature provided him with a center of gravity that allowed him to remain steady as he avoided one, two, three England players, all seeking revenge. He made it into the box, where surely it would've been best for the last defender to take him down, at least giving England the slimmest of chances that the penalty wouldn't be converted. Instead, he dodged that last defender and dribbled past Shilton, scoring quite possibly the best goal the world has ever seen.

In that moment, *El Pibe*, the kid from the poor side of the tracks, lifted up his entire country. His play was a fantasy for all of Argentina. Not only his play but his person redefined what football could mean. Rather than blistering and bruising, the game could be beautiful, filled with perfect passes and slippery, winding runs. Maradona's performance against England not only resulted in a win for one side, a loss for the other, but it also represented a country asserting themselves against another with far more power. As journalist Tim Vickery notes, the first goal was a triumph of the *pibe* (street kid), using his wits to overcome power. The second was Argentina saying, "Not only are we as good as you, but we are actually *better*."[32] That point was further driven home when Maradona, in probably his best performance of the tournament, scored two goals against Belgium to take Argentina to the final. There, he slipped through the perfect pass that led to the winning goal. At that point, Maradona became a god to most Argentines. And he was about to become one to Naples.

Even his sexual escapades couldn't keep the supporters from upholding Maradona as their god. In 1986, a local woman, Cristiana Sinagra, came forward to say her unborn child was a result of an affair with Diego. Paternity tests upheld Sinagra's claim, but Maradona refused to acknowledge his son. Suddenly the rest of the country had yet another reason to hate the superstar. This was Italy, and family was the cornerstone that held the country together. Again, his reputation took a dive among the media and fans of other clubs. Napoli supporters, however, stood strong behind their man.[33] In fact, they followed him around, traveling to distant stadiums and packing visitors' sections, ensuring chants of support for the *Partenopei* and jeers about the hosts and their cities would echo around the stands.

Maradona è il Napoli

In return, Maradona supported the supporters. Like Verona's "Welcome to Italy" banner, most of the North was also afflicted by geographism. Those in the South were almost universally looked down upon by Northerners, who labeled them *terroni* (an insult inferring that all they were capable of was working the land) and accused them of being poor, lazy, and uneducated. In the stadiums, the antagonism was dialed up and displayed for all to see, with Northern supporters making monkey noises at Southern players, and displaying banners like "Vesuvius, wash them with fire" and "Give us a present, Vesuvius." But the actions of the Northerners backfired. According to Maradona, "That whole North v South battle made me stronger and gave me a chance to do what I like best—fight for a cause. And if it's the cause of the poor, all the better."[34]

And what better cause for the poor than giving them a reason to be proud of their city, their club? Okay, perhaps there are many other causes to take up, but this was what Maradona could do—lift up Napoli so that they could look the Northerners in the eye and say, "We are as good as you. Our city has a proud, slightly outrageous history, one that should not be ignored. And by lifting a scudetto, everyone will realize that Naples has not slid into inexorable decline. Without help from the North, we built a club strong enough to top the Serie A table. Naples is still a proud kingdom, but one that has every right to be part of a united Italy." That's what Maradona gave to Naples in 1986–1987. At first, the city was deserted on that May afternoon. Then came the eruptions of noise from the San Paolo. The people poured into the streets, draped in blue, draping cars, balconies, and entire homes in blue. They danced on rooftops and in the middle of roads. The city held mock funerals for Juventus and Milan, rich and successful clubs that represented what Neapolitans despised about the North. They saw it as a matter of revenge; it was now time for Neapolitans to stand tall, rather than simply being regarded as being from a city filled with thieves. They were building a new empire, they proclaimed. The celebrations that followed Napoli's first scudetto must have thrilled Maradona; while he was hardly the only extremely talented player on the squad, Diego was the one who made the title possible. He brought joy to the South. He made good on his words. He fought for the poor of Naples.

And he remained loyal. Not only to the club, not only to the city, but to the players who made up the impressive squad that worked together to ensure that Napoli won their first title. A short time after Maradona arrived at Napoli, Giuseppe Bruscolotti, the team's captain from 1978 until 1985, passed the armband to the diminutive man who was already a superstar in the summer of 1985. Bruscolotti told Diego that he was now the captain, but he must keep his promise—Naples was waiting for their scudetto. Given that the first half of 1984–1985 was a struggle for both Napoli and Maradona, and Diego wasn't even sure that he had made the correct choice in going to the club, those words had a deep impact. When, in 1987, Napoli drew 1–1 with Fiorentina to win the title, Maradona did not forget Bruscolotti. Rather than celebrating himself, he pulled Bruscolotti in front of the cameras, hugged him, and told the world that although he was wearing the captain's armband when Napoli secured the scudetto, the real *Vesuviani* captain would always remain Beppe Bruscolotti.[35]

But for all his desire to help the poor of Italy's South, Maradona was certainly no saint, and his professionalism had not improved with the move from Barcelona, where he had already acquired his cocaine habit. In fact, according to *Goal*, the Argentine said, "One touch and I felt like superman. In Naples, drugs were everywhere. And I took more and more there."[36] He lived with his long-term girlfriend, Claudia (who *may* have been a saint) yet regularly consorted with prostitutes or simply beautiful women he met, such as Cristiana, the mother of his son. His odd-looking appearance did him no harm—it was his feet. Or perhaps more accurately, the fame and fortune that accompanied his feet. And the cocaine probably didn't hurt. It hurt the team, however. Maradona was starting to feel invincible at the same time that his body was beginning to feel the side effects of late-night parties flowing with booze, drugs, and women. He regularly showed up late to training, or missed it altogether. He often arrived days late after holidays and breaks, and was in poor shape to boot.

Although they had their break-ups, Diego and Claudia did eventually marry in 1989 after having two daughters together—Dalma and Gianinna. The marriage itself was a lavish affair, costing a reported 1 million pounds. Over a thousand guests attended, with Maradona flying in 200 of his Napoli teammates and their families on a Boeing 747. The venue featured a 40-foot waterfall, while the couple arrived at the reception in a 1937 Dodge convertible. In true Maradona fashion, the party lasted until 8am the following morning. (Claudia had learned about Cristiana's baby and other women had told her of their escapades with Diego, but she believed him when he said they were lying.)

Yet, even with the pain-relieving drugs, and the supplementary cocaine, and his occasional lack of fitness, Maradona could still captivate a crowd. He rarely scored "just" a goal; he sent in tremendous free kicks, went on long, dribbling runs, scored goals while lying on the ground. He could even silence a crowd with a gorgeous pass. One particularly memorable moment came in the semifinal of the 1988–1989 UEFA Cup. Napoli played their second leg at Bayern Munich, carrying a 2–0 lead into the match. Confidence rolled off Diego Maradona. He captivated the entire stadium for three minutes, the small dark figure in untied boots, by essentially dancing through his warm-up, flicking the ball, performing tricks, and juggling the ball along his body.[37] Napoli went on to draw with Bayern 2–2, Maradona guiding the team, but Careca scoring the two *Azzurri* goals. With a 4–2 aggregate victory, Napoli moved on to the final, in which they beat Stuttgart 5–4 over two legs.

During the 1990 World Cup, Maradona finally managed to alienate the few remaining Italian admirers and media personalities who still supported him outside Naples. The tournament was held in Italy, and Argentina played Cameroon in the opening match at the San Siro in Milan. Diego was hated in the North, viewed as all that was bad of the *Mezzogiorno*, a symbol of a Naples getting too big for their britches, winning two scudetti and daring to challenge Milan during the latest winning run. When the crowd of 80,000 saw Maradona's face on the big scream, they booed and whistled. Despite the racism that spurred on the formation of the Lega Nord, the crowd felt the short little man from Naples deserved their ire more; they elected to support Cameroon throughout.[38]

Maradona knew how to fire up a crowd, and he would do exactly that before the semifinal against Italy, at his own San Paolo stadium. Although Italy were the favored side, politicians and sportspeople still felt the need to convince the crowd to support their nation rather than their Diego. Maradona, meanwhile, had found an opening to speak to his beloved people, the ones who had taken him in and made Naples his home. "Neapolitans, you shouldn't forget that in Italy they do not consider you to be Italians." Perhaps surprisingly, one banner hung in the San Paolo during the match read, "Diego, we love you but at the end of the day we are Italians."[39] John Foot writes, "It is unclear quite how many Italians in the ground supported their national team that night. Some claim 'the majority' did, others are unsure." As for the game, it was a nervy one, in which Italy scored first, and Argentina scored with about 20 minutes to play. Or so most would assume, but the referee added on an unprecedented nine minutes of extra time. Seeing as Italy had loaded their team with defensive players, they couldn't pierce the Argentina backline. The game inevitably ended in penalties, and just like in a movie, it was Maradona who stepped up for the key penalty. He kept his head, strolling to the spot and simply sliding the ball in to give Argentina the win and kick Italy out of the World Cup. And as the crowd booed the Argentina national anthem before the final, when the camera paused on Maradona, he mouthed the phrase "*hijos de puta*," easily understood by millions around the world—"sons of bitches."

Hoisted on his own petard

By the time he was 30, Napoli were becoming tired of Maradona's hedonistic self-indulgence. His body could no longer recover quickly from his nightly excesses. He was no longer the player who helped guide his team to a second scudetto in 1989–1990. His sins were becoming more egregious, and the club would finally need to treat him like the others on the squad. Napoli were set to play Spartak Moscow on November 7, after a disappointing 0–0 draw in the first leg at the San Paolo. Maradona didn't travel with his teammates; as usual, he had partied during the beginning of the week, and he spent the night before once more ensconced in sex and drugs. Even while he was diving deeper into the cocaine and other self-destructive habits, Diego still had powerful friends who supported him, told him how wonderful he was.[40] One lent him a private jet so he could travel to Moscow, but before he met up with his teammates, Maradona, a socialist, took the opportunity to have a private tour of Lenin's mausoleum. For once, Napoli punished their star, banishing him to the bench. Though he came on during the second half, he was unable to break down the Spartak defense, and once again the game ended 0–0. Although Maradona scored his penalty, a miss by Marco Baroni in the 3rd round sent Spartak through on penalties, 5–3.[41]

After his antics in Russia, Napoli no longer tried to erase Maradona's misdeeds by covering up stories or paying fines. Diego could not hide from his detractors forever. The Italian media had been reporting on his escapades with women, wine, cocaine, illegal gambling, and ties to the Camorra for years. His behavior at the 1990 World Cup had only increased the vitriol. Finally, in 1991, he tested positive for cocaine and was banned from playing in Italy for 15 months.[42] A few months later, the media happily smeared his name across the football pages. They alleged that Diego was part of a drug-trafficking network involving the Camorra. Although his voice was audible in wiretapped conversations, he was cleared of trafficking, which would have landed him in prison. Instead, he was fined 5 million lire and handed a suspended sentence for possession. Then came the kiss-and-tells from the prostitutes Maradona had slept with—and partook of cocaine benders with him. Some even told of the Camorra fixers that found women and cocaine for him.[43] Yet for many Napoli supporters, Diego could still do no wrong; he was one of them, and they couldn't punish him for joy, spontaneity, or even his failures, not when these were embedded in the culture.

The media spurred on the Italian football association. Maradona had now found himself a vicious enemy, the FIGC. The Italian FA targeted Maradona for "random" drug testing. Maradona was repeatedly drug tested until, in March of 1991, the test administered after the Bari match came up positive. It would be the end of his career in Italy. The Camorra couldn't save him, and Napoli was no longer interested in protecting him. The FIGC handed him a 15-month ban, and Maradona walked (or rather flew) away, back to Argentina. He continued to insist that this was another North–South rivalry, and he was being tested because he had the temerity to lead his team to wins over Northern teams, and of course he had helped the squad win two scudetti.[44]

He returned to Spain for the 1992–1993 season to play for Sevilla. Aged 32, he played 26 matches and scored just 6 goals. Despite leaving Sevilla, he attempted a comeback at the 1994 World Cup in the United States, but it was cut short when once again he tested positive for banned substances after Argentina's 2–1 win over Nigeria. This time they found five different stimulants that were not allowed in football: ephedrine, norephedrine, metephedrine, normetephedrine, and pseudoephedrine. Experts said all of these substances are common ingredients in non-prescription cold, cough, and allergy medications, yet they also act as stimulants.[45]

Given his impoverished background, his dedication to helping the poor, and his desire to make clear how poorly the South was treated compared to the North in Italy, it is unsurprising that Maradona was known to hold tight to many left-wing ideologies. When he died and a columnist labeled him the "worst of the worst," one of the reasons was because he was a "communist." The critique was likely correct, given that Maradona made friends with Fidel Castro, and later had his face tattooed on his left leg, and Che Guevara's on his right arm. He led thousands of Argentinians in a protest of former US President George W. Bush when he visited the country, calling him "an idiot and a murderer."[46] He supported Bolivia's Evo Morales, Venezuela's Hugo Chavez and Nicolás Maduro, and the United States' Barack Obama. His work as a political activist backed up the many speeches he gave regarding the North/South divide while in Naples.

After officially retiring as a player in 1997, after two seasons with former club Boca Juniors, Maradona then took on a managerial career and other official roles with clubs and countries, but often was unable to maintain a job for more than a year. He began with a year at Deportivo Mandiyú in 1994 and then Racing Club in 1995, alongside his friend Carlos Fren, while still playing in Argentina. Maradona was the actual manager, but didn't yet have his coaching license, so Fren, who was licensed, was officially listed as the manager. Perhaps his longest role was at Argentine club Deportivo Riestra, where he served as "mental coach" from 2013–2017.[47] Meanwhile, his most infamous role was with the Argentina men's national team, to which he was appointed manager in October 2008. With the World Cup approaching in 2010, he needed to guide his team through the qualifiers. He barely managed to sneak them through. After a 6–1 defeat to Bolivia and with two matches left to play, Argentina were in 5th place in CONMEBOL qualifying. In the last match, an 85th minute goal by substitute Mario Bolatti put Argentina up 1–0 over 10-man Uruguay, punching Argentina's ticket to South Africa and forcing Uruguay into a playoff match against Costa Rica.[48]

But Diego's time as the Argentina coach is not remembered for his ability to get his team to the World Cup, but what happened after they arrived there. The team won all three of their group games, against Nigeria, South Korea, and Greece, to advance to the knockout rounds. It was clear that Maradona was not using his players in the most effective manner; in fact, that had been obvious from the time that he had named his 23-man squad, leaving out Javier Zanetti and Esteban Cambiasso, both of whom had just won the *tripleta* (the scudetto, the Coppa Italia, and the UEFA Champions League). In press conferences prior to the start of the tournament, he talked about his team being heroes and warriors, and that the rest of the squads should fear Argentina because they had done so well in their training preparations for the World Cup.[49] Later, he criticized the tournament's ball, saying it was impossible to control. While many had negative things to say about the ball, it seemed Maradona might be preparing the press for when his team failed to win. They conquered Mexico 3–1 in the Round of 16, after reprimanding the journalists who failed to take his team seriously. Then came Germany in the quarterfinal, a repeat of the last tournament, in which Germany beat Argentina on penalties. This time, however, Germany exploited the cracks in Maradona's system, and destroyed the side 4–0. When asked if it would be the end of his career, or at least his time with Argentina, Diego responded, "[The result] was like a smack in the face from Muhammad Ali. I am drained of strength, but will I leave? I really don't know. We will see what the future holds."[50]

Despite his "diminishing strength," Maradona did not leave the football world. He spent the next few seasons switching back and forth between the United Arab Emirates, Mexico, and Argentina. He was the coach of Argentine Primera División club Gimnasia de La Plata when he passed away in November 2020. Ironically, Maradona had spent his last decade trying his best to get healthy. His time in Cuba (where he met Castro) was an attempt to curb his cocaine addiction. He put on weight easily, as evidenced by watching the end of his playing career, so in 2005 he underwent gastric bypass surgery so he could slim down, given that at the time he weighed 280 pounds. Two years later, he was admitted to a psychiatric facility to treat his addiction to alcohol. When he came out of the clinic, he told the public that he had quit drinking and had not used drugs in over two years. However, a decade later, his displays at the 2018 World Cup convinced the world that he was once again using cocaine, though Maradona denied it by blaming his behavior on alcohol.[51]

The sanctified sinner

Diego Armando Maradona died from heart failure related to a recent brain surgery on November 25, 2020, just a month after his 60th birthday. An investigation into his death resulted in seven medical professionals in Argentina being charged with homicide, including his neurosurgeon and psychiatrist.[52] The primary reason was due to the fact that Maradona was being looked after in his home, rather than being transferred to a hospital for treatment of his heart condition.

A week after his death, mourners were still arriving at what was soon to become the Stadio Diego Maradona, "lighting candles and adding flowers, footballs and messages of devotion to a shrine that stretches halfway around the stadium."[53] Yet even 30 years after he left Italy, his time there still reflects the country's North–South divide. While Naples was in emotional tatters, many in the North were determined to remind the rest of the world that Maradona was no saint—even if the city still maintained miniature shrines to their idol and hung his photo in nearly every cafe and restaurant. Soon after his death, an *Il Giornale* editorial called him a "cheat, drug addict, alcoholic, violent sexist with women, tax evader and communist: the worst of the worst of the worst."[54]

While it can be argued that many of those descriptors are true, those who loved him never believed him to be without his flaws. Harvard professor and Argentina native Mariano Siskind, when speaking of Diego's legacy and coming to terms with his poor behavior, stated:

> I'm bothered by the reported cases of gender violence. I'm bothered by the fact that he had many sons and daughters that he only recognized late in his life. To answer your question about how one comes to terms with these aspects of

his life, we don't. Because there is no need. Maradona was the most imperfect of human gods. There's no need to reconcile the contradiction that our love for him creates in us; you just live with that contradiction the same way you live with contradictions in your own life. You don't come to terms with it. Morality and love don't go together.[55]

Napoli fans raised Maradona above those flaws, upholding him for the beauty with which he played the sport, honoring him for the two scudetti, Coppa Italia, and UEFA Cup he helped put in Napoli's trophy cabinet, and deifying him because for a few shining years, he was able to make Naples part of the national conversation, often able to drown out the voices of critics like the one in *Il Giornale* by showing the world that the South could, in fact, succeed. As one supporter told *Politico*, "The south has always been hungry and abandoned. When he arrived we had a thousand difficulties, but the moments of joy he brought us on the pitch made us forget them, like a love story."

Maradona left behind a legacy of pain and brokenness, but he also helped to change the entire world of football. Believe him or not, he thinks Lionel Messi would have had a much worse career had Maradona not helped referees realize that defenders executed harsher tackles on the best players, and they needed to be reined in. Such crunching tackles, which kept Maradona on pain medication and likely reinforced his coke habit, are not as common in modern-day football. Comparing how defenders treated both Argentines, it seems Diego had a point. Maradona was also the first player to have a full-time agent, working to ensure the clubs kept his best interests in mind, and, as an 18-year-old kid at Argentinos Juniors, he demonstrated his burgeoning power by demanding the players receive the payments they'd be promised for friendlies, and threatened to not play in those friendlies if it didn't happen. In general, he fought for the rights of players, from those looking for a decent contract to the conditions in Mexico 1986, when the participants were nearly passing out from the sweltering heat. He even rebelled against FIFA, before it was the cool stance to take. Diego's belief was that the attention of the public be focused on the players, not on the governing bodies. He wanted all players to get the respect he felt they deserved.[56]

Tim Vickery may have said it best: "His admirers thrived on the way he would fall down only to get back up again. It humanised a figure whose epic life was as mazy as one of his left-footed dribbles."

[1] Looch, C. (2019, June 10). How Maradona changed Naples forever. *The Culture Trip*. https://theculturetrip.com/europe/italy/articles/how-maradona-changed-naples-forever/

[2] Carter, J. (2012, March 9). Maradona brings success to Napoli. *ESPN*. https://www.espn.com/soccer/columns/story/_/id/1033038/rewind-to-1987:-maradona-brings-success-to-napoli

[3] Garganese, C. (n.d.). Diego Maradona: The god of Naples. *Goal*. https://www.goal.com/story/diegomaradonathegodofnaples/index.html

[4] Id.

[5] Looch.

[6] Garganese.

[7] Id.

[8] Id.

[9] Doidge, M. (2015). *Football Italia: Italian football in an age of globalization*. Bloomsbury Academic.

[10] Ludden, J. (2018). Once upon a time in Naples (3rd ed.). CreateSpace Independent Publishing Platform.

[11] Gladwell, D., & Tomas, M. (2020, October 30). The big read: Maradona, the FC Barcelona years. *FC Barcelona*. https://www.fcbarcelona.com/en/news/1615168/the-big-read-maradona-the-fc-barcelona-years

[12] Pastore, G. (2016, July 5). Operazione San Gennaro. *Sky Sport*. http://www.ultimouomo.com/operazione-san-gennaro/

[13] Vickery, T. (2020, November 25). Diego Maradona: How tormenting England made him an Argentine deity. *BBC*. https://www.bbc.com/sport/football/55074235

[14] Garganese.

[15] Biography.com Editors. (last updated 2020, November 25). Diego Maradona biography. *Biography.com*. https://www.biography.com/athlete/diego-maradona

[16] Id.

[17] Maradona: Así fue su debut. (2020, October 20). *El Gráfico*. https://www.elgrafico.com.ar/articulo/1088/33691/maradona-asi-empezo-todo

[18] Maradona e' nostro, annuncia il Boca. L'ha pagato 10 miliardi (un record). (1981, February 13). *Corriere della Sera*, 28.

[19] Gladwell & Tomás.

[20] Balague, G. (2020, November 26). Diego Maradona dies: Guillem Balague on "the magician, the cheat, the god, the flawed genius." *BBC*. https://www.bbc.com/sport/football/55084504

[21] Looch.

[22] Garganese.

[23] Pastore.

[24] Garganese.

[25] Id.

[26] Ludden.

[27] Garganese.

[28] Rispo, R., & Bell, H. Maradona era episode 1: 1984/85 season. *Far from Vesuvius*. Podcast audio. May 1, 2020.

[29] AP News. (2020, February 20). For Napoli fans, the team is a religion and Maradona is god. *GVWire*. https://gvwire.com/2020/02/20/for-napoli-fans-the-team-is-a-religion-and-maradona-is-god/

[30] Carter.

[31] Veltroni, W. (2016, December 6). Veltroni intervista Giordano: "I trionfi con Diego. Ora sogno la Lazio." *Corriere dello Sport*. https://www.corrieredellosport.it/news/calcio/2015/12/05-6469468/veltroni_intervista_giordano_i_trionfi_con_diego_ora_sogno_la_lazio

[32] Vickery.

[33] Looch.

[34] Maradona, D. (2005). *El Diego: The autobiography of the world's greatest footballer*. Yellow Jersey Press.

[35] Ciccarelli, L. (2012, October 24). La storia siete voi: "Pal 'e fierro" Bruscolotti. *Tutto Napoli*. https://www.tuttonapoli.net/rubriche/la-storia-siete-voi-pal-e-fierro-bruscolotti-120374

[36] Garganese.

[37] Carter.

[38] Foot.

[39] Id.

[40] Balague.

[41] Id.

[42] Looch.

43 Id.

44 Id.

45 Berkowitz, S. (1994, July 1). Maradona booted from the World Cup. *Washington Post*. https://www.washingtonpost.com/archive/sports/1994/07/01/maradona-booted-from-world-cup/edc8123d-5057-4449-a65d-24ef3f112955/

46 Taylor, C. (2005, November 6). A big hand. *Guardian*. https://www.theguardian.com/football/2005/nov/06/sport.argentina

47 Diego Maradona debuta hoy como asesor espiritual de Deportivo Riestra. (2013, August 19). *Telam*. https://www.telam.com.ar/notas/201308/29256-diego-maradona-debuta-hoy-como-asesor-espiritual-de-deportivo-riestra.html

48 Late winner puts Argentina in World Cup finals. (2009, October 15). *CNN*. http://edition.cnn.com/2009/SPORT/football/10/14/football.samerica/index.html

49 Fryer, R., & Hills, D. (2010, July 4). World Cup 2010: The best of Argentina manager Diego Maradona. *Guardian*. https://www.theguardian.com/football/2010/jul/04/diego-maradona-argentina-world-cup

50 Id.

51 Couzens, G. (2018, 27 June). Diego Maradona blames bizarre World Cup behaviour on white wine binge in VIP box. *Evening Standard*. https://www.standard.co.uk/sport/football/worldcup/diego-maradona-blames-bizarre-world-cup-behaviour-on-white-wine-binge-in-vip-box-a3873281.html?fallback=true

52 Church, B., & De La Fuente, H. (2021, May 21). Seven medical professionals charged with homicide after investigation into Diego Maradona's death. *CNN*. https://edition.cnn.com/2021/05/21/football/diego-maradona-death-homicide-charges-spt-intl/index.html

53 Roberts, H. (2020, December 4). Maradona divides Italy, in life and in death. *Politico*. https://www.politico.eu/article/maradona-death-naples-italy-divisions-league/

54 Id.

55 Mineo, L. (2020, December 4). Why Maradona matters. *Harvard Gazette*. https://news.harvard.edu/gazette/story/2020/12/harvard-professor-explains-why-diego-maradona-matters/

56 Balague.

CHAPTER 9

Down, Down in an Earlier Round

9

While the rest of *calcio* was delighted to see Maradona leave the game, taking his cheating ways and embarrassing habits back to Spain, the vast majority of Napoli supporters seemed to understand that their fandom would be forever changed. Although the club retained a strong squad without their superstar, after he left in March, they had only managed to finish 8th. Napoli fans had watched their side lift two *scudetti* and a UEFA Cup when the God of Naples had graced them with his fabulous feet. They now knew what it felt like to watch a superstar take the pitch, week in, week out. Nothing less than spectacle and success (preferably both) would appease them. Unfortunately, the club had spent beyond its means to acquire Diego and meet his demands. They could not afford another marquee name; they could not even afford to retain the excellent supporting cast they had assembled. The *Partenopei* were on the road to ruin. It turned out the narrow, pot-holed track dead-ended at a sharp cliff, and Napoli were moving much too fast to brake before they plunged over.

New-look Napoli

Just before the news that Maradona had tested positive for cocaine, Luciano Moggi packed up and returned to Turin, citing irreconcilable differences with Corrado Ferlaino. Not an uncommon problem, but Moggi had read the writing on the wall, and knew the team would struggle to find their way after their defining player left the squad. He wanted assurances from Ferlaino that there would be a plan for reconstruction, and that he would be more involved with the club.[1] When those weren't given, he decided his former club, Torino, would be a better fit. That left Giorgio Perinetti, the former head of Napoli's youth system who had just been promoted to sporting director, to tell Diego about his 15-month ban. It must have been a wonderful way for him to begin his new job.

The supporters wondered what the new Napoli would look like. The club worried about whether the fans would continue to turn up. It helped little that several starters were sold, while turnover also occurred at the upper levels of the SSC Napoli hierarchy. In the summer of 1991, Alberto Bigon left the club due to their poor finish in Serie A. Despite winning the *scudetto* and Supercoppa with Napoli, his career went downhill after his departure, coaching smaller clubs on the peninsula before looking outside the country for opportunities. He was replaced by Claudio Ranieri. Although now a legendary manager, Ranieri's biggest success at the time had been leading Cagliari to promotion to Serie A in 1989–1990. He still came to Napoli with demands, however: He would need squad reinforcements immediately. Ferlaino actually cracked open his wallet to bring in central defender Laurent Blanc from Montpellier, whom Ranieri used as a *libero*. (Alas, Blanc had yet to develop the leadership skills he became famed for and later harnessed for his impressive managerial career. Although he did well at Napoli, he was never quite comfortable in Italy and remained there for just one season.) Center-forward Michele Padovano was bought from Pisa in Serie B, defensive midfielder Stefano De Agostini from Reggiana in B, and left-back Massimo Tarantino returned from loan with Barletta in Serie C. Reinforcements Ranieri could have, but most had to be young and cheap.

Fortunately for Napoli, Maradona had left them with strict instructions: The Napoli management should focus on Gianfranco Zola's development. Diego said, "Napoli doesn't need to look for anyone to replace me, the team already has Zola."[2] Zola had arrived in 1989–1990, and the 23-year-old, viewed as an understudy to Maradona, played only a minor role in Napoli winning their second scudetto. But that didn't mean he wasn't learning. "I learned everything from Diego," said Zola. "I used to spy on him every time he trained and learned how to curl a free-kick just like him. After one year I had completely changed."[3] Soon, Zola didn't need to spy. He and Diego would practice free-kicks together, spending hours after team practice perfecting their shots.

In the same interview, the journalist inquired about his favorite memory of Maradona. Zola answered:

> Maradona always used to keep the No. 10 shirt for himself. However, once we played an Italian Cup game in Pisa and so I went to take the No. 9 shirt, but Maradona brought the No. 10 shirt over to me. He told me that he wanted to wear the No. 9 to pay homage to his friend Careca, saying he wanted to play once with that number, but I later found out that he gave it to me for the experience. I was very flattered. That was a great gesture. It gave me a lot of confidence.[4]

It's interesting to note, however, that it does not appear that Napoli faced Pisa, or played anywhere near Pisa, in the Coppa in the year and a half these two were at the club together. A misremembered city? An incorrect translation? A mere soundbite trotted out to strengthen the story that Maradona was his mentor? No, Zola simply got the tournament wrong: Napoli drew 1–1 in Pisa, Ciro Ferrara scoring the goal. Regardless, Zola's confidence did increase under Diego. And he was one of the last Napoli players to wear that honored No.10 on his back; Maradona's number was retired in 2000.

Zola did make one major contribution to the 1990 scudetto-winning squad—in February, he scored a significant goal at Genoa in injury time that led to a 2–1 Napoli win, essential to keep Napoli top of the table. After Maradona left in March 1991, Zola needed to quickly step into his boots. Increased confidence or not, the gaping hole at the center of the squad was an immense one for the 25-year-old to try to fill. Zola did as well as anyone at the time probably could have done, at a club that was in disarray and in a city that was in mourning. Partnered alongside Careca, the two led the team on an eight-match unbeaten run to end the season, including each scoring a goal in the final game, a 3–2 win over Bologna. They helped lift Napoli up the standings, but the side still missed out on European classification by a point.

Now, in the first season of the post-Diego era, it was Ranieri's job to bring Napoli back into competition with the great clubs of Italy. And do it within the restraints of Ferlaino's budget restrictions. Later in his career, Ranieri would be known for employing a rigid 4-4-2 formation, concentrating on defensive solidity rather than an attacking prowess, and excessively rotating his squad. In his first year at Napoli, however, he fielded the same players consistently, and the attack, led by Careca and Padovano, with Zola in a playmaking/attacking midfield role, was the third-best in the league. Careca led the squad with 15 goals while Zola scored 12; Blanc—as of 2022, still Montpellier's all-time leading goalscorer—was next with 6, not bad for a defender. But aside from their libero, the defense was not at their best. They shipped 40 goals, on level with Lazio who would finish 10th, and more than Atalanta (13th) and Cagliari (11th). That weakness forced Napoli, alongside Inter, into a role in which they could only pressure the 1st-placed side throughout the season; it was Milan and Juventus who would truly fight it out for the scudetto. The *Rossoneri* crushed Napoli's hopes of the title in the 15th round, beating them 5–0. Ultimately Milan would win the title at the San Paolo, two weeks before the end of the season, with a 1–1 draw in Naples. Napoli finished 14 points behind, but it was enough to take 4th and send them to the UEFA Cup the next season.

Although Careca, almost 31 at the start of the season, was blossoming once more, looking like he'd seized upon one last season of youthful energy, it was Zola who the team was constructed around. While he was occasionally inconsistent, he is considered one of the best Italian creative attackers of all time. He was everything one could ask for from a playmaker—imaginative, technically gifted, unpredictable, with the willingness to create chances and provide passes for his teammates rather than hogging the spotlight. It is nearly impossible to list all his technical abilities. In addition to his eye for goal, he was outstanding with the ball at his feet, able to dribble around opposing players while superbly controlling the ball. He had an eye for the perfect pass and the ability to execute it with either foot. Unsurprisingly, given his tutelage under Maradona, he was a penalty-kick and set-piece specialist, known especially for his accuracy at bending direct free-kicks. Finally, his overall vision of the game and ability to read it, alongside his technical intelligence, allowed him to orchestrate a match to get the best out of his teammates.

After the first season without Maradona resulted in a short title run and the return of European competition, Napoli fans felt optimistic about 1992–1993. That optimism was short-lived.

– HEROES –
Careca (1987–1993)

Napoli appearances: 164
Goals: 96

Due to his head of curly black hair, he was called "Careca," after the Brazilian clown *Carequinha*—but when it came to football, he wasn't kidding about.[5] Antônio de Oliveira Filho began playing professionally at age 17, with his local side Guarani, who won the Brazilian Championship for the first time in his first year. From the start, his impressive pace and astonishing ability to convert chances into goals (he scored 18 in 17 games in his last season with Guarani) drew the eye.

Five years later, in 1983, he signed with São Paulo. There, he scored 17 goals in his first season and 12 in his third (he recorded no appearances in Série A in his second, an absence for which there are no clear answers). In his third year with the *Tricolor*, they did the double and Careca came away with the *Bola de Ouro*, recognizing him as the year's best player.

It was more than enough to punch his ticket to Mexico 1986. Although he missed the 1982 World Cup (and Brazil missed him back, falling to Italy 3–2 in the 2nd round), he had become the center of the *Seleção* attack. It's no surprise he caught Napoli's attention there. Brazil dominated in the group stage; Careca scored in the 1–0 win over Algeria and knocked in a brace in the 3–0 against Northern Ireland.

Although he scored from the spot in the 4–0 triumph over Portugal and the only goal in open play against France, taking his total to five goals in the tournament, he was not chosen to take a penalty to resolve the 1–1 draw with *Les Bleus*. Sócrates and Júlio César missed, and Brazil were out. It's likely Careca would have made more of a splash in the international area had Diego Maradona not stolen the show with the "Hand of God" and "Goal of the Century."

When the Partenopei came for Careca, he needed little convincing from the team that had just been crowned champions of the best league in the world, who boasted the best player in the world at the time. Arriving at Napoli in 1987, he quickly became an integral part of the squad, one point on the fearsome MaGiCa trident, alongside Maradona and Bruno Giordano. Writer Gary Thacker puts it beautifully:

> Some may argue that it would be easy to play alongside a talent such as Maradona in his prime, others would argue the reverse; but the rapidity with which the Brazilian striker struck up a relationship with the Argentine genius argued very much for the former.[6]

The "Ca" of the magical trio scored 18 goals across all competitions that first season, and helped contribute to many of Maradona's 21. One of Careca's immense talents was being able to maneuver into space, and with opposing defenders distracted by Diego, he became even more proficient at making his way into scoring position. With the next year came Napoli's victory in the 1988–1989 UEFA Cup, and the Brazilian played a starring role. Giordano had left, leaving just Diego and Careca to take on the mighty defenses of Europe. They paired perfectly in the 2–0 semifinal win over Bayern Munich at the San Paolo, with Careca getting on the end of a plumptious pass from Maradona to net the opener. He did it again in the reverse fixture, slipping into space in the box to slide in a ball from his partner. His second strike was jaw-dropping, a goal sprung off the counter, sinking into the back of the Bayern net. The magic continued in the two-legged final against Stuttgart; the two each scored in the 2–1 win in Naples, and in the second, Careca flew past his defender to find yet another Maradona ball, clipping it over the keeper to sound the death knell for Stuttgart.

Careca continued to flourish alongside Maradona as Napoli won their second scudetto in 1989–1990, scoring 12 goals in all competitions. After *El Diego* left, he and Gianfranco Zola did their best to pick up the pieces. The *Azzurri* managed to finish 4th in 1991–1992, an improvement over the previous season's 8th-place finish, Careca leading the way with 15 goals. But despite the addition of Daniel Fonseca, who helped keep the attack lively, Napoli sank down to 11th, and both Zola and Careca said their last goodbyes.

Casting lines with barely a bite

The summer seemed to bring good news—Ranieri remained as manager, and together he and Giorgio Perinetti convinced Ferlaino to buy Uruguayan striker Daniel Fonseca from Cagliari, where he had done well but not particularly excelled. The two believed he'd flourish when played centrally, a position he was better suited to than the left, where he'd been consigned to while in Sardinia. Ferlaino spent 15 billion lire and sold the club's half of Vittorio Pusceddu's co-ownership deal to bring him to Naples.[7] Ranieri moved him back to center-forward, and he scored 24 goals across all competitions. The only striking departure was that of Laurent Blanc, who returned to France, flourishing once more at Saint-Étienne.

The Napoli attack remained strong, despite the veteran Careca, who maintained his spot in the starting XI, starting to deteriorate. He scored just seven goals in the league that year. Meanwhile, Zola was still creating and scoring, with 12 goals to his name in Serie A over the season. Roberto Policano, brought in to strengthen the midfield, also scored seven. Again, the defense was an issue. Defenders Giancarlo Corradini (age 32) and Giovanni Francini (29), both of whom had been with Napoli for quite a few seasons, were starting to look a little rusty. But the real issue was the goalkeeper, Giovanni Galli, now 34. At that age a keeper typically is still rather dependable, but Galli had always had a few difficulties in the area, particularly with hard or tricky shots. That season he let in 50 goals in the league, while the squad scored 49. Perhaps it didn't help that FIFA had changed the rules of the game applicable to goalkeepers. No longer were they permitted to touch a back pass from one of their players with their hands, unless that pass was a header.[8]

The season began with a boring home draw to Brescia, which some fans would later look back on with fondness as the season progressed. But in the next round Napoli beat Foggia 4–2, with goals from Careca and Zola as well as a brace from their new forward Fonseca. A few days later, Napoli would learn just how magnificent the Uruguayan known as *El Castor* (the beaver), due to his impressive buck teeth, truly could be. Fonseca was a speedy striker with an accurate, powerful shot from distance. He also excelled in the air. Fonseca and Zola shared many of the same talents, including excellent dribbling skills and an innate ability to read the game that resulted in the confidence that the ball at their feet would make it into the penalty box. Like Zola, he wasn't too proud to also create chances for his teammates. One quirky difference between the two of them was that at times, Fonseca switched out his team socks for those of Uruguay's, just a bit lighter than Napoli blue, as he believed they brought him good luck.[9]

The old videos make it impossible to tell if the socks Fonseca was wearing on September 16, 1992 were of a lighter hue—but it would certainly make sense if they were. The squad had flown to Spain to face their first UEFA Cup opponent, Valencia, who had also finished 4th the previous season. In the 20th minute, Fonseca, racing up the left side of the pitch, latched on to a long ball from near the halfway line, ran a few steps, then used his left foot to slot it into the bottom right corner of the goal. Napoli went into the dressing room 1–0 up, feeling quite pleased to be leading away from home.

Valencia briefly shattered Napoli's confidence when Francesco Roberto equalized in the 56th. In the same minute, Quique Sánchez Flores was shown a red card. Fonseca turned the dial up to 11. In the 60th minute he repeated almost the same trick, racing up the left and slipping a low ball into the opposite corner of an empty net, prompting the tiny Zola to jump up and wrap himself around Fonseca's body in celebration. Four minutes later, Fonseca was once again running up the left, this time to pounce on a loose ball. He threaded his shot through a number of defenders to pick up his third goal. For his fourth, in the 87th minute, a perfect Zola pass led to a perfect Giovanni Francini pass to El Castor, in a perfect position for a one-on-one with Valencia José Manuel Sempere. He easily beat the keeper. Finally, in the 90th, Fonseca put a cherry on top of what was already a decisive Napoli victory when a long ball from the back deflected off a Valencia player to fall at his feet as he, once again, raced down the left. He beat the last defender and then tapped the ball past Sempere's outstretched arms. The noise from the small Napoli away section echoed throughout the stadium. With a 5–1 victory and five away goals to fall back on, it was almost certain Napoli would advance to the next round.

Upon their return home, they were greeted by a visit from Inter, who would eventually finish 2nd. The *Nerazzurri* scored two goals in quick succession in the second half, and Fonseca could only manage a consolation goal in the 84th. Although their brilliant Uruguayan would score early in the next match, at Ancona (who were playing their first Serie A season), the newly promoted side equalized in the second half, giving Napoli just one point in the two games before their return leg with Valencia. And Juventus would be waiting in the next Serie A round.

When *Los Murciélagos* came to town, the San Paolo was ready to witness another exciting spectacle. Disappointingly for the hometown crowd, the game was much calmer, and Fonseca needed just one goal to win it for Napoli, 1–0. But at least they were on to the next round of the UEFA Cup, to be played against Paris Saint-Germain in October. For the next three weeks, they could focus on the league. The match against Juventus at the San Paolo went about as expected. The Old Lady were up 3–0 in the 80th minute, and the fans were demonstrably upset. Then El Castor struck in the 84th, a shot that felt like nothing more than another consolation goal. However, Zola scored two minutes later, and suddenly the crowd was roaring, screaming for their side to do the impossible. Unfortunately, they were unable to find an equalizer, and Juventus held on for a 3–2 win. The next match was what had the supporters truly upset, as Napoli lost 2–0 to little Udinese, a team that would avoid relegation only by goal difference, a team coached by their old manager Alberto Bigon. Three points from their 2–1 over Roma at the San Paolo—the goals coming from Fonseca and Careca—did much to increase both the supporters' and the squad's confidence ahead of the next European match.

Later that week, the fans prepared for a visit from PSG. Surely this time they'd be able to witness the same sort of routing that took place in the first match at Valencia. They were sorely disappointed; their side couldn't make much of anything, while PSG scored two goals. Yet they remained optimistic for their visit to Atalanta, as surely Napoli wouldn't lose again to another "small club." But they did, 3–2, with another last-gasp attempt at rescuing a point, this time the goal from Ciro Ferrara in the 90th.

Annoyingly, Napoli then went to Paris and picked up nothing; they barely produced a chance, much less a goal in the 0–0 draw. On that November night they were dumped out of the UEFA Cup, and came home to ready themselves for a visit from Fabio Capello's Milan.

It did not go well.

Last season's victors were in the middle of an eventual 58-game unbeaten run. *Gli invincibili* would not be felled until March 21st, when Parma beat them 1–0 in Milan. Even in the midst of this incredible run, the game at the San Paolo stood out. The visitors ran circles around Napoli. The Rossoneri danced on their graves, almost literally, for it was this match that put Napoli in a hole they were almost unable to dig themselves out of. Milan star Marco van Basten scored four goals (the first and only time in his career he would do so), and Stefano Eranio added a fifth. Once again Napoli scored late, but Zola's 83rd minute goal barely made a dent in their humiliation. And it came far too late to save Claudio Ranieri.

The writing had been on the wall after PSG kicked Napoli out of European competition, but the 5–1 loss to Milan was the final straw. Ferlaino sacked Ranieri and brought back Ottavio Bianchi. Soon after, sporting director Perinetti, who already had issues with Ferlaino (not a rare problem in Naples), resigned in protest of the managerial change.

Bianchi failed to revive a sagging Napoli. The team managed to collect a few points against smaller sides, but they were no longer able to beat the big clubs. This was exemplified by Napoli's visit to Juventus, in which the *Bianconeri* went into the break up 2–0. Zola scored in the 51st minute, and Ferrara put in the equalizer in the 71st. Immediately after, Fabrizio Ravanelli put Juventus ahead once more. An 80th minute penalty, converted by Fonseca, put the sides back on level terms. But Napoli didn't have the strength, mentally or physically, to hang on. Andreas Möller exploited their defensive weaknesses in the 87th, giving Juventus a 4–3 win.

In the end, though, Napoli *were* able to claw their way up. They mustered three wins and five draws to save themselves and finish the season in 11th place.

This was insufficient for Ferlaino, who was also worried about the club's deficit, which was around 80 billion lire at the time.[10] He passed on control of the club to director Ellenio Gallo, but not before installing Bianchi as general manager. Bianchi then put Marcello Lippi, a man known for his one-year stints at clubs, on the bench for the 1993–1994 season.

Thus began a period of austerity for Napoli. The first casualty of the new financial measures was Gianfranco Zola, who was sold to Parma for over 12 billion lire.[11] This angered the fans, who accused him of being a traitor. Yet it was Ferlaino who had put the club in such a position with the money he'd spent on big signings, and Zola made it clear he had no choice but to accept the transfer. Careca moved to Japan to finish his career. Napoli also sold Giovanni Galli to Torino, bringing back Giuseppe "Pino" Taglialatela from his latest loan stint at Serie B side Bari. Inserting two young Fabios into the squad was the best decision the club made that year. Center-back Fabio Cannavaro, who would later become a hero with both Juventus and the Italian men's national team, was brought up from the youth team and immediately put into the starting XI. Fabio Pecchia, a central midfielder from the Avellino youth side, joined him.

Despite these measures, Napoli nearly defaulted on their loans three times, and the players went without their salaries for months.[12] Yet they managed to lift themselves up five places in the table, finishing 6th, the last qualifying spot for the UEFA Cup. Pecchia distinguished himself as an irreplaceable cog in Lippi's system, a quick player who understood his manager's vision for fast-moving forward play, and even at a young age could direct others to follow it. But without Zola, who scored 18 league goals at Parma that season, Napoli's goalscoring abilities weakened considerably. Fonseca finished the season with 15; the next highest scorer was Paolo Di Canio (on loan to Napoli from Juventus) with 5. On the other hand, the defense had been strengthened with the collection of journeyman players that wore the Azzurri shirt that year; at the same time, other than Cannavaro, there was nothing truly remarkable about the players filling those roles. Napoli managed to finish in the European places thanks to their performances in the final six rounds, when the side seemed to finally click. They beat Milan (1st), drew with Juventus (2nd), and scored two unanswered against Parma (5th), to ensure they finished 6th, one point above Roma.

In 1994–1995, President Gallo continued to operate the club conservatively. Marcello Lippi moved on to Juventus, replaced by Vincenzo Guerini, who as coach of Ancona had been the one to bring them to the top flight, then right back down to Serie B. Along with the clear plan of how the team should play, the casualties of that summer included those who could execute it: Ciro Ferrara, who followed Lippi to Turin, and Daniel Fonseca, who took his goals to Roma for 17.5 billion lire, plus the remainder of the co-ownership deal for Benny Carbone. The five goals of Di Canio were lost when his loan ended. Both Fonseca and Di Canio had found their rhythm in Lippi's squad in the final third of the previous season. In an attempt to reinforce the side without spending much money, they picked up Brazilian defender André Cruz, a set-piece specialist, on a free from Standard Liege.

Nine other players joined the squad that year, including forward Massimo Agostini, who Guerini knew from Ancona, and midfielder Freddy Rincón from Brazilian club Palmeiras. As evidenced in the previous season by Napoli's performances in the spring, a big change is often difficult for the coaching team to deal with, as the players come in from teams with different tactics, and the new squad need time to learn each other's personalities and approaches to the game. New manager Vincenzo Guerini lasted just six rounds, in which Napoli picked up a win and two draws, and lost 5–1 to Lazio in his last game.

Vujadin Boškov replaced Guerini, and suddenly Napoli were chasing a European spot rather than fighting against relegation. Again the team finished on a high, with four consecutive wins at the end of the season—including a win over Milan, last year's champions—but the side's earlier struggles were reflected in their 7th-place finish, one position and one point out of European qualification. It was a particularly distressing outcome, as Napoli had made it to the 3rd round of the UEFA Cup that year and were ready to challenge the teams of Europe once more.

The format of the UEFA Cup changed in the 1994–1995 season. Many new teams were added, as the UEFA Champions League demoted the winners of smaller countries' domestic tournaments to the UEFA Cup (this change lasted three years). In order to address the influx of new teams, the tournament added a preliminary round, which subsequently reduced the number of clubs involved from 91 to the usual 64.

Napoli joined in the 1st round, one of five Italian clubs, including Inter, who had barely escaped relegation the season before but had made up for that embarrassment by lifting the UEFA Cup. The Partenopei faced Skonto, a rather new Latvian club that had formed in 1991. In the first leg, Benny Carbone scored both goals, unanswered, one in each half. In the second, in Riga, midfielder Renato Buso (another player who had hit his stride at the end of the previous season) claimed the only goal. With a 3–0 win on aggregate, Napoli moved on to the 2nd round. Boavista, a Portuguese side, proved slightly more of a challenge. In Porto, Napoli came back from losing 1–0 at the halftime whistle to draw 1–1, again thanks to Carbone. At the San Paolo for the second leg, Agostino took over the scoring, putting Napoli up 2–0 before the break. Boavista attempted a comeback in the 76th minute to make it 2–1. Napoli fans chewed their fingernails, cheered their team on, and yelled helpful suggestions, sometimes all three at once. They knew if the visitors scored again, the *Vesuviani* would be knocked out, and Boavista would go through on away goals. Fortunately for the hearts of many Napoli supporters, the home side hunkered down and refused to allow another goal to land in Pino Taglialatela's goal.

In the 3rd round, Napoli traveled to Germany to face Eintracht Frankfurt, managed by Jupp Heynckes. The previous week's 5–2 loss to Fiorentina at the San Paolo, with own goals by both André Cruz and Fabio Cannavaro, was still fresh in their minds. Just before the first half ended in Frankfurt, Cannavaro was shown a second yellow, and Napoli were down a man. Then, in the 55th minute, Tony Yeboah pounced on a ball inside the crowded penalty area. As Napoli frantically tried to defend their goal, a low shot slipped through Taglialatela's fingers and into the net. Yeboah joyously celebrated with his teammates, but the goal would go down as an own goal. It turned out Buso had gotten his toes on the end of the ball, trying to nudge it away from Yeboah, and instead kicked it straight past Pino. The second leg wasn't nearly as dramatic: a 54th minute goal this time, legitimately scored by Eintracht midfielder Ralf Falkenmayer, in front of the home crowd at the San Paolo. The Napoli supporters witnessed the German side win 1–0, for an aggregate score of 2–0. Napoli were out of the UEFA Cup, and (spoiler alert!) would not return to European competition until 2008.

Napoli may not have reached the two-legged final, but it was an all-Italian affair, and for the third year in a row, a Serie A side emerged victorious. In 1995, Parma beat Juventus, 2–1, both goals from Dino Baggio, to lift the trophy for the first time; Gianfranco Zola scored five times during the tournament. The 1990s showcased Italian clubs' dominance in Europe, winning new calcio fans throughout the world. Prior to their victory over the Bianconeri, Parma had won the European Cup Winners' Cup in 1993 and lost the final to Arsenal the next year. Meanwhile, Juventus held the 1990 and 1993 UEFA Cups and went on to beat Ajax the next year in the European Cup, also reaching that tournament's final in 1996 and 1997. Milan won the same tournament in 1990 and 1994, and finished runners-up in 1991 and 1993. Even Sampdoria, who won their only scudetto in the 1990–1991 season, won the Cup Winners' Cup in 1990 and made it to the European Cup final in 1992 while under Vujadin Boškov, who would become Napoli's manager a few years later. Lazio also triumphed in the Cup Winners' Cup, clinching the tournament's final trophy in 1999, the same year Parma lifted another UEFA Cup. Finally, Fiorentina made it to the UEFA Cup final in 1990, losing to Juve, whose city rivals, Torino, had also faced Ajax, losing the 1992 final.

The decline and fall of the Italian football empire

I beg for Napoli fans' patience as I take a broader view of Italy in the 1990s and 2000s, and how the country's economy affected Serie A clubs. As the decade began, like much of the world, the Italian government was attempting to liberalize the economy, lower its debt, and convert state-owned businesses into private companies. It appeared they had succeeded; by the mid-1990s, Italy's GDP had nearly caught up with that of Belgium, France, Germany, and the Netherlands.[13] In turn, this allowed more money to fall into the hands of football club owners, who were typically high-ranking businessmen. However, Italy decided to enter the Eurozone, which required fiscal austerity and structural reforms. As a result, output and incomes decreased, while unemployment rose. This caused a decline in owners' fortunes at the same time as other countries' clubs became richer, triggering rising transfer costs and increased player salaries.

European competition involves prize money and, in Italy, success results in a larger share of the money from TV rights. It was natural that owners would chase the opportunity to play midweek football. But without sufficient investment, clubs began to topple like dominoes, sliding down the table, falling into Serie B, even dissolving into nothingness. According to the *Guardian*, during 2002–2018, "153 Italian clubs have refounded, merged with other clubs or disappeared altogether."[14] Chievo Verona, who had spent much of the last two decades in Serie A and even qualified for the UEFA Cup in 2002–2003 and the Champions League in 2006–2007, were expelled from Serie B in 2021 and, at the time of writing, have yet to return to league football. Little wonder, then, that Italian football is often viewed with derision, as many still remember the glory of the 1990s and its contrast with today's results; Serie A last lifted a European trophy in 2010, when José Mourinho's park-the-bus Inter produced effective yet soporific results. In addition, economic difficulties often lead to a rise in insular and nationalistic attitudes, which are reflected in Italian supporters' continued use of racial slurs and geographist chants, further alienating potential fans and owners (and their money).

Conquering Europe broke nearly every Italian club who attempted it, and even those who merely pursued the chance. For example, Genoa reached the UEFA Cup semifinals in 1991–1992 but were relegated in 1995, after which a number of poor decisions kept them from reaching Serie A again until 2007. Almost every Italian club who reached a final later went bankrupt, were involved in a scandal, or both. Only Inter, UEFA Cup winners in 1991, 1994, and 1998, and runners-up in 1997, emerged unscathed. Udinese provided a rare model for smaller clubs who wished to compete in Europe: They played in the UEFA Cup in 1996–1997, 1997–1998, and 1998–1999, ending the decade with qualification to the Intertoto Cup. Although they slipped in 2001, when Luciano Spalletti returned to manage the *Zebrette* in 2002, he found the club had quickly instilled an organized, disciplined ethic, enabling them to reach the UEFA Cup once more.

Others would have done well to follow the *Friulani* example. The first sign that success would lead to greater ambition, which in turn could lead to a club's downfall, came from Florence. In the summer of 1990, Fiorentina were battling against relegation while simultaneously preparing to face Juventus in the UEFA Cup final. On the day of the match, owner Flavio Pontello sold beloved young star Roberto Baggio to their opponents, infuriating the *Viola* supporters. He had sold other promising players over the previous years, such as Daniel Bertoni, Sócrates, and Daniel Passarella, but was now holding a yard sale, his financial difficulties forcing him to sell much of the squad. The loss to Juventus, combined with the sale of Baggio, was the last straw for Fiore fans, who took to the streets in protest, forcing out Pontello. Under filmmaker Mario Cecchi Gori, they were relegated in 1993. Their instant bounce back up to Serie A did not provide a cautionary enough tale; they finished 4th in 1996. Fiorentina's league play disappointed in 1996–1997, but they made it to the Cup Winners' Cup semifinal, losing to Barcelona. Fiorentina finished the decade in 3rd, qualifying for the Champions League. But in June 2000, after a disappointing performance in both Serie A and Europe, they were forced to sell their revered Gabriel Batistuta to Roma for the then-highest price ever paid for a player over age 30. This kept them in the black and even helped the Viola secure a bank loan, but the damage was done.[15] Fiorentina were revealed to have debts of approximately 50 million US dollars and were unable to play their players' wages. The club entered administration in June 2002 and were denied a place in Serie B for the upcoming season. Entrepreneur Diego Della Valle re-established the club as Associazione Calcio Fiorentina e Florentia Viola, and they entered Serie C2, Italy's fourth tier.

Torino behaved in a similar manner, also failing to heed their own warning signs. In the early 1990s, they experienced their best years since the days of *il Grande Torino*. After knocking out Real Madrid in the 1991–1992 UEFA Cup semifinal, they faced Dennis Bergkamp's Ajax in the two-legged final, finishing runners-up only by virtue of the away goals rule. The next season, the away goals rule allowed the *Granata* to beat rivals Juventus in the semis of the Coppa Italia, after which they drew 5–5 with Roma. This allowed 9th-placed Torino back into Europe through the Cup Winners' Cup, where Arsenal knocked them out in the quarterfinals. After offloading almost their entire starting XI in hopes of bringing in funds, they finished 11th; Torino were relegated the next year in 1996. After a few years of yo-yoing between Series A and B, Torino were back, and had qualified for the 2002–2003 Intertoto Cup, where they lost to Villarreal on penalties. That year was their worst in Serie A, but they secured promotion immediately. Unfortunately for such a historic club, President Francesco Cimminelli had accumulated serious debts, and instead of moving to the top division, they filed for bankruptcy in August 2005.

While Sampdoria never found themselves facing dissolution and reformation, trying to stay relevant—in other words, securing European football—resulted in suffering and heartbreak. The winners of the 1990–1991 scudetto and the 1991 Supercoppa, and the runners-up of the 1992 European Cup, were relegated in summer 1999, bookending a glorious and strange decade for Samp. Clarence Seedorf and Juan Sebastián Verón had both played for the club, and after returning to Serie A in 2003, they acquired Antonio Cassano to help Walter Mazzarri's side reach the European positions in 2007–2008. With the help of Giampaolo Pazzini, whom Samp paid 9 million euros for in the January transfer window, they succeeded in qualifying for the UEFA Cup; two seasons later, they finished 4th and moved up to the Champions League playoffs. It was all too much. Sampdoria failed to qualify, were kicked down to Europa League where they didn't make it out of the group stage, and then finished 18th, meaning they would play the 2011–2012 season in Serie B. They may have survived had Riccardo Garrone chosen to purchase more players rather than stretching the squad so thin, but he was one of the few owners who refused to bury himself in debt; consequently, Sampdoria were able to bounce right back up to Serie A.

Parma, arguably the hardest hit of the clubs playing in Europe in the 1990s, continues to feel the ramifications of their attempt to keep pace with the big names. In 1996–1997, Parma clinched 2nd place in Serie A, their highest-ever finish in Italy's top division. The next season, they quietly bowed out of the group stage of their first Champions League appearance and finished 6th in the league, but this didn't deter ownership from continuing to pursue glory. Under Alberto Malesani and guided by stars such as Hernán Crespo, Juan Sebastián Verón, Fabio Cannavaro, and Gigi Buffon, Parma finished 4th, won the Coppa Italia, and surprised much of the world by flying to a 3–0 victory over Marseille in the UEFA Cup final. Verón left that summer, Crespo followed him to Lazio the next, and Buffon left after the 2000–2001 season, the last time Parma would qualify for Europe.

The butter hit the fan in 2003, when it was revealed where owner Calisto Tanzi had found the millions of euros he had spent on young stars such as Adrian Mutu, Adriano, and Alberto Gilardino: he had been embezzling from his dairy corporation, Parmalat. Parma declared bankruptcy in the middle of the season, finished 5th again, and went into administration until January 2007. Somewhat ironically, they were then relegated from Serie A at the end of the 2007–2008 season. The *Crociati* bounced right back up, spending six seasons aiming at a return to Europe, but when they finally qualified for Europe in 2014, their failure to pay their taxes meant they were unable to compete in the Europa League and were docked seven points at the start of the season. They finished dead last in 2014–2015 and declared bankruptcy again. Refounded as S.S.D. Parma Calcio 1913, they began anew in Serie D. Astonishingly, the new Parma, bought by Chinese business tycoon Jiang Lizhang, achieved back-to-back-to-back promotions to arrive in Serie A at the start of 2018–2019, but after selling the club to the Krause family in 2020, Parma again fell to 20th, and began 2021–2022 in Serie B.

But even Parma's woes could not compare to the biggest scandal to hit Italian football: *Calciopoli*. This match-fixing ring stretched over two years and two leagues, involving six clubs, touching at least nine others, and ultimately saw nine club officials, five FIGC representatives, and one referee punished. Calciopoli did not involve traditional match-fixing; in other words, no players were encouraged to play poorly or allow their opponents to pick up points. Instead, throughout the course of the 2004–2005 season, Juventus general manager (and former Napoli sporting director) Luciano Moggi—and to a lesser extent, club officials at Milan, Fiorentina, Lazio, Reggina, and Arezzo—had been working with the FIGC to secure favorable referee appointments and influence those referees who were appointed to oversee matches. Prosecutors uncovered the scandal while recording conversations on telephone lines at GEA World, a sports agency associated with Moggi and his son. By May 2006, the FIGC were ready to hand down the first set of punishments (most of which were later reduced).

Lazio got off lightly. The *Biancocelesti* had defeated Manchester United in 1999 to win the UEFA Super Cup, after which their chase for further success resulted in a scandal similar to what Parma would experience the next year. They sold their stars and parted with their beloved captain Alessandro Nesta. New owner Claudio Lotito aimed for Europe once more and succeeded in 2006, when the club qualified for the 2006–2007 UEFA Cup. Their involvement in Calciopoli meant disqualification from the tournament, but because their initial punishment of relegation to Serie B was reduced to an 11-point reduction in Serie A, and further adjusted to take just 3 points away, Lazio managed to finish 3rd in 2006–2007 Serie A and qualify for the Champions League 3rd qualifying round.

Like Lazio, Fiorentina were relegated to Serie B when the ruling was first announced, but upon appeal, their points reduction was reduced to 30 points, keeping them in 9th and thus in the top division. They were, however, kicked out of the next season's Champions League. The Viola began the 2006–2007 season with -15 points, yet still managed to finish in 6th, qualifying for the UEFA Cup.

Milan, too, were fortunate. The Italian giants had won Europe's top competition in 1990, 1994, and 2003, while also appearing in the 1993, 1995, and 2005 finals. Perhaps it was this accumulation of hardware that prompted the FIGC to spare the Rossoneri; after all, they had been playing in the Champions League semifinals just two months before the first set of punishments were handed down. Originally, Milan escaped relegation, but were docked 44 points from the 2005–2006 season, 15 points for the 2006–2007 season, and were banned from Europe. On appeal, the punishment was reduced to 30 points for the previous season, enabling them to stay 3rd and enter the 2006–2007 Champions League—which they then proceeded to win. Milan also finished 3rd again that season, despite an 8-point deduction.

Hardware, however, could not spare Juventus, not when Moggi was at the center of everything. The Old Lady fell and fell hard. They were stripped of the 2004–2005 title, which has never been awarded, and watched as their nemesis, Inter, added the 2005–2006 scudetto to their trophy cabinet, after Juventus were unceremoniously dropped from first all the way down to Serie B. The Bianconeri have proven themselves nearly impossible to keep down, and—despite a nine-point reduction to start the season—conquered the league in 2006–2007, bouncing right back up to Serie A. Juventus fans remain bitter about Calciopoli, arguing that many clubs (such as Inter) should have been punished but weren't, and counting those two scudetti as part of their tally.

It was Napoli that Juventus beat in the 2006–2007 Serie B season, and Napoli who would come up alongside them that year. First, though, came the Vesuviani fall.

The depths of despair

It was in 1995–1996 that Napoli truly began their long, dark slide. They would briefly poke their noses up for air in 2000, but it would be a decade before the fans truly regained faith in their side.

Again, the season started with a new coach, as Vujadin Boškov shifted to technical director, and Aldo Sensibile was brought in to take his place. The latter hadn't coached in several years, and really functioned more as Boškov's assistant. Again, the matter of costs hovered over the heads of everyone associated with the club, this time the clouds much darker and stormier. The club went into liquidation and was near bankruptcy when Corrado Ferlaino was convinced to come back to the helm.[16] In order to keep Napoli afloat, Ferlaino had to make difficult decisions, such as selling Benny Carbone to Inter and, especially, selling Fabio Cannavaro to Parma; he would go on to become one of the best central defenders in Italy's history.

Despite making few helpful additions to the squad, Napoli actually started off the season fairly well—11 goals from 5 matches, including a 2–1 win over Inter and a 1–1 draw with Juventus. But they experienced a dearth of firepower even more paralyzing than that of the last season, when Massimo Agostini had scored just nine goals in the league. Arturo Di Napoli, the most impressive of the players the club brought in despite coming from Serie C side Gualdo, had the most goals, with just five. However, three of them were crucial—1–0 wins against Bari, Lazio, and Sampdoria; those nine points rather than three points kept them from fighting a relegation battle.

The squad relied on draws (and Di Napoli's vital goals) to keep up, but in the second half of the season the defense began to fall apart. That left them with no way of stopping the opposition *and* no means of attacking to cover up the defensive flaws. Over the course of the season, Napoli failed to score in 15 matches. Supporters began to wonder; given the club's dire financial circumstances, and their questionable personnel choices, could Napoli ever be competitive again?

Fans paying attention at the beginning of the 1996–1997 season believed the answer to that question was a definitive yes. After all, Napoli, under Luigi Simoni, were in 2nd at the time of the Christmas break, behind Juventus. Overall, the side looked much better. Alfredo Aglietti, who arrived from Reggina in Serie B, scored eight league goals, despite the fact he was brought in as a reserve striker. Fabio Pecchia had improved enough in midfield to wear the captain's armband, and Alain Boghossian was looking like the player Napoli had hoped he'd be several years back. In addition, Pino Taglialatela was now not only stopping penalties, but taking them. On the other hand, Nicola Caccia, on loan from Piacenza, and Beto, who came on a free transfer from Brazilian club Botafogo, did not perform as expected.

Simoni set up his side in a 4-4-2 that was aggressive when it needed to be, with a midfield not only ready to go on the attack but also willing to shield the defense when required.[17] André Cruz, particularly, was putting in a better year at defense in front of the talented Argentine sweeper Roberto Ayala. Ironically, the only player they'd paid any real amount of money for, Caio, for whom they gave Inter 5 million euros, barely saw the pitch.

In the *ritorno*, Napoli lost the little amount of firepower they began the season with, and sank into a quick decline thanks to their sheer inability to win; after beating Parma in the first match of the second half of the season, they went 11 long games without a win. The excitement the fans and the squad had felt in the first half of the season disappeared, replaced with a depression that drifted through the ranks. It didn't help that Gigi Simoni had already signed a contract to coach Inter the next season; he no longer cared much what Napoli did or didn't do. Between that and the quick slide down the table, Ferlaino sacked him and brought in youth coach

Vincenzo Montefusco for the last month. He didn't fare much better.

The one truly bright spot in the season was Napoli's performance in the Coppa Italia, where they made it all the way to the final, beating Monza, Pescara, Lazio, and Inter. In Milan, the first leg ended 1–1. It was a *"Notte alla Maradona,"* according to *Gazzetta*,[18] where 50,000 fans showed up 2 hours before kick-off in Naples; 70,000 groaned in dismay when Javier Zanetti gave the visitors an early lead in the 11th minute. Maurizio Ganz helped level the playing field by hitting Francesco Colonnese, putting the Nerazzurri down a man ten minutes before the break. At the 77th, Beto equalized. An electrical current whipped through the stadium, pressing Napoli into attacks again and again and ultimately sending the match into extra time. With no way through, Napoli relied on goalkeeper Giuseppe Taglialatela, who made a miraculous save to deny Ivan Zamorano and take the game to penalties. Tagliatela played the hero again, saving Massimo Paganin's penalty. Every Napoli player scored, clinching the 6–4 victory that sent them into the final.

The two-legged final was against Vicenza, the side Napoli would play in the last round of the league. In the first leg, at the San Paolo, the hosts won 1–0, a goal from Pecchia in the first half sufficient for the victory. The second leg, however, was much more dramatic. At Vicenza, their midfielder Giampiero Maini scored in the exact same minute as Pecchia had, the 21st. But when 90 minutes were up, the score was 1–1 on aggregate, and the game moved into extra time. The Partenopei made it through the first 15 minutes holding off Vicenza, perhaps hoping to make it to penalties as they had done with Inter. But then came the second 15-minute half. That's when the Napoli defense finally crumbled. Maurizio Rossi scored in the 118th minute, and Alessandro Iannuzzi 2 minutes later. Vicenza won 3–1 to lift the Coppa, while Napoli settled for beating them in the final Serie A round, 1–0 once again with Beto scoring a rare goal in the 4th minute. The three points lifted the Azzurri into 13th, four points above the drop.

They didn't know it then, but Napoli fans needed to cling to the team's every small victory, whether it was reaching the Coppa final or beating Vicenza in the league to avenge themselves. Much worse was headed their way.

Napoli had been hoping the money from winning the Coppa Italia would help them survive financially. Without it, in order to stave off bankruptcy, the club had to sell off five starters prior to the 1997–1998 season: Alain Boghossian, André Cruz, Francesco Colonnese, Mauro Milanese, and Fabio Pecchia. Beto, never a success, went back to Brazil on a free transfer. The players sold brought in very little money; Juve paid a little less than 6 billion lire for Pecchia, almost all of which Napoli used to buy José Luis Calderón from Argentina's Independiente. The center-forward made six appearances and scored exactly zero times, demonstrating yet again that Napoli were not identifying cheap players that could actually improve their squad. The only standout was 22-year-old Claudio Bellucci from Sampdoria, in his first season with the club, who scored 10 goals.

The season went so badly that Napoli fans might want to skip this paragraph. It was the worst played by Napoli in the top flight, excluding their first season in 1926–1927 when they didn't win a single game. They managed to win two games in 1997–1998 but could only collect 14 points, 12 fewer than Lecce who finished 17th, and 22 points from safety. They conceded 76 goals and scored just 25, finishing with a goal difference of -51. The club plowed through four coaches—Bortolo Mutti, Carlo Mazzone, Giovanni Galeone, and Vincenzo Montefusco (again). Napoli were officially relegated on April 11, after a 3–1 loss to Parma. The season ended on May 16, five joyless rounds later.

Napoli fans were dejected, to put it mildly. It was the first time the club had been relegated in 33 years. Just eight years ago, they had been on top of the world, celebrating having won their second scudetto. Unfortunately, that ambition had left the club broken, tapped out of every available resource. Fortunately, Napoli supporters were not the type to abandon their club just because of a little thing like dropping down a division.

Ferlaino was determined that the club would go straight back up to Serie A (never mind the financial situation) so he brought Antonio Juliano back on board as sporting director for the 1998–1999 season. For his new manager, he selected Renzo Ulivieri, who had done well at Bologna for four seasons; however, he had lasted barely four months the previous season with Reggina, before being sacked for poor results. Interestingly, only 9 players left the club, while more than 15 were brought in. Again, this was by a team in serious financial trouble. Yet they didn't spend all that much; Juliano and Ferlaino focused on players who had been good at one point, but were at the end of their careers. Predictably, they did little to help the side.

After the *andata*, Napoli had 22 points, but most were from draws, and they never came close to the 4th place spot needed for promotion. The side worsened in the second half of the season and Ferlaino fired Ulivieri, bringing back Montefusco yet again. The mediocre manager wasn't much better than the man he replaced, but at least he managed to guide Napoli away from relegation and into 9th place. Such a finish, 13 points away from the promotion spots, was certainly not what Ferlaino had in mind. Something had to change.

Because they were so in debt, Ferlaino allowed new parties to buy shares in Napoli. This would ultimately lead to their demise, as Ferlaino now held the same number of shares as Giorgio Corbelli, creating a co-presidency that only caused trouble. The trouble began before Corbelli was officially on board: he wanted to buy the entire club, while Ferlaino wanted to keep the majority of shares; they settled on a 50–50 split after months of negotiations.[19] Just a year later, Corbelli would decide to leave Napoli; two months after that, he demanded Ferlaino leave instead. When Ferlaino decided to leave football entirely, Corbelli bought the rest of his shares, becoming the sole owner of Napoli in November 2001. Two weeks later, Ferlaino sued, arguing that Corbelli had not transferred the shares correctly.[20]

But before all that, Napoli had a title to win.

Amazingly enough, the club brought in a coach who stayed for the entire season. Walter Novellino had worked wonders to keep Vicenza in Serie A the season before, and they hoped he could work similar magic for Napoli. They were in even more of a mess financially due to not getting the money that accompanies promotion, but managed to put together a decent team, even though Juliano was no longer at the club to help. Napoli sold 18-year-old Paolo Cannavaro, Fabio's younger brother and also a central defender, to Parma for 1 million euros. They also sold goalkeeper Pino Taglialatela to Fiorentina for an undisclosed amount. In return they picked up young players like forward Roberto Stellone from Lecce and convinced Milan to loan out right-back Massimo Oddo. In addition, they also made sure they hung on to Stefan Schwoch, the 29-year-old striker who'd transferred from Venezia the previous January.

Like previous seasons, Napoli got off to a decent start, but never had a steady run unmarred by losses. Their longest unbeaten run that season was just five games. But they kept pushing through, their defense and their attack both equally acceptable. That said, it was Schwoch who led Napoli to 2nd in the league and promotion back to Serie A, thanks to his 22 goals scored. He'd made the difference in both games against Sampdoria, essentially ensuring that Napoli had a chance at promotion (despite the fact that the game occurred two months before the end of the season). That April win guaranteed the Azzurri would finish even on points with Brescia and Atalanta, on 63, just 1 point above Samp. Officially, Napoli clinched promotion in the penultimate round with a 1–0 win over Pistoiese. Of course, it was Schwoch who scored the lone goal in that match. In the final round, Napoli held back and let their *gemellaggio* friends Genoa win 3–1, but it was too little, too late, and the *Grifoni* finished 6th, five points behind their city rivals Sampdoria.

Napoli also made it rather far in the Coppa Italia that season, especially for a Serie B team (although, having won their first Coppa while in Serie B, that history might have boosted their confidence). All Serie B and Serie C teams entered the tournament in the group stage when it began. Napoli topped their group, just edging out local rivals Salernitana; both were even on 13 points, but Napoli had a +9 goal difference, while Salernitana's was +8. Next came the 2nd round, a knockout stage where the eight lowest finishing Serie A teams from the previous season join the cup. The Partenopei came up against Bari, a Serie A club. The first leg was played at the San Paolo, and where Napoli scraped by with a 1–0 victory, thanks to a goal by midfielder Cristiano Scapolo in the 92nd minute. In the return leg, Bari struck first when Daniel Andersson scored in the 56th minute. But this Napoli team remained strong and focused, and in the 79th minute, Anselmo Robbiati scored the equalizer. Napoli moved on to the next round thanks to a 2–1 aggregate win.

Unfortunately for the Vesuviani, they drew Juventus in the 3rd round. The Bianconeri came to town on December 1 and punished Napoli for believing they could lift another Coppa while in Serie B, scoring three goals to Napoli's one, a penalty converted by Francesco Turrini. When Napoli visited Turin two weeks later, Juventus barely seemed to need to work at all. After Juan Esnáider scored in the 26th minute, the hosts simply made sure the Napoli attack, strong enough for Serie B but not fierce enough to pierce the Juventus defense, was stymied at every turn.

They may have failed in their pursuit of the Coppa, but at least Napoli were back in Serie A for the 2000–2001 season. Soon, however, they likely wished that they had remained in the second division, given that the season began with a 2–1 loss to Juventus at the San Paolo, a 3–1 defeat at Inter, and, most humiliating of all, a 5–1 loss to Bologna in Naples. Napoli did not win a match until the 9th round, a 1–0 victory over Bari, who ultimately finished last in Serie A with just 20 points.

At the beginning of the season, Giorgio Corbelli had officially taken on the mantle of president, while Corrado Ferlaino remained on the Napoli board. Corbelli proved just as bad as Ferlaino at making personnel decisions, as well as financial ones. The new president had made grand promises about Napoli shortly returning to European football.[21] Corbelli then turned around and sold top scorer Stefan Schwoch to Torino for an undisclosed sum, a wildly unpopular move. He created almost an entirely new team for Serie A, overpaying for players he considered "names," whom even the most diehard Napoli fans would fail to remember, unless it was due to outrage over how much had been spent to acquire them. For example, the club bought attacker David Sesa from Lecce for 7.25 million euros, but he made only 16 appearances and scored just 1 goal.

Another name was that of Zdeněk Zeman, who had been at Fenerbahçe the previous season. Zeman is a bit of a cult icon in calcio, not only for his chain-smoking on the touchline but also for the aggressively attacking approach his teams take. Corbelli chose Zeman specifically because of his attacking principles, in order to entertain the fans (and himself). However, for a Zeman team to work effectively, the squad must be constructed carefully in order to fit his tactics; this Napoli squad was put together in a rather haphazard manner in which very few quality attackers were brought in. In addition, Zeman needs time with his squad to perfect the way he will attack. With no consideration of this need, Napoli fired him after six rounds, in which Napoli had picked up just two points. Corbelli believed that bringing in Edmundo from Fiorentina in January would excite the supporters, but one should always approach any player nicknamed "The Animal" cautiously. Fans were rightfully wary considering the Brazilian had left Fiorentina in the 1998–1999 season while they were chasing the scudetto; he never returned from *Carnevale* and the Viola lost out on the title. They never really experienced his volatile temper, though, because he was injured in his debut and could do little to help the club evade relegation after that.

But even with Emiliano Mondonico taking Zeman's spot on the bench, Napoli spent just 9 of 34 weeks out of the direct relegation zone (they never climbed higher than 15th; 15th and 16th entered a playoff with the 3rd- and 4th-placed Serie B sides to determine who would participate in Serie A the next season). According to Corbelli, his Napoli would have remained in Serie A if not for a scandal created by Parma. In the penultimate game of the season, the Crociati, who had secured a Champions League spot for the next season, lost to Hellas Verona 2–1. Given the crowd at the bottom of the table—Napoli and Vicenza both with 36 points, Reggina, Verona, and Lecce with 37—final placements were determined based on head-to-head record. Without the extra three points from the Parma win, Verona would have been relegated. Corbelli alleged that Verona was under the control of the president of Parma, Callisto Tanzi. In 2010, those suspicions were confirmed when Tanzi admitted Verona had been his puppet club from 1998 until 2004.[22] Though important in the overall scheme of Italian football, Corbelli must not have known how to accurately do the sums needed to determine whether the match-fixing had any effect on Napoli. If he had, he would have realized that even taking those three points away from Verona would not have saved his club.

So once again Napoli fell back down to Serie B, There's little to say about the 2001–2002 season that hasn't already been stated—a new coach, Luigi De Canio, an entire squad rehaul in which any decent players, such as Fabio Pecchia and Nicola Amoruso, bailed so they could remain in Serie A, Giorgio Corbelli and sporting director Luigi Pavarese compiling a squad that could barely be competitive in Serie B, much less bring them back up to the first division. As though the club had angered the calcio gods, September brought significant rainfall, and holes began to open in the San Paolo's concrete. This forced them to play their home games in other areas of the South, in much smaller stadiums.[23]

About all that can be said for this lackluster Napoli season is that the club spent hardly any money, so when they finished 5th, six points away from the promotion spots, it didn't feel as though they had spiraled even further downward.

Yet somehow they had.

Napoli's 2002–2003 Serie B season wasn't horrible; at least, not in the context of what followed. Salvatore Naldi had acquired 10% ownership of the club in December 2001, then, after a legal battle, he liquidated the shares owned by Ferlaino in February 2002. By May, he had taken over the club completely. But, as Napoli supporters had learned by this point, simply being rich didn't make someone a good calcio owner. Not only was the squad dramatically changed before the season began, but five players left and seven came in during the January transfer window. Franco Colomba began the season as coach and finished in the same way. In between, his second in command, Sergio Buso, took over for one match, after Napoli lost 3–1 to Genoa. Then Franco Scoglio stepped in, making it all the way until March 11, the day after his side lost 2–0 to Sampdoria. Colomba returned, guiding the side to 16th in Serie B.

Even through all this, most Napoli supporters stood by their club. They had some of the largest crowds in Serie B. While some clubs, even a few in Serie A, were drawing under 10,000 fans to their matches, 64,500 turned up to watch them against Vicenza at the end of March, a strange time as it was too early in the season to be truly worried about promotion or relegation. But according to *Gazzetta*, the fans poured into the stadium both out of pride, showing that they did not want to see their club in the relegation zone, and out of protest, telling Naldi "not to take Neapolitans for fools."[24]

That being said, Napoli fans *were* worried about their club. They understood how poorly it was being run. There was no way to avoid seeing just how badly the squad was performing, and that there were no players talented enough to turn it around. Some fans threatened the club's board members and officers. Some went after the players, attacking them for playing so poorly. (These two groups were not mutually exclusive.) One group cornered and threatened Renato Olive, a fringe player who had recently received a red card and suspension, with a knife.[25]

For the penultimate game of the season and the final home match, 52,500 showed up to the San Paolo to try to lift their team away from relegation by cheering them on against Ternana. It worked—the visitors scored an own goal and Napoli won 1–0. In the end, the supporters had little to worry about anyway, as Serie B would be expanded to include 24 teams. In the end only Cosenza lost their place, as they filed for bankruptcy and dropped down to Serie D.

It was a timely warning, foreshadowing the fate of Italy's fourth-most beloved club. That year, Naldi brought in Andrea Agostinelli, most recently coach of Piacenza in Serie A, who was fired after just three rounds, the victim of a disaster of a match at Messina. Despite the hosts missing several players, and thus appearing to barely try to win the match, Napoli couldn't eke out even a point. It's said the referee did everything he could to stop Napoli from winning, including awarding a seemingly ridiculous late penalty to Messina.

Gigi Simoni returned to coach the side for the remainder of the season, a season in which high squad turnover once again couldn't make up for the lack of talent the club were able to bring in on a limited budget. Two weeks later, he, Naldi, and the rest of the Napoli staff would need to face the aftermath of a fanbase that had been pushed to their limits during a derby with one of their southern rivals, Avellino. The two sides had not met since 1988. The Napoli ultras first clashed with the police, leading to many serious accidents, and then invaded the pitch. Their raw anger spurred a tragedy—in all the chaos, a young Napoli fan, Sergio Ercolano, fell 20 yards, a drop that resulted in head trauma and serious internal injuries. As a result, he suffered cardiac arrest and died in the hospital in Avellino. The fighting resulted in hundreds of injuries to the fans and at least 30 among the police and *carabinieri*, while the Stadio Partenio sustained extensive damage. The game was called off and Avellino awarded a 3–0 win, while Napoli were forced to play five games behind closed doors and at a neutral ground in Campobasso. In the end, Ercolano's mother Carmen was incorrect when she said, "*Non si può morire così, non si può morire per una partita di calcio*" (You can't die like this—you can't die for a football match); it is clear there are fans who would lay down anything, even their lives, for the sport. But that doesn't make it okay to sacrifice the lives of others alongside them.

While unable to win games or score goals, Simoni's magic resulted in 26 draws, enough for Napoli to finish in 14th, five points above the relegation playoff spot. Their season ended on June 12 with a goalless draw at home against AlbinoLeffe. Only 10,500 fans came out to the San Paolo.

Less than two weeks later, SSC Napoli were in a much worse position.

Owner Salvatore Naldi, along with Napoli's entire board, resigned on June 22. This departure should have pleased Napoli supporters, because President Naldi had deceived the fans and made promises he could not keep, and surrounded himself with people who were unable to help him manage a football club. However, Naldi was unable to recapitalize and bring new shareholders into the club, who held a debt of 70 million euros and were required to pay 8.6 million euros immediately to Morelli, a notary who was the go-between between the club and the banks, and who was tasked with finding even more money. When unable to do so, Napoli decided they had to give up.[26] Without repayment, the club would be liquidated at the end of the month, giving potential new owners just eight days to decide whether to risk buying a club that had managed to get themselves into such a huge amount of trouble. The players began to flee as soon as possible, quickly accepting offers from other teams. It was understandable, as the club had managed to pay them just twice over the course of the season, and in their contracts they had agreed to wait for 30% of their salaries over the following seasons—money they would now never see. A new owner needed to be found immediately. Should the club slide into bankruptcy, it was likely they would need to begin again in Serie D, the highest amateur level in calcio.

[1] Moggi al Torino—Con Ferlaino e' rottura. (1991, March 15). *La Repubblica*. https://ricerca. repubblica.it/repubblica/archivio/repubblica/1991/03/15/moggi-al-torino-con-ferlaino-rottura.html

[2] Made in Italy: Gianfranco Zola. (2018, May 15). *Chelsea FC*. https://www.chelseafc.com/en/ news/2018/5/15/made-in-italy

[3] Id.

[4] Bird, J. (n.d.). Searching for Maradona. *Mundial*. https://mundialmag.com/blogs/articles/diego-maradona-naples

[5] Thacker, G. (2017, November 15). The glory of Careca at Napoli. *These Football Times*. https:// thesefootballtimes.co/2017/11/15/the-glory-of-careca-at-napoli/

[6] Id.

[7] Calciatori—La raccolta completa Panini 1961–2012, vol. 9 (1992–1993). (2012, 2 July). Ed. speciale per *La Gazzetta dello Sport*.

[8] Passaggio al portiere, nuova norma Fifa? (1992, April 24). *La Repubblica*, 38.

[9] Bedeschi, S. (2013, September 18). Gli eroi in bianconero: Daniel Fonseca. *TuttoJuve*. https://www. tuttojuve.com/gli-eroi-bianconeri/gli-eroi-in-bianconero-daniel-fonseca-62877

[10] Sannucci, C. (1996, December 26). Napoli & Simoni "all" Italiana, questo e' calcio. *La Repubblica*. https://ricerca.repubblica.it/repubblica/archivio/repubblica/1996/12/27/napoli-simoni-all-italiana-questo.html

[11] Zola saluta 'ceduto per pagare I debiti.' (1993, June 29). La Repubblica. https://ricerca.repubblica.it/ repubblica/archivio/repubblica/1993/06/29/zola-saluta-ceduto-per-pagare-debiti.html

[12] Sannucci.

[13] Storm, S. (2019, April 10). How to ruin a country in three decades. *Institute for New Economic Thinking*. https://www.ineteconomics.org/perspectives/blog/how-to-ruin-a-country-in-three-decades

[14] Morris, M. (2018, July 23). More historic Italian clubs go bust...while Juventus sign Cristiano Ronaldo. *Guardian*. https://www.theguardian.com/football/the-gentleman-ultra/2018/jul/23/cristiano-ronaldo-juventus-italian-clubs-bust-bari

[15] Primo passo contenere gli ingaggi. (2002, June 9). *La Repubblica*. https://ricerca.repubblica.it/repubblica/archivio/repubblica/2000/06/09/primo-passo-contenere-gli-ingaggi.html?

[16] Sannucci.

[17] Carratelli, M. (2007). *La grande storia del Napoli*. Gianni Marchesini Editore.

[18] Maradei, L. (1997, February 27). Napoli da impazzire. Gazzetta dello Sport. http://archiviostorico.gazzetta.it/1997/febbraio/27/Napoli_impazzire_ga_0_97022710470.shtml

[19] Diciotto mesi di inutili trattative. (30 November 2001). *La Repubblica*. https://ricerca.repubblica.it/repubblica/archivio/repubblica/2001/11/30/diciotto-mesi-di-inutili-trattative.html

[20] Azzi, M. (2001, December 14). Corbelli e Ferlaino in tribunale. *La Repubblica*. https://ricerca.repubblica.it/repubblica/archivio/repubblica/2001/12/14/corbelli-ferlaino-in-tribunale.html

[21] Hodges-Ramon, L. (2019, August 15).The decline of Napoli post-Maradona: From paradiso to inferno. *These Football Times*. https://thesefootballtimes.co/2019/08/15/the-cecline-of-napoli-post-maradona-from-paradiso-to-inferno

[22] "Tanzi controllava il Verona e fece retrocedere il Napoli." (2010, November 21). *La Repubblica*. https://parma.repubblica.it/cronaca/2010/11/21/news/tanzi_controllava_il_verona_e_fece_retrocedere_il_napoli-9349653/

[23] Hodges-Ramon.

[24] Malfitano, M. (2003, April 1). Il Napoli torna a galla. *Gazzetta dello Sport*. http://archiviostorico.gazzetta.it//2003/aprile/01/Napoli_torna_galla_ga_0_0304013897.shtml

[25] Hodges-Ramon.

[26] Mimmo, M. (2004, June 23). Naldi scappa Napoli a picco. *La Gazzetta dello Sport*. http://archiviostorico.gazzetta.it/2004/giugno/23/Naldi_scappa_Napoli_picco_ga_10_0406237586.shtml

CHAPTER 10

How Am I Going
to Be an Optimist About This?

10

Many consider the heyday of Italian football to be the 1990s. The country built on their success in Italia 1990 and were now in a position to attract more world-class players. Napoli could lay claim to their involvement in promoting *calcio* with their second *scudetto*, lifted in 1990, and their star, Diego Maradona. But while the *Partenopei* were one of the first to cause those outside the peninsula to pay attention to Italian soccer, they were also the first to fall. They set aside fiscal responsibility in favor of bringing glory to the South, first pouring money into catering to Maradona and cleaning up his messes, then spending the rest of the decade trying to recapture that glory by shelling out for players destined to be stars, only to have to sell them a year or two later to avoid bankruptcy. The final fall occurred in August 2004, when Napoli Soccer had to begin again in Serie C1. By spring 2008, SSC Napoli had returned to Serie A. It's tempting to think that a miracle along the lines of San Gennaro's blood liquefying had transformed the team. Instead, it was down to a little ingenuity, a lot of hard work, ever-present support from the fans, and an owner who finally knew how to balance the books.

The shame spiral, summarized

During the 1990s, Serie A was often called the most beautiful league in the world. With all eyes on Italy, most teams fell under the enchantment of this siren's song; they purchased flashy players and trotted out squads teeming with stars. The 1995 Bosman ruling helped, as European leagues could no longer restrict the number of foreign European Union players on each squad, allowing for easier freedom of movement. So, too, did the media, with Silvio Berlusconi's Mediaset and other corporations buying up television rights and promoting the league. Berlusconi also supported the shift from the European Cup to the UEFA Champions League, which added a round-robin group stage to the competition. From 1997 on, the best leagues could also have up to four teams in the tournament, a decision that not only aided Berlusconi's plan to establish a more elite competition in which Milan would not be eliminated in the 1st round,[1] but also tempted more clubs into securing better and better players as they sought to finish high in their domestic leagues.

Despite the fact that both the new Champions League and the Bosman ruling made it easier for the richest leagues in the world (England, Spain, Germany, and Italy) to become even more successful, and thus even richer, Serie A clubs remained relatively poor compared to the others. The media deals were growing, but not at the rate they were elsewhere. In addition, Italy refused to court the global market; in fact, until

2010 they restricted media access by selling TV rights individually, unlike the Premier League, which sold contracts to international television companies, helping promote the league's *global* brand. The lack of privatized stadiums made it impossible for Italian teams to capitalize on ticket revenue, and attendance rates were falling, while England and Germany were seeing growth in the number of fans showing up at the grounds. Attempting to bridge the gulf, clubs sought instant gratification by spending money they didn't have, borrowing or using more unscrupulous methods to get their hands on enough cash. It made Serie A the most exciting league to watch at the time, and Italy the favored destination of the game's best players.

In addition to Napoli, the siren slayed a number of clubs—Fiorentina, Lazio, Parma, and Torino all fell, while other teams were ensnared in the scandals of the 2000s (for more information, see the previous chapter). Two decades later, calcio has yet to rebuild their reputation and win back the adoration of the majority of the world's fans.

After their 13th place finish in Serie B 2003–2004, Napoli looked secure. Until they had a system in place for long-term planning, they likely wouldn't be flaunting calcio's beauty, but they'd be able to move forward. But then, with debts of around 70 million euros, Napoli were denied a Serie B license and declared bankrupt. Fortunately, Napoli's flagrant disregard for fiscal responsibility would not result in the demise of football in Naples. Under the *Lodo Petrucci*, part of the FIGC's rules governing Italian football clubs, the Federation could award the old club's name to a new company governing the team, even without that company clearing the previous debts. The club would drop down a division, but would continue on.

A phoenix rising

Four companies were established, each one hoping to convince the city that they could recreate their beloved club. The presidents of Perugia, Luciano Gaucci, and Siena, Paolo De Luca, were both in the running, but neither had the sufficient funds demanded by the Naples city council.

However, the FIGC did not permit the mere transfer of "SSC Napoli" under the Lodo law. In August, they declared that SSC Napoli would need to accept amateur status and work their way back up the ranks of Italian football, or they would need to dissolve the club and start a new franchise in the city that could compete in Serie C1, the third division.[2] That club would be a successor to SSC Napoli, absorbing their records, titles, and history. Enter Aurelio De Laurentiis, the prolific Italian filmmaker, entrepreneur, and resident of Naples. Unwilling to permit the extinction of professional calcio in his city, he paid the courts 30 million euros for the club. Renaming it "Napoli Soccer," De Laurentiis was determined to prove that a football club from Naples could be fiscally responsible while still flourishing on the pitch. De Laurentiis, or ADL as the fans call him, is sometimes brilliant, sometimes highly erratic, and often eccentric, but there's no denying he is a good businessman.

ADL wanted, and continues to want, to do more to fix football in Italy. As almost everyone who pays attention to calcio knows, "Football in Italy today . . . seems plagued by entrenched problems of racism and hooliganism, by financial irregularity and by a relative lack of modernization of the sport's key structures and facilities."[3] De Laurentiis is aware of how the country and the leagues are holding back Italian football, having never truly capitalized on their 2006 World Cup victory to attract more fans, to refurbish crumbling stadiums, or to entirely replace elements of football's infrastructure, as other winners of major tournaments have done.[4] He knows the way ultras behave and, as Napoli owner, has had to confront both racism and geographism; i.e., the vitriol directed at those from the South. In his over 15 years in charge of Napoli, he has spoken out against these issues, and has long campaigned to have a new stadium built in Naples, a modern, privately owned structure that would bring in more revenue for the club.

But before De Laurentiis could begin to take on the entire culture of calcio, he had to first get his new team into a position where they were able to capture the spotlight. And that meant dragging them up to Serie A as soon as possible—without going broke while doing so.

In terms of saving money, having players spend their own cash to go out and buy vests and footballs for the team to use while practicing was a good first step, albeit likely not an intentional one. By the time Napoli Soccer was formed, there were only two weeks to go before the 2004–2005 season began. The new club would have to build a team from scratch in that amount of time,[5] in addition to getting them trained and accustomed to the tactics of coach Gian Piero Ventura. Ventura had finished the previous season coaching Cagliari to an 8th-place finish in Serie B. Perhaps more importantly, ADL convinced sporting director Pierpaolo Marino (who had been charged with helping build a strong squad around Diego Maradona back in the 1980s) to leave Udinese to join his new club.

When Napoli went bankrupt, 26 players left the club. In order to keep the bills down while picking up players who possessed enough ability to help Napoli Soccer win the Serie C1 title, most were brought in to the new squad on one-year loans. Two particularly significant players emerged from the vast number acquired— Francesco Montervino, the club captain, who would come to be seen as a symbol of rebirth for the city, and Roberto Sosa, who would help the club return to its former glories in any way he could, including joining with Montervino to buy the aforementioned balls and vests.[6] Sosa, originally brought in from Udinese, where he'd moved from Argentina thanks to the sharp eye of then-sporting director Marino, opted to follow Marino back to Napoli, rather than play in Serie A with Messina, a side he'd helped guide to the top division.

Serie C1 was a new challenge for Napoli Soccer. While many of the newly signed players had been there before, for others the format was yet another unexpected thing they had to get used to. At the time, the third division was divided into two groups based on geographic location, Serie C1/A and Serie C1/B. The top team in each group automatically moved up to Serie B, while those who finished 2nd–5th entered an eight-team playoff. (The current third division, now called simply "Serie C," is composed of three geographical groups of 20 each, with the winner of each group automatically promoted to Serie B, as well as the winner of the promotion playoff involving 27 teams, placed 2nd–10th in each group, as well as the Coppa Serie C champions . . . or the squad that finished 11th in the group containing the runners-up, assuming they, too, have already been promoted or received one of the playoff slots. No wonder some long for the "good old days.")

The squad De Laurentiis had put together was fairly strong, and Napoli finished 3rd in Group B, nine points back of winners Rimini. In the semifinals of the group playoffs, Napoli faced Sambenedettese, from San Benedetto del Tronto in the province of Marche, on the Adriatic Sea. There, Napoli drew with the home side, 1–1. When Sambenedettese traveled to the San Paolo—the immense stadium was still being used to hold large crowds of Napoli fans loyal to their club, regardless of how far they'd fallen—the Partenopei won 2–0. Napoli then came up against Avellino, who had beaten Reggiana, in their final bid for promotion. A couple of seasons back, Napoli fans' appalling behavior toward their Southern rivals had caused the game to be called off. More importantly, the derby brought back memories of the young fan killed during the altercations between police and supporters, falling to his death from the stands at the Avellino stadium. The first leg, in Naples, ended in a goalless draw, so the squad headed to Avellino. There, the *Irpini* won 2–1, consigning Napoli to another season in the third division.

A hop, skip, and a jump away

The Partenopei faithful spent the summer hoping they'd be elevated to Serie B, not by virtue of their talent or wits, but through administrative decisions. They were holding out hope that another Southern side would fold: Messina had surprised in Serie A, finishing 7th, but were in dire financial straits. Had they folded, a Serie B club would have moved up, allowing Napoli to take their place in the higher division.

But Messina survived and Napoli moved into another squad rebuild, retaining only a few important individuals, mostly those brought in midseason. Coach Edy Reja, who had taken over from Gian Piero Ventura when he was sacked in January, was persuaded to stay, although he was worried about finding decent players.[7] Gianluca Grava, a solid defender who was one of the loyal players, also remained. (In the end, Grava stayed with the club until 2013. His loyalty was rewarded and he is now the technical director of Napoli's youth squads.) They also kept Piá, a Brazilian forward who was brought to Italy by Atalanta, but scored just 1 goal in 23 games for them in Serie A. The club lent him to Serie B side Ascoli for a season, where he was much more successful, and then arranged for a co-ownership deal that saw Piá transfer to Napoli in the winter of 2005. Emanuele Calaiò had come up through the youth system at Torino, but rarely played at the senior level. After failing to impress in two seasons with the *Granata*, he was sent on loan to Serie B clubs Ternana and Messina before being bought by Pescara. Joining the team midseason, he helped the club win promotion in 2003, and was the club's top scorer in 2003–2004, with 21 goals. But Pescara was set to go right back down to Serie C1 until the banking irregularities by Napoli, as well as Ancona, gave them a lucky break. With six goals at the beginning of the 2004–2005 Serie B season, it's little wonder that he caught Marino's eye, and Napoli secured a co-ownership deal with the *Delfini* in January 2005, ultimately buying him outright at the end of the season.

Before the 2005–2006 season began, ADL and Marino shed 15 players, only one of whom, Ignazio Abate, went on to become a strong presence in a Serie A side. He returned to his boyhood club, Milan, and spent a decade with the *Rossoneri*. They went on to spend a decent chunk of money on acquiring new players; however, the most spent on a single talent was 750,000 euros, paid to Defensor Sporting in Uruguay for central midfielder Nicolás Amodio. They also brought in less flashy but talented players such as goalkeeper Gennaro Iezzo and defender Rubén Maldonado, both coming in on free transfers, and another Uruguayan midfielder, Mariano Bogliacino, for whom they offered a ridiculously low fee to Plaza Colonia. The club philosophy was to find young players and raise them up with the club's values, favoring the culture of the team rather than the individual.[8] Napoli had spent money, but had spent it wisely, and doing so paid off. They did have to spend 2.85 million overall to make Emanuele Calaiò's transfer permanent, but he'd scored 18 goals in just a half season with the club, certainly enough to justify the expense. The left-footed striker went on to lead Napoli to the Serie C1 title and lifted the division's *Capocannoniere* award.

Even before the Serie C1 season began, Napoli had already proved they would be *the* team to overcome. They began in the 1st round of the Coppa Italia by beating Serie B side Pescara, with Calaiò scoring against his former team late in the first half and an insurance goal from Roberto Sosa in the 79th minute. Next, Marco Capparella played the hero, scoring the only goal to beat Reggina, a Serie A side, and advance Napoli to the 3rd round. At this point the fans, who had numbered no more than 10,000 for the first two matches, had become rather excited about this squad, and many more came out for the 3rd round against Serie B side Piacenza. Almost 29,000 turned out to watch Sosa score a last-minute goal to take Napoli, a Serie C1 side, into what most consider the "real" Coppa Italia, the round in which the best Serie A sides enter the competition. But first it was time for league play.

The 2005–2006 season was a rather typical one for the third division of Italian football. The league was again divided into two regional groupings, C1/A and C1/B. In addition, the season was divided into two parts—the regular season, in which the teams topping each group automatically qualify for promotion, and the playoff season. The winner of each playoff final joined the outright winners of Group A and Group B in Serie B. The playoff also featured a *playout* final, where the teams placed 14th–17th competed against each other to avoid being relegated to Serie C2; the two 18th-placed teams were relegated automatically. In other words, four teams are meant to move up to Serie B, and four down to Serie C2. But as often happens in the third division, things were a little more complicated than that. In the south, Gela, who finished 12th, and Sassari Torres, who finished an impressive 3rd, were declared bankrupt and relegated to Serie C2. Meanwhile, Acireale and Chieti, both in the relegation playoff spots, were also declared bankrupt and relegated to Serie D. That left two holes to fill, so Pro Sesto in Group A and Massese in Group B were given a reprieve.

When Napoli kicked off the season against little Acireale, a side from Sicily, they'd already shown themselves to be a squad able to take on teams in the upper divisions and come away with wins. To start the 2005–2006 league season, Napoli went on a 13-game streak in which they refused to lose, falling only to Perugia at the beginning of December. They may have been affected by their upcoming game—a two-legged Coppa Italia match against Roma, in Serie A. Napoli felt lucky to draw Roma, who'd placed just 8th in Serie A the season before. On the other hand, the club had made it to the Coppa Italia final in 2005 and were out to avenge their decisive loss to Inter. Sure enough, just days after the Perugia match, Napoli lost 3–0 to the *Giallorossi*. They then returned to league play, traveling to Frosinone to meet their fiercest competitors for the C1/B title. Thanks to two goals from Bogliacino, with one from Piá sandwiched between, Napoli easily swept away their competition, 3–1.

The *ritorno* began on January 8 with a 1–0 victory over Acireale at the San Paolo, the goal again coming from Sosa, boosting Napoli's confidence prior to the midweek match in Rome. The second leg of the Coppa match began with a first-half goal by Roma midfielder Alberto Aquilani, followed by a goal by Brazilian midfielder Mancini just before the break. A consolation goal in extra time by Amodio wasn't nearly enough; Roma won 5–1 over the two legs, and Napoli were out of the Coppa Italia. Roma made it to the final once more, where again Inter won, 4–2 on aggregate.

Napoli turned to the league, although they played rather less effectively than they had in the first half of the season. Players knew that they'd basically already won the C1 title; they began slacking off a bit. They lost their next three games on the road to Massese, who were ultimately relegated; Torres, who came 3rd but were sent down due to bankruptcy; and Juve Stabia, their closest geographical rivals, who scored three against them. Despite their confidence, ADL was not pleased with Napoli's road performances, and took it out on coach Reja, saying, "This team will get to Serie B, but are playing like a Serie C team."[9] He knew it was important to continue to attract supporters by playing matches that were at least moderately entertaining. However, even in the third division, around 20,000 Napoli fans were showing up to each match at the San Paolo, fairly secure in the knowledge their club would win promotion.

From the Juve Stabia loss on out, Napoli didn't lose a single game; they also never lost at home during the season. The number of fans attending steadily increased. Napoli were ranked 10th in Italy that season for the average number of spectators.[10] On April 15, 42,681 Napoli fans turned up at the San Paolo, by far the most that Serie

C1 season, and likely one of the largest crowds ever to take in a third-division game. They were there to see Napoli mathematically clinch a place in Serie B for the next season. This time, the Partenopei were able to best Perugia, with goals from Calaiò and Capparella giving the hosts a 2–0 victory. Capparella's spectacular volley in the 67th minute nearly brought down the house, the fans thrilled that at this time next season, they'd be watching Serie B matches. Some accounts say 51,000 actually attended. It's possible that extra fans snuck in to celebrate, setting off fireworks, not even waiting for the final home match to begin the party properly. Thanks to ADL's business trip to Los Angeles keeping him away from Naples, Napoli fans elsewhere were able to cheer their side as they moved up; the owner allowed the television station RAI Internazionale to show the match for free.[11]

Next, Napoli went to Grosseto, drew 2–2, then came home to draw 1–1 with Frosinone in front of an even bigger crowd, ready to welcome back their revived team to the bigger leagues. The *Vesuviani* finished off their season with a goalless draw at Lanciano. The squad may have gone out with an exhausted whimper, but the fans certainly created a bang! back in Naples, and across the world.

After topping Group B and being promoted to Serie B, Napoli had one more challenge to face—the two-legged Supercoppa di Lega di Serie C1, against Spezia, who had won Group A. In the first match, 8,000 at Spezia's Stadio Alberto watched the two teams play out a goalless draw. A week later, at the San Paolo, only 7,000 fans turned out to see Bogliacino convert a penalty to draw 1–1 with Spezia; however, the latter lifted the trophy due to having scored an away goal. The low turnout at both games was due to the fact that Italian fans often do not turn out to support their clubs in what they consider "minor" tournaments. (It should be noted that the Coppa Italia itself is not considered "minor" by most Napoli supporters, who are proud of the six trophies in their cabinet; this is why so many turned up prior to the official start of the 2005–2006 season to watch Napoli play in the early rounds.)

– HEROES –
Roberto Sosa (2004–2008)

Napoli appearances: 131
Goals: 30

Roberto Sosa, originally brought in from Udinese, had moved from Argentina thanks to the sharp eye of sporting director Pierpaolo Marino, who oversaw his move from Argentina's Gimnasia La Plata and subsequently brought Sosa with him to Naples. As an Argentine, he was immediately appreciated and quickly became an idol for millions of Napoli fans. From his very first season at the club, the supporters displayed a banner reading "*Habemus Pampa*," a play on "*Habemus Papam*," the phrase used when a new pope is appointed. In other words, the fans were proclaiming that "*Pampa*" Sosa was their new pope, their new religion.

"Loyalty" is perhaps the best word to describe the attacker. What stands out about Sosa is that although he led another Southern club, Messina, back to the top flight, rather than play in Serie A he opted to follow Marino back to Napoli, then in the third division. Pampa was a defining figure in Napoli's rise to Serie A. When he arrived at the club, Napoli Soccer had almost nothing to call their own—not even a squad. Along with club captain Francesco Montervino, one of the few who had stayed on through the drop, the two bought vests and balls for the developing squad.[12]

Looking at his stats, it's easy to think Sosa meant little to Napoli's promotion bid in 2005–2006; after all, his average goals decreased from ten to six per season. But coach Edy Reja had Sosa playing in a different role—rather than starting every game, he would often come in off the bench, disrupt play, and find a way for Napoli to grab points. He was a powerful forward and could execute an especially strong header, which came in handy on the often poor pitches of Serie C1. Sosa would get on the end of a long ball from the back, then head it on to a teammate who found himself in a good goal-scoring position. Sosa is one of many cases in Napoli history in which goals scored, even as an attacker, do not tell the full story.

Sosa was the last player to wear the No. 10 shirt, after which it would be reserved for Maradona and Maradona only. Prior to this time, players didn't wear their names on the back of their shirts, only numbers, so after Diego quite a few had worn the No. 10 shirt while playing that position—but it still certainly seemed to mean something to Sosa to have this honor. When he left in 2008, Sosa ripped off the shirt to reveal another, reading "He who loves does not forget." His career may have faded toward the end, prompting a move back to his beloved Gimnasia La Plata, but he was the last player to score wearing the precious Neapolitan *Azzurri* No. 10,[13] and neither he nor the fans will forget what he did for their beloved Napoli. In fact, when he became a pundit for Sky Sports after his playing career ended, many accused him of being too biased toward the Partenopei.[14]

Promotion vaulted the club that had once again become "Società Sportiva Calcio Napoli" into what was likely the most competitive Serie B season ever. (ADL had purchased both the old name and the old crest at an auction in 2006.[15]) Genoa, the giant of early Italian football, had been promoted from Serie C1 with Napoli, through the Group A playoffs. Brescia had won Serie B three times and came in second four times. Bologna held seven Serie A titles. But the biggest threat came from Juventus, who had been about to be crowned champions of Serie A once again when the *Calciopoli* scandal broke. They were relegated to Serie B and had nine points docked right from the start—actually letting them off comparatively lightly, given that the first punishment handed down was a 30-point deduction.

Calciopoli may not have involved Napoli directly, but its tentacles squirmed into every crack and fissure in Italian football. Even 15 years on, it continues to lead to heated arguments over what actually occurred, and if the right teams were punished.

News of the Calciopoli scandal broke on May 2, 2008. Wiretaps put in place by prosecutors in Turin and Naples incriminated a number of major clubs—including Juventus, Milan, Lazio, and Fiorentina—and associated individuals such as owners and sporting directors, as well as referees. More than 7,500 transcribed pages of these conversations were combed through from the 2004–2005 and 2005–2006 seasons. Those involved were accused of "illicit sporting activities," a polite way of saying they bribed officials in order to have certain referees, friendly to the clubs involved, oversee their matches. At the same time, in Rome, investigators examined the potential illegal activities of GEA World, an Italian sporting agency involved with the managing and the buying and selling of a large number of players. In addition, it uncovered an illegal gambling ring, which implicated 21 professional players, including Vincenzo Iaquinta, a member of the Italy 2006 World Cup-winning squad.[16]

In June, FIGC Attorney Stefano Palazzi called for Juventus, Milan, Lazio, and Fiorentina to be punished for violating the sporting justice code. He alleged that Juventus (and former Napoli) sporting director Luciano Moggi and general manager Antonio Giraudo, Fiorentina owners, brothers Diego and Andrea Della Valle, Lazio owner Claudio Lotito, and a number of referees were involved. In one particularly damning telephone conversation, Moggi accused referees Pierluigi Collina and Roberto Rosetti of being "too objective" and asked them to be "punished." *La Repubblica* printed a number of other incriminating statements pulled from the transcripts.[17]

Palazzi wanted to punish those involved harshly, likely to ensure the same problem wouldn't arise again. He demanded that Milan, Lazio, and Fiorentina be relegated to Serie B, and Juventus sent to the third division, at minimum.[18] It didn't quite work out that way. The original punishments were handed down by a Rome tribunal just five days after Italy won the World Cup.[19] Juventus were stripped of the 2004–2005 as well as the 2005–2006 title. They were also the only one of the four clubs relegated to Serie B. Initially, they were meant to start the 2006–2007 season at -30 points, which was reduced down to -17, and then down to -9. Palazini wanted the other three clubs dropped to the final three positions in Serie A, which would send them down a division, and for points to be deducted the next season. He argued the three clubs should be barred from playing in Europe the next season as well—Milan and Fiorentina from the Champions League, and Lazio from the UEFA Cup. When the dust settled, only Juve's relegation was upheld. Milan, Lazio, and Fiorentina all had 30 points deducted from their 2005–2006 season, and Fiorentina were docked 15 points, Milan 8 points, and Lazio 3 points for 2006–2007. Milan were permitted to play in the 2006–2007 Champions League as even with their points deduction they remained in the top three teams, but Fiorentina and Lazio were expelled from European competition. A wide range of individuals, including Moggi and Giraudo, received bans from football ranging from three months to five years. Finally, once the appeals process was over and the clubs involved docked points, Inter were awarded the 2005–2006 scudetto as they finished highest in the revised table.

The purpose of the original punishments—relegations combined with docked points for the next season—was to ensure that the clubs involved in illegal sporting behavior would have a difficult time climbing back to the first division. But of the three clubs that were not relegated, Lazio and Milan finished the 2006–2007 Serie A season in the Champions League places (having gone from potentially being disqualified to trophy winners, Milan would have been automatically entered in the group stage regardless), and Fiorentina finished 6th, qualifying for the UEFA Cup. Taking away nine points from Juventus in Serie B had almost no effect; the Old Lady topped the league with 86 points, 6 more than 2nd-place Napoli, and had a goal difference of +58. Although a number of important players had jumped ship when Juventus were relegated, quite a few stayed on, demonstrating their loyalty. These players included Mauro Camoranesi and Alessandro Del Piero as well as Gianluigi Buffon, Giorgio Chiellini, Pavel Nedvěd, and David Trezeguet. With such a strong squad, it's little wonder that Juve overcame their points deduction.

Napoli were slightly handicapped in that regard. Although many more players stuck around after the promotion to Serie B, ADL knew they'd need reinforcements to field a squad that could compete in this exceedingly tough season. Still, he was playing the long game; he wanted to show that he could build a squad to conquer Serie B, then sustain themselves in Serie A, without taking the club into another financial crisis. De Laurentiis knew the supporters would be furious if their Napoli became a yo-yo club, bouncing back and forth between the top two divisions. Cristian Bucchi was the player they spent the most on, paying 4 million euros to Modena, followed by Roberto De Zerbi, who came from Catania for 2.5 million. Both proved their worth by scoring crucial goals, as well as providing support to Calaiò, who again scored the most goals for the Azzurri. The others the club picked up tended to be free agents or returning from loans, supplements for the season but not intended to shine in Serie A.

Yet the Partenopei held their own. They kicked off the season at the San Paolo with a 4–2 victory over Treviso in front of nearly 40,000 fans, with two goals coming from the newly purchased Bucchi. In the 10th round, Napoli came back to draw 1–1 against Juventus, with a goal from Mariano Bogliacino (who would remain with the club until 2010), kicking off a streak of 19 games unbeaten in the league. In November, they climbed their way to the top of the table. De Laurentiis, ever the pragmatist, then revealed he was not ready to deem the season a success, instead reminding fans that they were just at the "beginning of shooting a film" and that the season would be long and exhausting for the players.[20] (Such a tradition is upheld by Napoli fans to this day, helped along by the fact that they consistently collapse come April.) He also noted that the supporters should not hold their breath in anticipation of another Maradona arriving (another tradition he seems to repeat during every transfer window). Instead, he mentioned Napoli's five-year plan, discussing the club's emphasis on youth, and the fact that he would rather raise the new Diego rather than go out and buy one. He then praised sporting director Marino for his ability to spot talent before others, allowing Napoli to buy players on the cheap. Reja also received accolations for being serious, balanced, and prepared, with ADL naming him as his ideal coach.

In the Coppa Italia, Napoli bumped up against Juve in the 3rd round, before the season began. The game went into extra time, and a goal from Alessandro Del Piero at the 120th minute looked to have won it for Juventus. However, Paolo Cannavaro scored the equalizer just a minute later, which took the game to penalties. A number of missed shots, including the Bianconeri's opener by Gigi Buffon, forced each team to send out seven shot-takers. Juve defender Federico Balzaretti failed to make the lucky seventh shot, and Napoli moved on to the next round. Napoli hosted the first game of the two-legged match against Parma, and won 1–0 thanks to a penalty converted by Cristian Bucchi. But in their own stadium, the Serie A side dominated Napoli, scoring three goals, the first from the spot. Napoli's Samuel Della Bona managed a consolation goal in the 83rd. Parma won 3–2 on aggregate, kicking Napoli out of the cup. De Laurentiis' words had proved prophetic: After the highs of fall, his squad bookended their Coppa crash with one-all draws against Frosione and Cesena.

On the final day of the season, Juventus had already claimed the title. However, Napoli's status in 2nd place was not secure. Serie B had a rule that as long as the 3rd-place team finished with 10 or more points than the 4th-place team, there would be no playoff—the 3rd-place team would automatically move into Serie A. Napoli walked a risky line, seemingly not playing their strongest against Genoa. Should Genoa have won, and Piacenza beat lowly Triestina at home, Napoli would have gone to the playoff round with the latter. If, however, Genoa managed just a point, they would avoid the playoff round if Piacenza failed to win. While Napoli were not about to allow Genoa a win, they seemed glad enough to play out a goalless draw. Meanwhile, Piacenza picked up just a point. Genoa finished with a ten-point lead, causing dramatic celebrations at the final whistle in the Marassi stadium.[21] The draw further cemented the *gemellaggio* between Napoli and Genoa, the longest-running in Serie A.

Napoli's back-to-back promotions—from Serie C1 to Serie B, then immediately to Serie A—demonstrated that ADL could run the club both successfully and sustainably. For a time, it endeared the owner to the supporters. After all, it had taken just four seasons for a club in ruins to find their way back to the top of the Italian league system. It was time for the new Napoli to prove themselves in the top division.

The Napoli mind, the Napoli heart, the Napoli *way*

De Laurentiis claimed he wanted to buy young and train the players up in the Napoli way; however, during the years in the lower divisions, potentially valuable squad members seemed to be brought in at 25 or 26 years old, typically at the peak of an attacker's performance. Midfielder Samuele Dalla Bona was 26, as was defender Cannavaro, while his fellow defender Maurizio Domizzi was 26. It can be argued that the latter two were a bit young for defenders, but certainly not young enough to be immediately trained in the "Napoli way" that De Laurentiis sought (although Cannavaro had trained at Napoli in his youth). Christian Bucchi, who'd scored those essential goals in the first match and put away three in the Coppa Italia, was 29. Piá, only 25, was instrumental in Napoli's Serie C1 title victory, scoring seven goals, and he continued his success in Serie B with goals in crucial victories against Treviso and Brescia. However, once the club reached Serie A, his loyalty was rewarded by sending him on loan back to Serie B, with Treviso, and ultimately selling him for 300,000 euros in the January transfer window. György Garics was just 22, but he had only 11 appearances that season—would he be worth the time, energy, and resources needed to create a *Napulitano* Serie A defender?

When surveying the squad set to rise up from Serie B to Serie A, many wondered: Would De Laurentiis remain committed to the plan he'd set out in November, a five-year steady progression focused on younger players who could be bought on the cheap, or would he throw that away as Serie A called, demanding a more competitive squad and throwing around the cash needed to obtain it?

Together with Marino, De Laurentiis looked to keep his word. Now that the club were back in the top division, ADL could really focus on his mission to train younger players rather than rely on aging veterans who could be purchased cheaply. The club let go of most players 30 or older. Bucchi, along with 32-year-old David Giubilato and 31-year-old Ivano Trotta, were out. Piá and Nicolás Amodio, both in their mid-20s, were sent out on loan. The 26-year-old Cannavaro took the captain's armband. Marino went on the hunt for young players they could train up in the Napoli way. Six new players were signed, all of whom played in the typical starting XI. Of these, the oldest were Marcelo Zalayeta, a striker; Manuele Blasi, a midfielder; and Matteo Contini, a defender.

Ultimately, the club spent 21.65 million euros on signings for the 2007–2008 season and bought out the rest of Domizzi's contract from Sampdoria. They earned just 1.2 million euros selling off those they felt could no longer contribute to the squad. Such an investment may seem low now, when teams are paying 100 million euros for one player, but at the time it was both rather high and rather necessary—at least, for a side that planned to stay up and become a competitive force in Serie A. At the same time the club managed to balance its books, a vital step given that fans lived in fear that another crash could occur.[22] Such a balance was something they would continue to be sure happened under ADL's leadership.

This season established a marked shift in the way SSC Napoli not only handled business, but how they conducted themselves on the pitch. The most important change involved bringing two new, young, exciting players into the club. The first was Ezequiel Lavezzi, a 22-year-old Argentine attacker. Napoli paid San Lorenzo 5.6 million euros for him, and in his first season he scored the most goals in all competitions. The second was Marek Hamšík, just 20, from Brescia. The Slovakian attacking midfielder scored the most goals in the league, nine, for the price of 5.5 million euros. Buying defensive midfielder Walter Gargano, 23, for 3.2 million euros from Danubio in Uruguay, helped protect the back line—even if his position held him back from attracting as much attention.

Many of the players who started most often for that 2007–2008 side remained with the club for at least a few years, giving it a steady spine and the willingness to trust one another when tactics or style changed. Looking back, it was the 20-year-old from Brescia, Hamšík, who would have the most impact on the squad, but at the time Lavezzi was the signing that excited the fans the most. In my interviews with fans who had stuck through the hard times and were thrilled that the club was back in Serie A, they all mentioned that the city of Naples was eagerly awaiting the man they hoped would be their new star. Neapolitans are almost always convinced it would be another attacker from Argentina who would elevate the side to another title win. Lavezzi was a young forward who quickly moved through the Rosario youth side to join Estudiantes for a year before being bought by Genoa in 2004 for 1 million euros. It certainly seemed like the pacy, dribbly Argentine could wind up being the new Napoli's Diego Maradona. His nickname even fitted with Maradona's style—*El Pocho* (The chubby one).

Genoa, however, did not seem to know what a player they had on their hands. They immediately loaned him to another Argentine club, San Lorenzo. Lavezzi was prepared to play for Genoa in the 2005–2006 season, to the point where he had been presented to the media as a Genoa player and he'd chosen 77 for his shirt number. However, Genoa allegedly fixed a match between themselves and Venezia on the last day of the season, to ensure Genoa would win the Serie B title. The FIGC determined they should be sent to the bottom of the table and thus relegated to Serie C1. The fans, obviously unhappy about the decision, protested loudly with fire and flares.[23] This did nothing to change the federation's position, however, and they were sent down—ironically finding themselves in competition with Lavezzi's later club, Napoli. Due to financial constraints, Genoa were forced to sell him to San Lorenzo for just 1.2 million euros, earning a mere 200,000 euros on their investment.[24]

As Lavezzi was helping San Lorenzo to the Clausura 2007 title, beating Boca Juniors by an impressive six points, Napoli and Genoa were celebrating their rise to Serie A. Both would need much stronger squads, but it was Napoli who swooped in for Lavezzi first, paying about 5 million euros more than Genoa had originally. He quickly proved his worth, scoring a hattrick, with two goals in extra time, to eliminate the Serie B side from the Coppa Italia on August 18. He ultimately scored 11 goals in all competitions that season, tied with defender Domizzi, as surprising as that may seem—but Domizzi was the squad's penalty taker, scoring 6 in Serie A and 2 in Coppa Italia.

Most calcio fans remember the 2007–2008 season due to its dramatic conclusion, in which Inter and Roma were separated by just one point and, in the last round of league matches, were set to play Parma and Catania respectively, both trying to avoid relegation. Roma could only draw at Catania, which secured the Sicilian club's place in the top division for another year, while Inter won their match, and the title, with a win at Parma, relegating the latter. Napoli did not experience nearly as much drama in their first season back, although their 2nd round 5–0 win at Udinese, which included Ezequiel Lavezzi's first Serie A goal, was rather impressive, and their back-and-forth 4–4 draw with Roma, in which a late Marcelo Zalayeta goal saw the Azzurri take a point away from the capital, was thrilling to watch. The team also managed a tantalizing win at home against Juventus. Del Piero scored early in the second half, but Gargano equalized after just a few minutes. Then the visitors conceded two penalties within two minutes, both of which Domizzi put away.

But for Napoli supporters, the season was a turning point. Napoli finished 8th, 35 points behind champions Inter, and qualified for the Intertoto Cup. The Azzurri had not only announced their return to Serie A, but also demonstrated that the club were willing to put in the hard work to stay there. And, indeed, they have not been relegated since. The system wasn't perfect, but they would have a good time working out the kinks. Being a Napoli supporter was about to get really, really fun.

[1] Doidge, M. (2015). *Football Italia: Italian football in an age of globalization*. Bloomsbury Academic. http://dx.doi.org/10.5040/9781472519221.0006

[2] Napoli declared bankrupt says ANSA. (2004, August 2). *CNN*. http://edition.cnn.com/2004/SPORT/football/08/02/italy.napoli/

[3] Gould, D., & Williams J. (2011). After Heysel: How Italy lost the soccer 'peace'. *Soccer & Society*, *12*(5), 586–601, 587.

[4] Doidge.

[5] Napoli's success story. (2011, April 21). *The Swiss Ramble*. http://swissramble.blogspot.com/2011/04/napolis-success-story.html

[6] Ciccarelli, L. (2012, December 12). La storia siete voi: il Pampa Sosa. *TuttoNapoli*. https://www.tuttonapoli.net/rubriche/la-storia-siete-voi-il-pampa-sosa-127531

[7] Capella, F. (2005, June 21). Squadra: si riparte da Reja e Calaiò ma il dg Marino può perdere Abate. *La Repubblica*. https://ricerca.repubblica.it/repubblica/archivio/repubblica/2005/06/21/squadra-si-

riparte-da-reja-calaio-ma.html

[8] De Laurentiis: "Il mio Napoli tra la grande." (2006, November 20). *SoloNapoli*. http://www.solonapoli.com/leggi_news.asp?Id=12814#

[9] Centi, F. (2006, April 15). Napoli in festa: promosso in B. Gazzetta dello Sport. https://www.gazzetta.it/Calcio/Altro_Calcio/Primo_Piano/2006/04_Aprile/15/Napoli.shtml

[10] Germano, B. (2006, April 16). NAPOLI Rieccolo! *Gazzetta dello Sport*. http://archiviostorico.gazzetta.it//2006/aprile/16/NAPOLI_Rieccolo__ga_10_0604168050.shtml

[11] Ore 16.40, il Napoli in B si scatena la festa dei tifosi. (2006, April 15). *La Repubblica*. https://www.repubblica.it/2006/04/sezioni/sport/calcio/napoli-in-b/napoli-in-b/napoli-in-b.html

[12] Ciccarelli, L. (2012, December 12). La storia siete voi: il Pampa Sosa. *TuttoNapoli*. https://www.tuttonapoli.net/rubriche/la-storia-siete-voi-il-pampa-sosa-127531

[13] Id.

[14] Materazzo, G., & Sarnataro, D. (2014). *I campioni che hanno fatto grande il Napoli*. Newton Compton.

[15] Napoli: Ritorna il vecchio marchio (2006, May 26). *Gazzetta dello Sport*. http://archiviostorico.gazzetta.it//2006/maggio/24/Napoli_ritorna_vecchio_marchio_ga_10_060524072.shtml

[16] Baroncelli, A. (2007). Calciopoli: Reasons and scenarios for the soccer scandal. *Italian Politics, 22* (1). 226–248. https://www.berghahnjournals.com/view/journals/italian-politics/22/1/ip220113.xml

[17] "Consigli" agli arbitri e minacce: Le telefonate dei potenti del calcio. (2006, May 12). *La Repubblica*. https://www.repubblica.it/2006/05/sezioni/sport/calcio/inchieste-intercettazioni/inchieste-intercettazioni/inchieste-intercettazioni.html

[18] Calciopoli, la cronistoria. (2006, October 27). *Corriere della Sera*. https://www.corriere.it/Primo_Piano/Cronache/2006/07_Luglio/14/cronocalcio.html

[19] Punishments cut for Italian clubs. (2006, July 26). *BBC*. http://news.bbc.co.uk/sport2/hi/football/europe/5215178.stm

[20] De Laurentiis: "Il mio Napoli tra la grande." (2006, November 20). *SoloNapoli*. https://www.solonapoli.com/leggi_news.asp?Id=12814#

[21] Meadows, M. (2007, June 10). Soccer—Napoli and Genoa promoted to Serie A. *Reuters*. https://uk.reuters.com/article/ba-soccer-italy-serieb-idUKL1021893120070610

[22] The Swiss Rambler.

[23] Calcio, Genoa retrocesso in C1: Tifosi furibondi, proteste in città (2005, July 27). *La Repubblica*. https://www.repubblica.it/2005/g/sezioni/sport/calcio/genovac/genovac/genovac.html

[24] Genoa rue Lavezzi sale. (2012, February 28). *Football Italia*. https://www.football-italia.net/node/16137

CHAPTER 11

Our Stars Were Never Aligned

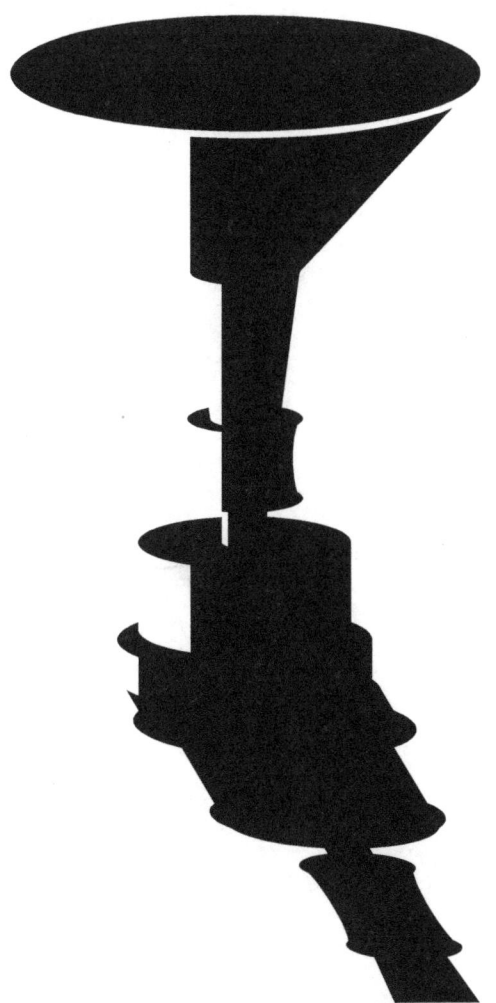

11

The Napoli culture that Aurelio De Laurentiis was trying to achieve would not spring up overnight. Time was needed to instill the club's goals and purpose into the young players being brought in, and even the older ones who had stuck around through the club's years in the lower leagues. ADL knew the Napoli supporters wanted what they had demanded for nearly 100 years—not only for their club to challenge for trophies, but also to put on exciting *calcio* performances. Most understood that the next *scudetto* would not be lifted at the end of the first season back in Serie A; however, given the performances witnessed over the 15 years of top-tier play under De Laurentiis, many were astonished that the third (and fourth, and fifth...) titles were not bestowed upon Napoli football club. Those same fans now hold little hope that their once-beloved owner, yet another savior in *Azzurri*, will assemble a squad that will finally claim their third title. Perhaps, in this time of modern football, a club that spend within their means and do not splash out on the biggest names cannot add more hardware to their trophy cabinet.

Sparkle and fade

First, though, to 2008–2009, which began with Napoli's participation in the Intertoto Cup. This rather silly named cup existed for a rather silly purpose—primarily, to allow betting on football matches to continue through the summer months. At first, an outright winner was declared after a playoff between group winners. That lasted through the tournament's first five years. The Intertoto Cup then abandoned the playoffs, declaring each of the group winners champions. In other words, a tournament created for the purpose of betting ended up as a tournament in which no one could bet on a single winner of the contest.[1]

By the time Napoli entered the Intertoto Cup in the summer of 2008 (its last iteration), the tournament held a bit more importance. It had been taken over by UEFA and now served as a way to allow lower-placed sides from a number of leagues to qualify for the UEFA Cup, the second-most prestigious tournament on the continent. Fifty teams entered the competition in three rounds, based on their 2007 team coefficients. The 1st round included 28 teams from the lowest-ranked leagues, playing two-legged matches to determine who would progress. The 14 winners were joined by the teams that qualified from the associations ranked 9–22. Again, two-legged matches decided the winners, and those 14 moved on to join the 8 highest-ranked teams. The 11 winners of those matches would enter the 2nd qualifying round of the UEFA Cup. Napoli entered the 3rd round of the Intertoto Cup and saw off Greek side Paniōnios 1–0, thanks to a goal from Mariano Bogliacino. Attendance figures for the 3rd round varied from 3,000 to 12,000, with Napoli the extreme outlier. They attracted 54,137 fans to the reverse fixture, supporters thrilled to see their team back in Europe, ecstatic to watch Marek Hamšík score for another 1–0 win and a 2–0 aggregate victory over Paniōnios, advancing them into the more prestigious UEFA Cup.

In the 2nd qualifying round of the UEFA Cup, Napoli saw off Albanian club Vllaznia 8–0 on aggregate, with three goals from Piá, two from Rinaudo, and one each from Germán Denis, Ezequiel Lavezzi, and Marek Hamšík. Alas, they then fell to Benfica 4–3 in the first "official" round, at which point 16 losers from the UEFA Champions League third qualifying round entered the lower tournament. The *Partenopei* won 3–2 at home, thanks to a goal from Luigi Vitale that quickly answered David Suazo's opener; Christian Maggio and Denis also scored, ensuring the one netted by Luisão, at the hour mark, kept Benfica at bay. However, Napoli could not manage to find the net in Lisbon, and so were eliminated in a 2–0 loss. But the Azzurri can "brag" that they were 1 of the 11 cowinners of the final Intertoto Cup, although another Portuguese side, Braga, are considered the overall winners of the competition, having progressed furthest in the UEFA Cup (they lost to PSG in the Round of 16).

Had Napoli known their fate for the 2008–2009 Serie A season, they likely would've celebrated finishing 8th in their first campaign all the more. The *Vesuviani* were quite active in the transfer market, hoping for a better year. The club let their joint top-scorer, Maurizio Domizzi, go to Marino's former club Udinese for 4.5 million euros. Emanuele Calaiò, who had scored 44 goals in all competitions in the past 3 seasons for Napoli, was deemed surplus to requirements after Ezequiel Lavezzi and Marcelo Zalayeta arrived, and was offered to Siena on a co-ownership deal for 2.3 million euros; the Tuscan club purchased him outright for 1.25 million euros the next summer.[2] Samuele Dalla Bona, so essential in the lower leagues, was let go for nothing, as was the once-beloved Roberto Sosa. The club made only 2.5 million on the sale of György Garics to Atalanta. Their total sales came to less than 11 million.

In hopes of a better finish than a season before, they had purchased Christian Maggio from Sampdoria for 8 million euros. The acquisition of the Italian national would prove vital; after he signed with Serie B side Benevento in 2018, Napoli consistently found themselves mired in difficulties finding the right fit at right-back. They also purchased yet another Argentinian forward, Germán Denis from Independiente, for 7.5 million, after he scored 15 goals in the first 13 games of the 2007 *Apertura* and 27 overall in 2007–2008. While Denis impressed in pre-season friendlies (and, obviously, in Europe), in 2 seasons with Napoli he was unable to reach his high-water mark of 15 goals, scoring just 13 during that time, 8 of which came in the first season. (Denis and Duván Zapata, at Napoli from 2013–2015, would go on to successful careers at Atalanta; unsurprisingly, Pierpaolo Marino had become the sporting director for *La Dea* in June 2011.) Meanwhile, the Vesuviani seemingly wasted 5.5 million on Leandro Rinaudo from Palermo, who had 37 appearances in 5 seasons and was loaned out twice, and 6.5 million on Jesús Dátolo from Boca Juniors in January; the attacking winger would last a year and scored just once in 22 appearances. Finally, they bought Salvatore Aronica from Reggina; the defender proved to be a player that fans had to choose whether to laugh or cry about.

Given Napoli's expenditures far outpaced their income, was De Laurentiis reneging on his promise to keep the club financially stable? Or was he counting on this squad to finish high in the table, in a position where he could reap financial rewards? If his strategy was the latter, it failed to come to fruition. Their second season back in Serie A was rather turbulent. Whether Marino had slipped up in the transfer market, or Reja was unable to figure out just how the squad would best work together, or Napoli was simply experiencing a sophomore slump, is difficult to determine.

At first the 2008–2009 season looked like it would be one in which Napoli would build off the previous season's 8th place finish. The season began with a 1–1 draw against Roma, and Napoli went on to lose just once in the first nine games. Marek Hamšík scored four over that period, while a hattrick against Reggina helped Germán Denis score five. Napoli even beat Juventus at the San Paolo once again, coming back from a 1–0 deficit almost immediately, with Hamšík scoring three minutes after Juve's Amauri, and Ezequiel Lavezzi sealing the 2–1 victory.

The team then struggled for the next few rounds, but as Christmas approached, another decent handful of results gave fans hope—a 2–0 win over Siena, a 3–0 victory over extreme underdogs Lecce, and a narrow 1–0 triumph over Catania. In addition to Lecce finishing dead last in Serie A, Siena and Catania would both fall below Napoli in the final table. During this time Napoli also lost to Torino, who were relegated at the end of the season. Then came the *ritorno*. The Partenopei fall actually began on January 18, 2009, the last round of the first half of the season, at newly promoted Chievo Verona, where Napoli came away with a 2–1 loss. This kicked off a period in which Napoli failed to win a single game for 14 long rounds (and a period of 10 years in which Napoli also often failed to pick up points against the *Gialloblu*, no matter how poorly Chievo were doing that season).

While Aurelio De Laurentiis had previously pledged his loyalty to Edy Reja, and the manager had made a fine showing during the club's first season back in Serie A, ADL couldn't keep the promise he had made when Napoli were in Serie B to keep Reja at the helm. Prior to the start of the second half of that second Serie A season, Napoli had seemed destined for at least 4th place in the league, and another trip to Europe. But after collecting just two points in nine games, they fell to 11th, and Reja was summarily dismissed after a 2–0 loss to Lazio on March 8. In his place came Roberto Donadoni, the manager of the Italian national team from 2006–2008.[3]

Donadoni had failed spectacularly at the 2008 European Championships, where his World Cup-holding side suffered their most significant defeat in 25 years, after which they exited during the quarterfinals. Unable to shake off that cloud of doom, he hadn't managed a team in nearly a year, and few were surprised that he wasn't able to turn Napoli around immediately. He did, however, succeed at pressing his squad to play for draws rather than wins; in his first four games they drew with Reggina, Milan, Sampdoria, and Atalanta, slipping down to 12th in the league. Then came a loss at Cagliari, which caused Napoli to fall to 14th. De Laurentiis directed his anger toward the players, stating that with the exception of Hamšík and Lavezzi, no one was safe from being sold.[4] Although the owner was not thrilled with Donadoni, he was willing to be patient, keeping the hope alive that bringing in new players over the summer would improve results.

Initially, the threats worked. Napoli beat Inter—who had been top since Round 11 and ultimately ran away with the title—1–0 with a goal by Marcelo Zalayeta, raising the fans' hopes. Up next came the struggling teams that Napoli had actually managed to beat in the first half of the season, so surely the side could climb the table. Unfortunately for the *tifosi*, they were forced to watch Napoli drop points against Siena, Lecce, and Catania, and once again lose to Torino, who slid into the relegation zone the next round. And at the time of their 1–1 draw, Lecce were dead last. By the time Napoli beat Chievo 3–0 at home to finish the season, the fans likely felt more relieved than celebratory. In fact, by that point the fans were so angry with the Napoli performance that they took to protesting against the club, and many of the 27,000 who attended the final match left early to show their disgust with their team.[5]

What the supporters didn't know was that the team that would show up in the next decade was one that they would take pride in. An exciting new talent was about to join the squad, creating a front three that could challenge any team in Europe. And a special *Mister* would be brought in to work his magic, resulting in a thrilling style of play that finally made those outside Naples sit up and pay attention. But first they had to get through 2009.

A slight shimmer of success

Pierpaolo Marino performed a few brilliant maneuvers over the summer of 2009. On the very first day of the transfer window, Napoli brought in Fabio Quagliarella. The 26-year-old had scored 21 goals in all appearances for Udinese the previous season, including 8 in 11 appearances in the UEFA Cup, where he led his team to the quarterfinals. Napoli paid 18.75 million euros to secure the striker for five years, and Marino described the signing as one of the most prestigious of his life. Meanwhile Quagliarella, a native of the Naples area, said that playing for Napoli had always been his dream.[6]

The next significant summer signing was Luca Cigarini from Atalanta. Napoli paid around 11 million for the defensive midfielder, and it was one of the bigger missteps

Marino made. Cigarini's value was only around 7 million,[7] and he never consistently made the starting XI with Napoli that season. In summer 2010, he was sent on loan to Sevilla, after just one year in Naples. Attacking midfielder Juan Zúñiga, purchased from Siena for 8.5 million euros, turned out to be a much better investment. On the same day Napoli brought in Zúñiga, they announced the 7 million euro signing of Hugo Campagnaro from Sampdoria, who became a mainstay in Napoli's back line for four seasons. Perhaps most importantly, Marino managed to woo experienced goalkeeper Morgan De Sanctis from Sevilla for less than 2 million euros, although he was worth at least 5 million. Meanwhile, the only real money the club brought in was from a co-ownership agreement with Sampdoria for Daniele Mannini, for 3.5 million. Napoli had again spent far more than they made in the transfer window. The club had few other revenue streams, as Naples owned the San Paolo, TV rights and resulting monies were based on success and popularity, and they were not in European competition. At some point the club would need to restock their bank account if they wanted to avoid sliding back into financial issues.

Despite the signings, almost all of whom would blossom into quality players, Napoli had a rollercoaster of a start to 2009–2010. Before the league started, they faced rivals Salernitana in a *Derby della Campania* at the San Paolo in the 3rd round of the Coppa Italia. The Partenopei easily dispatched the Serie B side, 3–0, with goals from Christian Maggio, Ezequiel Lavezzi, and new signing Erwin Hoffer, from Rapid Vienna. But in Serie A, the side's only wins in the first seven games came against Siena and Livorno, both of whom would be relegated at the end of the season. The fans were frustrated, but not as much so as owner Aurelio De Laurentiis. He had lavished a significant amount of money on new players, more than most clubs in Italy had spent that transfer window, in order to assemble a squad that was capable of competing for a place in Europe. Not known for his subtlety, ADL also was said to be unhappy with a few of the players who came in, as well as a few of those who had left.[8] He attacked Morgan De Sanctis, blaming him for Napoli's 3–1 loss at Inter in Week 5.[9] De Laurentiis then went on to crucify Marino on live TV the next week, after a 2–1 victory over Siena, criticizing the squad's poor results. The next day, by mutual agreement, Pierpaolo Marino and Napoli parted ways, despite his having signed a new contract only a few months earlier. The club insisted he and De Laurentiis remained friends, even after concluding it was best for Marino to leave. The five-year rebuilding project was said to have been completed, and the club now needed more innovation for the next phase of the mission.[10]

Meanwhile, Roberto Donadoni, faced with a number of new signings, was unable to settle on a starting XI, and the constant squad rotation, despite the club not playing multiple games per week, caused many to scratch their heads in confusion. Though his tactical formation, a 3-5-2, remained the same, he could not settle on a starting back three, making it difficult for the relationship between his defenders and new goalkeeper De Sanctis to solidify. He also tried a number of different players in the more advanced midfield roles, but again, the constant changes made it hard for the two consistent midfielders, Walter Gargano and Marek Hamšík (interestingly enough, the two would later become brothers-in-law), to form the strong connections needed for the 3-5-2 to work; at the same time, they often did not know who the two most advanced attackers would be. Although after his club lost 2–1 to Roma on October 4, De Laurentiis complimented Donadoni, calling him "adorable" and reassuring the press that everyone makes mistakes, he refused to commit to keeping his manager.[11] In the end seven points from seven games and 15th position in the table was not enough for the president. It did not help that the players were dissatisfied with the tactics Donadoni imposed, and it was rumored that a few were dissatisfied with their manager overall.[12] De Laurentiis' lack of trust in Donadoni, confirmed by his goalkeeping coach Sergio Buso, and his constant questioning of Donadoni's managerial decisions[13] further heightened the tension in the locker room, contributing to the poor displays on the field. On October 6, ADL fired Donadoni, and brought in Walter Mazzarri to take his place.

Mazzarri had last held the helm at Sampdoria, where in 2008 they'd finished 6th. Perhaps more importantly to the players under Donadoni—and the fans who'd had to watch the painful matches at the beginning of the season—the new Mister's football was fast, attractive, and often effective. At Sampdoria, much of his success had relied on the front two, Italian internationals Antonio Cassano and Giampaolo Pazzini. That success ran to ground in the 2008–2009 season, when Sampdoria finished 13th. Despite an exciting Coppa Italia campaign that led them to the final, where they lost to Lazio on penalties, Mazzarri announced he'd be leaving the club after the last game of the season. Sampdoria felt the Coppa final was enough to keep Mazzarri around;[14] according to Walter, he and the club had come to a mutual agreement regarding his resignation.

In many ways, Mazzarri was the perfect manager to serve under De Laurentiis. Like the movie director, he has a flair for the dramatic. He paces the touchline, windmilling his arms and even kicking his feet to convey messages to his players. He'll clutch his head or pull out his hair, pick up empty Lete water bottles with his teeth (a perfect image for the Napoli sponsor), or get down on his knees and tear the grass from the field.[15] He would often smoke as he moved up and down the field, and many suspected him of having whisky stashed in the dugout—he would certainly drink when banned from a match and forced to watch from the owner's box, while the delighted TV cameras watched him right back.

Given Mazzarri's personality and his previous stylish showings of play at Sampdoria, things at Napoli were finally looking up. In Mazzarri's first game in charge, Napoli lined up in the same 3-5-2 formation, with many of the same players who had lost to Roma in the previous round. In the 15th minute, the fans who'd excitedly welcomed Mazzarri to the pitch such a short time ago were stunned into silence when Bologna striker Adaílton put in an amazing free kick from outside the area, striking the top of the net and falling behind Morgan De Sanctis. Afterwards, Napoli certainly didn't play like they were cowed—just unlucky. Matteo Contini sent up a beautiful free kick that Quagliarella danced around to contain, but Jesús Dátolo missed the target. The Bologna goalkeeper got a hand on an Ezequiel Lavezzi free kick that was just as stunning as Adaílton's. The side continued to press forward, trying to work their way through a strong Bologna defense. The break came late in the second half, when a corner from Napoli finally managed to befuddle Bologna. The ball bounced around before landing at the feet of Quagliarella, who was near the far post. He slotted it in for the equalizer— and kissed the badge of his childhood club. In the 79th, De Sanctis elegantly denied Marco Di Vaio during a one-on-one situation. Then, Lavezzi reminded the supporters of the reasons they were thrilled at his arrival. He gathered the ball in the other half of the pitch, blew past a Bologna player by toeing the ball over his head, then outran the rest of the defenders to go one-on-one with the Bologna keeper . . . but was denied. Soon after, when a ball fell to him after a corner, he faked out the defender to slot a perfect pass into the scramble in front of goal, where Christian Maggio was waiting to slot into goal. With that extra-time shot, Napoli came from behind to beat Bologna 2–1.

The Bologna game deserves attention because it helped foreshadow what was to come. First, Napoli went 15 games unbeaten before finally being felled by Udinese in February. By that time, Napoli had moved up to 4th—11 positions higher than when Mazzarri had taken over. Second, it revealed what would come to be known as "Mazzarri time." Similar to the *Zona Cesarini*, Mazzarri's squad would rarely accept a defeat, and kept playing strong until the final whistle, with many a goal coming after the 85th minute and a surprising number happening after the 90th. (After the Bologna match, Napoli came from behind to take at least a point against Milan, Juventus, Roma, Bari, and Cagliari, and scored a late winner the next week against Fiorentina.) Third, it showed off the fast breaks that would come to define the team. Yet one factor was still missing.

After the loss to Udinese, Napoli ran out of steam, going seven games without a win and falling to 8th in the table. But the Partenopei still knew how to prepare themselves for their strongest of opponents, and when Juventus came to town, they convinced even those already beginning to doubt Mazzarri that he was, in fact, a special Mister (as he'd come to be known). In the 2009–2010 season, Napoli did the double over Juventus, coming back from behind once again to score three in the second half, with goals from Hamšík, Quagliarella, and Lavezzi. For fans of a team that had rarely reached the European positions over the course of their long history, and who had been disbanded and reformed only five years ago, this was nearly as good as winning the scudetto. That season they did reach the European positions, finishing 6th—four points over Juventus, and their best finish since 1993–1994—and earning entry into the Europa League (formerly the UEFA

Cup) qualifying round. Marek Hamšík was the club's *Capocannoniere* that season, with 12 goals.

If the 2009–2010 season showed Napoli's potential, the 2010–2011 season demonstrated their power. While many owners would have been content to make few changes to their squad, instead waiting to see how the current side would gel under a full season under Mazzarri, De Laurentiis would not rest until his club regularly made European football. More precisely, Champions League football.

The Holy Trinity

Mazzarri had a plan for how this team would function, and it included just one "true" striker. ADL and Mazzarri put their heads together and determined that the path to the Champions League would be guided by the best forward they could convince to join the club, someone they hoped would finish near the top of the league's scoring charts. They looked south to Palermo, where the 23-year-old Edinson Cavani had formed a terrific partnership with Fabrizio Miccoli. The latter had scored 19 goals, while Cavani added 13 more. Together they ensured Palermo finished 5th (6 points above Napoli in 6th) and qualified for the Europa League playoff round. Cavani fit the profile of what the club were looking for—young, hardworking, and clinical in front of goal. They quickly sent Riccardo Bigon, who had taken over as sporting director from Pierpaolo Marino the previous season, down to Sicily to secure the Uruguayan's transfer. Surprisingly, he came rather cheaply— Palermo agreed to loan Cavani for 5 million euros, with an obligation to buy for an additional 12 million.[*] It may seem like mere pennies today, but it was the most Napoli had ever spent on a player.

Interestingly enough, the Cavani deal required extra sweetening. He wanted the #7 jersey, at the time worn by Ezequiel Lavezzi. Although quite a rivalry exists between Lavezzi's Argentina and Cavani's Uruguay, Lavezzi had no problem giving up his shirt. "He is a great striker and he will be important for all of us here at Napoli which is getting stronger," he told the press. "The number 7? I can give it to him. For me what counts is being a protagonist on the pitch. Now it's down to Mazzarri in terms of who plays."[16] Lavezzi also made the point that the club's primary goal was to secure Champions League qualification, and Cavani's talent would help them reach that goal. For a player often accused of being too selfish on the pitch, Lavezzi was truly gracious about the arrival of an attacker with the potential to steal the limelight.

Others, however, were not as pleased, about both the money spent on the Cavani transfer and the implication that last season's squad was not strong enough to

[*] UEFA introduced financial fair play in 2009 and implemented it at the start of the 2011–2012 season. This prompted many Italian clubs to lean on the nifty "borrow to buy" trick, which allows payments to be spread across two seasons, one being made for the loan, the next to complete the purchase. By dividing their expenses between two fiscal years, they reduce their risk of being punished for spending beyond their means.

qualify for a Champions League place. To meet Mazzarri's demands of how he felt a team should function, Napoli sold striker Germán Denis for just 2 million euros (he was worth at least double that amount), while fellow forward Marcelo Zalayeta left on a free. After the first Europa League match, before the transfer window closed, they sent Fabio Quagliarella, the Neapolitan who'd spent just one season at the club, to Juventus on loan.

Prior to the start of the 2010–2011 season, Napoli introduced the Club Azzurro Card (now the "Fan Stadium Card"), in compliance with government orders. A fan of a Serie A club must have such a card to buy a season ticket and to sit with the away side when attending matches elsewhere. The primary reason for implementing the system was given as being able to keep the away fans in one area of the stadium, reducing clashes with the home fans. However, many protested, including politicians and club officials, as it was obvious that the police were looking for a way to keep closer tabs on fans they believed might be troublemakers.

The season began for Napoli on August 19 when they hosted Swedish club Elfsborg in the Europa League playoff round. Over 35,000 were in attendance at the San Paolo that day, likely hoping to catch a glimpse of Cavani, the man they hoped would be their next superstar. Cavani came in at the 62nd minute, but it was Lavezzi who was the hero of the day, scoring just before halftime. Napoli took their 1–0 lead to Sweden, where they parlayed it into a 3–0 aggregate win. This time Cavani was in the lineup from the start, and scored two goals in the first half.

A few days later, on August 29, Napoli opened their Serie A account with a 1–1 draw at Fiorentina. Cavani scored in the seventh minute, but the visitors' defense gave way to allow a goal from *Viola* midfielder Gaetano D'Agostino early in the second half. They next drew 2–2 with Bari, with yet another goal from Cavani, as well as one from Paolo Cannavaro.

Next they were off to the Netherlands for the Europa League group stage. Napoli played out a goalless draw with Utrecht, and then a 3–3 draw with Steaua Bucharest, points they would later regret dropping in a competition that often paired group winners with tamer opposition in the first knockout round. Mazzarri time was working overtime in Bucharest, however, when Cavani scored in the 97th minute to seal the draw. Next it was another goalless draw, at the San Paolo against Liverpool; in the reverse fixture at Anfield they lost 3–1. Napoli then drew, *again*, a 3–3 with Utrecht, Cavani scoring all three of the Azzurri goals. Finally, in the last round, they beat Steaua 1–0, again with a Mazzarri time goal by Cavani, in the 93rd. This gave Napoli a one point advantage over the Romanian side, and so they advanced to the knockout stages. An apt term, as they were quickly knocked out by Villarreal. Napoli once more drew 0–0 at the San Paolo, then lost 2–1 in Spain. Mazzarri time could only take them so far. Meanwhile, group winners Liverpool advanced after dispensing with Sparta Prague. Still, the Partenopei were the only Italian team to make it out of the group stage.

– NUMBERS –
UEFA coefficients and calcio's further downfall

By 1980, UEFA had implemented a coefficient system to rank the men's football associations of Europe to determine the number of teams from each that would participate in tournaments they hosted, as well as where they would enter. As of 2022, these club tournaments are the Champions League, the Europa League, and the UEFA Europa Conference League.

Determining each association's coefficient is a complicated process involving the results of each club's performance in each of the three tournaments over the past five years. Each win equals two points, each draw equals one; however, these are halved for each qualifying round that occurs before a tournament's group stage. In addition, associations are awarded bonus points for each club that reaches the group stage, with more added for each subsequent round. Extra points are given for the more prestigious tournaments (i.e., +2 points for advancing in the Champions League, +1 for advancing in the Europa League). For instance, although in 2021–2022 Roma played 11 games and triumphed in the Conference League final, they collected only 25 points, as compared to Liverpool's 33 points over 11 games that ultimately resulted in a loss in the Champions League final.[17] Finally, the coefficients compiled at the end of each season are used to determine association rankings for the tournaments occurring two years in the future; the 2021–2022 numbers will affect the 2023–2024 events.

In 2018, UEFA changed the rules regarding how clubs could directly qualify for the Champions League group stage: in addition to the previous year's Champions and Europa League winners, each of the top four associations would send the teams that finished 1st–4th in the previous season (at the time, those were Spain, Germany, England, Italy).[18] But in the previous years, when Napoli were aiming for a steady presence in the top tournament, things weren't quite so simple.

After Italian clubs had rose to prominence in the 1990s, Italy had grown used to sending four teams to the Champions League in the first decade of the 21st century; while the 3rd- and 4th-placed teams entered prior to the group stage, the "official" start of the tournament, at least three Italian sides advanced each year. But in 2012, Italy dropped to 4th in the rankings, meaning the 4th-placed side entered the Europa League and the team in 3rd started in the playoff round. In 2014, Italy's decline became even more evident, when Juventus lost to Benfica in the Europa League semifinals, and Portugal began to close in on the top four. Fortunately for Serie A fans, a strong performance in the 2017 Champions League group stage boosted their pride, and by 2018, UEFA's rule change meant each of Italy's top four went straight to the group stage. And as the UEFA rankings appear designed to allow the top associations to remain in position, it is likely Italian clubs can remain confident that a 4th-place finish is sufficient for a trip to the Champions League group stage.

But Mazzarri time certainly affected their performances and, ultimately, their final place in Serie A. After losing 2–1 to Milan in Round 8, Napoli began to tear up the league. In the next 11 matches, Napoli emerged victorious in 8 of them. Three— Cagliari, Palermo, and Lecce—were won after 90 minutes had passed, with Lavezzi, Maggio, and Cavani the scorers in Mazzarri time. Napoli's 4–3 victory over Lazio in Round 31 perfectly encapsulates Mazzarri's approach to the game, and how the players fit into his system. At the hour mark, the visiting *Biancocelesti* led 2–0, despite Napoli taking the game to them from the whistle. Attempts by Cavani, Lavezzi, and Hamšík were all stifled, but at the other end, Stefano Mauri threaded his way through the defense to score Lazio's first, while André Dias snuck behind the backline to score the second with a toe-poke off a free kick. Minutes later, Andrea Dossena scored his first of the season, heading on a set-piece from *El Pocho*. Then, in the 62nd minute, it was Cavani's turn to snap in a sharp header to set things level. Napoli's hopes were crushed in the most dramatic of fashions six minutes later, when Salvatore Aronica slid a perfect ball into the net . . . of Morgan De Sanctis. This is the incident that caused Walter Mazzarri to infamously start pulling up the grass on the touchline and shredding it into smaller pieces. Fortunately for the Mister's health, El Matador came to the rescue once more. The hosts were awarded a penalty in the 82nd minute after Cavani was felled in the box. He stepped to the spot himself, once again putting Napoli on equal terms with a cool right-footed strike. As the fans bit their nails and watched the clock run down, Giuseppe Mascara headed the ball to an onrushing Cavani. Fernando Muslera got a hand to it, but the Lazio keeper only succeeded in helping the lobbed shot find its target. The fans went wild, filling the San Paolo with chants of "Ca-van-i, Ca-van-i" and celebrating another win in Mazzarri time, this one seemingly miraculous, temporarily lifting them above Inter into 2nd.

This game remains firmly etched in my mind as a defining moment in Napoli history, one in which I had to stifle my screams throughout, given that it occurred in the pre-dawn hours on the West Coast. The tension, the agony, the drama, the beauty have ensured it is one most supporters remember, a guidepost marking the offense Napoli fans would come to expect, and a defense they would spend the next decade arguing about how it could best be reinforced.

The squad finally began to slow down in the final six rounds, picking up just five points, with no useful goals after the 85th minute mark (Giuseppe Mascara scored in the 96th minute against Udinese, but the *Friulani* won 2–1, thanks to goals from future Napoli midfielder Gökhan Inler and ex-Napoli forward Germán Denis). Despite scoring few points in the last few games, Napoli still managed to secure 3rd place in the penultimate round thanks to a goal by Juan Zúñiga to draw level with Inter, 1–1. They were able to celebrate their place in the Champions League in front of an exuberant home crowd at the San Paolo. They put a nice bow on the end of the season by drawing 2–2 with Juventus, who finished 7th. The number of Napoli victories, 21, equaled that of the wins they'd recorded in 1989–1990, their second scudetto-winning season. Cavani, with 26 goals in the league, finished 2nd in the Serie A Capocannoniere race (behind Udinese's Antonio Di Natale) and, with an additional 7 goals in the Europa League, broke Antonio Vojak's record to become Napoli's top scorer in a top division season. Goalkeeper Morgan De Sanctis also made history by going 798 consecutive minutes without allowing a goal at the San Paolo.

– THREE TENORS –
Edinson Cavani (2010–2013)
Napoli appearances/goals: 138/104
Uruguay appearances/goals: 130/54

Marek Hamšík (2007–2019)
Napoli appearances/goals: 520/121
Slovakia appearances/goals: 135/26

Ezequiel Lavezzi (2007–2012)
Napoli appearances/goals: 188/48
Argentina appearances/goals: 51/9

For two years, Napoli boasted the most dynamic, the most entertaining, the most thrilling three-pronged attack in European football. Chris Nee, a football journalist with an exceedingly keen eye, notes, the Three Tenors "helped Napoli to treat their achievements together in the blue shirt as a springboard, not cause for celebration in their own right."[19] But while the club may not herald their achievements, most Vesuviani speak passionately about their squad's style from 2010–2012, and long to see such flair consistently demonstrated on the pitch once more.

Edinson Cavani arrived in the summer of 2010, the center prong of the trident that featured Ezequiel Lavezzi racing up the left and Marek Hamšík coolly advancing down the right. Such a simple description, however, does not convey the incredible alchemy the three achieved on the football field. From their first start together, the second leg of the Europa League playoff match against Elfsborg, the chemistry was clear. Cavani built off the 1–0 victory Lavezzi had orchestrated in the first leg, scoring twice in the first half for a 2–0 win, both goals reflecting El Pocho and Marekiaro's awareness of where central midfielder Walter Gargano was and their ability to predict where Cavani would pop up next.

The Three Tenors were similar in some ways. They often executed overlapping runs, switching positions in a way that would later be perfected by Hamšík working with Dries Mertens on the left and José Callejón on the right. All three were capable of tracking back, and it was most impressive to watch Cavani, with his classic center-forward build, slide in to prevent the ball from crossing the line. And all were able to predict, practically perfectly, where the other two would be when the ball was in play.

It was when the three worked in tandem that their differences came to the forefront, highlighting their best traits and downplaying their few weaknesses. Cavani, the new superstar, was the poacher, scoring 26 times in the league during his first season, breaking Antonio Vojak's nearly 80-year record of 22 league goals when he scored his impressive hattrick against Lazio in April. *El Matador* went on to score 23 league goals the next, and an incredible 29 times in 34 outings in 2012–2013, the first from Napoli to secure the league Capocannoniere title since Diego Maradona.

While Lavezzi, much like his compatriot *El Diego*, could score amazing goals from tricky, dribbling runs, after Cavani arrived, he became more of a workhorse—albeit one who was impossible to look away from. Contrary to his nickname, El Pocho had pace aplenty, and flew up and down the sides of the pitch. His seemingly boundless supply of energy, combined with his ability to shake off defenders, made Lavezzi seem indispensable. With his dramatic flair and model-like looks, Lavezzi simply oozed charisma, and was a joy to watch. Even better, his synchronicity with Cavani and Hamšík looked as though it would propel the Azzurri to even greater heights. It seemed Napoli had finally found another Argentine who could help lift them to the scudetto once more. Alas, El Pocho flitted off to PSG in 2012, leaving a trail of broken hearts in his wake.

Surprisingly, perhaps, for those who watched Napoli in l'era Mazzariana, it is Hamšík who went on to become a true legend at the club. The Slovakian does not have Cavani's goalscoring abilities or Lavezzi's flair, making it difficult for outsiders to understand exactly why the Partenopei faithful adore him. Those who watched Napoli often enough could appreciate his skill, of course. He is a true maestro, orchestrating play, mapping out moves five steps ahead, as though taking on a chess grandmaster. His passing is impeccable and his use of space nearly incomparable. None of this, however, is evident at first glance. Marekiaro, like Naples itself, revealed himself slowly, layer upon layer, giving people reason after reason to fall in love. Hamšík gave that love right back, admiring and adoring the city and its people. More than that, he became a symbol of the new Napoli during his nearly 12 seasons at the club. Marek Hamšík, more than any other player or manager, even more than the owner, is the one who ensured that, since Napoli's re-entry into Serie A in 2007, they could fight for the scudetto, lift the Coppa Italias, and battle it out in Europe—and look damn good as they did so.

The 2010–2011 season was indisputably a fantastic one for a team that had so recently rebuilt themselves. Now, the challenge was to put together a squad that could not only continue to dominate in Serie A, but also to progress in the Champions League.

Napoli threw caution into the wind and leaned into the approximately 40 million euros[20] they'd earn from making it to the top European championship to bring in a host of new players, along with personal investment by Aurelio De Laurentiis. Having sold Fabio Quagliarella to Juventus for 10.5 million euros (a rather low figure, given he went on to score 23 goals in 84 appearances for the Old Lady) and Víctor Ruiz, (a rare misstep they'd brought in the previous January for 8.5 million, and who played only six times in a Napoli shirt) to Valencia for 8 million, Napoli did have some funds to work with. But certainly not enough to suit the big dreams of Mazzarri, Bigon, and De Laurentiis. In fact, the amount earned from players sold barely covered the 12 million euros owed to Palermo to officially purchase Edinson Cavani.

But no matter. The squad needed to be reinforced, and Napoli were going to need to spend some funds to do so.

The midfield needed an overhaul, as several players who often started matches—Michele Pazienza, Hassan Yebda, and José Sosa—were sold, left on a free transfer, or reached the end of their loan period. Luca Cigarini (who had just returned from a loan to Sevilla) and Raffaele Maiello were loaned to Atalanta and Crotone, respectively. Napoli replaced them by purchasing Gökhan Inler from Udinese and Blerim Džemaili from Parma while picking up both Marco Donadel and Mario Santana from Fiorentina on free transfers. The Inler signing showed ADL's theatrics. After signing him from Udinese for 13 million euros, the club arranged a press conference where they brought out a player wearing a full lion mask (keep in mind the lion is not a Napoli mascot). Then they turned him to reveal the number 88, which Inler had worn at Udinese, after which they allowed him to take off the mask to raucous cheers.[21] According to the Swiss international, the mask was a spontaneous decision that showed off De Laurentiis' sense of humor:

> We were standing around after I had completed my medical and he was wondering out loud what kind of presentation we should have . . . De Laurentiis turned around and saw a lion's mask in the changing rooms. His face lit up and he said: 'That's our presentation. That is our surprise for the fans.' To begin with I didn't want to do it at all. I am quite a reserved person and don't want to make a big deal of it. But he is very infectious and in the end I went ahead with it. The fans loved it.[22]

While Napoli brought in several other players that season, no one else received the full-lion treatment. Center-backs Miguel Britos, purchased from Bologna, and Federico Fernández, from Argentine club Estudiantes, were meant to shore up the defense and provide extra cover when needed. Eduardo Vargas and Goran Pandev were both brought in as "second strikers," but came nowhere close to Cavani's talent or his chemistry with Lavezzi and Hamšík. Pandev, at least, had a poacher's instinct when coming off the bench. Vargas, however, was a hasty error committed by a club desperate to pull in more talent. Purchased from Universidad de Chile for 13.5 million euros, he made only 19 appearances in 3 seasons with the club; the majority of his time was spent out on loan.

Entering at the group stages of the Champions League and the knockout phase of the Coppa Italia meant that Napoli's season began with Serie A play (although the start of the championship was delayed a week as players went on strike due to a dispute over their collective bargaining agreement[23]). The Vesuviani kicked off with a 3–1 victory at Cesena before beating reigning champions Milan by the same scoreline at the San Paolo, a Cavani hattrick setting off excitement in the crowd about the new season. In between, Napoli had traveled to Manchester and came away with an impressive 1–1 draw against City in the first Champions League game, the goal courtesy of Cavani again. Napoli then lost to Chievo Verona away, 1–0, and played host to goalless draw with Fiorentina a few days later, Then came a 2–0 victory over Villarreal in the Champions League at the San Paolo, a 3–0 win at Inter, a 2–1 loss at Parma, and a 1–1 draw at home to Bayern Munich. In the incomparable words of Google Translate, Napoli were "inconsistent due to insufficient incisiveness." Something was lacking. The team sat 7th at the end of the *andata*, with seven wins and eight draws. Meanwhile, Juventus hadn't lost a game all season.

However, Napoli had made it through to the Champions League knockout stages. To finish out the group stage, they had lost 3–2 to Bayern in the return fixture (the brace came in the surprising form of Federico Fernández) but beaten both Manchester City (2–1, both goals coming from Cavani this time) and Villarreal (2–0, goals from Inler and Hamšík). This put them 2nd in their group with 11 points, two points behind Bayern and one above City.

In the league, Napoli went on a nine-game unbeaten run after losing to Genoa 3–2 to kick off the ritorno. They were only stopped by the unbeatable Juventus. In time, Napoli would have their revenge. Unfortunately, they would have no way to get revenge on Chelsea for an exceedingly painful Champions League loss. Napoli surprised much of the world when they beat Chelsea 3–1 in the first leg, with a brace from Lavezzi bracketing a Cavani goal. Chelsea returned the favor in London with a 3–1 victory. Gökhan Inler had rescued Napoli with his goal, leaving the teams tied 4–4 and each with one away goal. The game moved into extra time, where a goal from Branislav Ivanović in the 105th minute broke the hearts of Napoli fans worldwide. Chelsea ended up lifting the trophy that year.

So Napoli were out of the Champions League (although their performances were commendable) and had crushed supporters' dreams in the league by finishing 5th, a single point behind Lazio and three behind Udinese, against whom they had the head-to-head advantage. Turning just one loss into a win could have put them in the Champions League again. It had now become more difficult for an Italian club to enter the tournament proper; a drop in Italy's UEFA coefficient ratings meant that, beginning in 2011–2012, only the top two clubs entered at the Champions League group stage, while the third entered at the playoff round prior to the groups. The 4th-placed team would now enter the Europa League.

What Napoli had was the Coppa Italia. They positively breezed through to the final, beating Cesena 2–1 and Inter 2–0 in the first two knockout stages. They met Siena, a side who had battled relegation for most of the season, in the semifinals. Surprisingly enough, Napoli lost in Tuscany 2–1, with Napoli's goal actually an own goal by Siena defender Emanuele Pesoli. Pesoli would really come to rue that mistake. He wouldn't be the only one. In Naples, in front of a crowd of 56,000—more than 10 times the number who had turned out to watch Siena—another Siena player slipped up. This time it was Simone Vergassola, who put Napoli ahead 1–0 in the 10th minute.

A goal by Edinson Cavani 20 minutes later wrapped up the scoring, ending Siena's dreams of a Coppa final. They had own-goaled their way out of the tournament, resulting in a 3–2 Napoli victory.

Napoli's reward was a final against Juventus. The Old Lady finished their unbeaten, title-winning season in Napoli's house with a 2–0 victory over the hosts. The Coppa Italia final took place in Rome a week later, and Napoli were out for blood. The Olimpico was packed solid with supporters excited to watch the rivals face off. Everyone expected the match to be a tense but exciting affair, with both teams seeking to attack from the start. Juventus would win, of course, but Napoli would make it difficult for them.

In fact, it was the opposite: The Three Tenors—Lavezzi, Hamšík, and Cavani—made life difficult for the *Bianconeri* from the off. The Napoli defense held firm, springing the Napoli counterattack, whose pacey maneuvers worked the ball to one of the front three. Napoli missed a few good opportunities in the first half, but Juventus also gave them a scare once they realized targeting Salvatore Aronica would lead to chances on goal. In the second half, Lavezzi latched on to a loose ball in the penalty area and was felled by Juventus goalkeeper Marco Storari. Cavani stepped to the spot and neatly converted to give Napoli the 1–0 lead in the 63rd minute.

Juventus surged forward, determined to find an equalizer, but Morgan De Sanctis made the Napoli goal his personal fortress. This provided the Partenopei with the opportunity to quickly move the ball forward once again, and it ultimately fell at Hamšík's feet inside the area. The attacker coolly slotted the ball past Storari and into the bottom right corner to make it 2–0 in the 83rd. Former SB Nation editor Graham MacAree wryly observed, "There was no doubt about the victory at that point, and Juventus made their loss even more embarrassing when Fabio Quagliarella was sent off for an elbow to Aronica's face."[24] Before the whistle blew, Walter Mazzarri and others were already celebrating on the sidelines. Napoli had done what was assumed to be impossible—defeated the mighty Juventus, on neutral turf, during a season in which the Bianconeri had never lost a league match. In the process, they lifted their first trophy in 20 years.

Part of the pantheon

While MacAree predicted a breakup of the Napoli squad prior to the start of the 2012–2013 season, the team remained mostly intact. The biggest loss was Ezequiel Lavezzi, part of the famous attacking trident. He was sold to PSG for 30 million euros, amidst rumors his partner wanted to leave Naples after the two of them were robbed (unfortunately, Napoli stars are all-too-commonly victims of crime in Naples). Walter Gargano, an important cog in Napoli's midfield, was loaned to Inter for 3.25 million, while Federico Fernández was loaned to Getafe in Spain. Napoli bought Valon Behrami from Fiorentina for 8 million, to replace Gargano and make the midfield "more rough and physical." Napoli also worked out a loan deal with Udinese for left-back Pablo Armero. The two other midfielders bought to strengthen the center of the pitch never quite worked out; fortunately, Napoli had spent just 3.2 million to bring in Omar El Kaddouri from Brescia and only 1.5 million to snag Giandomenico Mesto from Genoa. So while the squad did experience some turnover that summer, as is to be expected, the Vesuviani were not picked apart and scattered among big-name clubs.

Perhaps most importantly, Lorenzo Insigne came back from his loan at Pescara, where he and Ciro Immobile had won the Serie B title the season before (Marco Verratti lent assistance, operating as a *regista* in this system rather than a *trequartista*). Pescara coach Zdeněk Zeman encouraged them to go all out in attack, resulting in 18 goals from Insigne, who was "regularly cutting in from the left to bend right-footed shots into the far corner."[25] However, while Lorenzo had 37 appearances in all competitions, he scored just 5 times and made the hometown fans unsure of his homegrown talent.

The 2012–2013 season began with the Supercoppa Italia between league winners Juve and Coppa winners Napoli. It was held on August 11 in Beijing, a move to try to garner more Asian fans; since 2011, the Supercoppa has been played overseas (exempting 2020, when Napoli and Juventus met again, in Reggio Emilia, and 2021, when Juventus visited Inter at their home stadium, the San Siro in Milan—naturally, the home side won—both due to the global pandemic).

Napoli more or less made fools of themselves in China. It wasn't so much the game itself, but the actions surrounding it that made Napoli look bad (or passionate, depending on whether one supported the club). The scoring began with a goal from Cavani in the 27th minute. The Napoli lead was canceled out ten minutes later thanks to a shot from Kwadwo Asamoah. Under the cover of pouring rain, Goran Pandev put Napoli ahead 2–1 going into the break, thanks to a delicious little lob that fooled both Leonardo Bonucci and goalkeeper Gigi Buffon. The communication between Bonucci and Buffon seemed broken, strange considering Bonucci had joined the Juventus defense two seasons before. But luck was on the Old Lady's side. The referee spotted contact between Juve forward Mirko Vučinić, who had come on at the break, and Napoli defender Federico Fernández, who had taken the place of Paolo Cannavaro barely ten minutes earlier. Arturo Vidal stepped to the spot and equalized for Juventus yet again, 2–2. The 74th minute goal pushed the game into extra time, and Napoli into exhaustion. They were down to just nine men. Pandev was shown a straight red in the 85th, for saying "one word too many." Juan Zúñiga received a second yellow in the 93rd for a "blatant and naive foul" on Sebastian Giovinco. The Partenopei was also without a coach, as Walter Mazzarri was shown a red for protesting Zúñiga's expulsion. The weakened team was unable to hold back the Juventus attack, consisting of Giovinco and Alessandro Matri, with Claudio Marchisio directly behind, and only Giovinco worried about being shown a second yellow. Disaster struck in the 97th minute, when Christian Maggio got on the end of a Juventus free kick by Andrea Pirlo and headed the ball past Morgan De Sanctis for a spectacular own goal. Though Juventus were up a goal, the attacks kept coming. Marchisio neatly threaded a pass through to Vučinić, whose 102nd-minute goal sealed the win for Juventus, 4–2.

In protest of the refereeing by Paolo Mazzoleni, Napoli refused to attend the award ceremony and announced a press blackout.[26] The FIGC punished the club, but only financially—they would need to pay 20,000 euros, while owner De Laurentiis was fined an additional 25,000.

Napoli began the Serie A season with a seven-game unbeaten streak, recording just one draw, when they were kept goalless by Southern rivals Catania. With 19 points by Round 8, Napoli were feeling confident going into their match in Turin, despite the Juventus win in the Supercoppa. Alas, the meeting resulted in Napoli's first loss of the season, thanks to late goals from Martín Cáceres and Paul Pogba. The Partenopei beat their bogey team, Chievo Verona, in the next round, but were beaten in Bergamo by a briefly surging Atalanta side, 1–0. Napoli finished the andata with eight unbeaten; the only team they lost to was Inter, 2–1 at the San Siro, just a week after beating Pescara 5–1. Mazzarri time mattered little in the first half of the season, with late goals coming only in matches where Napoli already held the lead, such as the first match, a 3–0 win at Palermo, when Cavani scored against his former team in the 88th minute, or the last match, a 4–1 triumph over Roma, when Maggio scored in the 90th. In the ritorno, they went 8 matches unbeaten (11 when the first part of the season is included) and even drew 1–1 with Juventus before coming up against the bogies and losing 2–0 to Chievo, with Cyril Théréau, a thorn in Napoli's side for years, scoring the second. Napoli responded with a nine-match unbeaten run, with last-minute goals from Cavani against Parma (2–1), Hugo Campagnaro at Lazio (1–1), and Lorenzo Insigne in front of the home crowd against Cagliari in the 94th minute (3–2). Their final match, a 2–1 loss at Roma, mattered little. Juventus had already clinched the scudetto, and Napoli, in 2nd, could not be touched by Milan.

A 2nd-place league finish meant that Napoli did not put much effort into secondary tournaments. The defending champions lost their first Coppa Italia match 2–1 to Bologna, with Mazzarri time working against them, as Panagiōtīs Kone scored the winner for the *Rossoblù* in the 91st minute. In the Europa League, the *Ciucciarelli* took 2nd in their group, behind Dnipro Dnipropetrovsk, despite the fact that they had beaten the Ukrainian side 4–2 at the San Paolo, all four goals coming from Cavani; the final two were scored after the 85th minute. But Mazzarri time couldn't save them against Viktoria Plzeň. The Czech club scored three at the San Paolo unanswered, and then came away with a 2–0 win at home, knocking Napoli out of the tournament with a 5–0 aggregate score.

But Napoli had secured Champions League play once more, and now it was time to see how they would do without Walter Mazzarri. Crucially, the fans would also learn what the club would do without star striker Edinson Cavani.

For the 2013–2014 season, De Laurentiis hired Rafael Benítez to help Napoli win the Serie A title. At minimum, it was expected he would ensure Napoli win at least one trophy. Almost 30 players left when he was hired, although not out of protest;

most had come to the end of their contracts and were able to leave on a free transfer. The lack of money coming in from those players leaving would have hurt Napoli if it weren't for Paris Saint-Germain buying Cavani for 64 million euros, paying off a release clause Napoli believed would never be met.[27] Yet Benítez, De Laurentiis, and the rest of the management team spent 107.5 million bringing in new players. The departures of Morgan De Sanctis to Roma for a mere 500,000 euros and Andrea Dossena to Sunderland and Hugo Campagnaro to Inter on free transfers did nothing to bridge that gap.

But this was a team that desperately wanted its first scudetto in nearly 25 years, so De Laurentiis opened the coffers. The club purchased big names like Gonzalo Higuaín, José Callejón, Dries Mertens (attackers), Raúl Albiol (defender), and Pepe Reina (goalkeeper), many of whom Benítez had encountered in Spain. Jorginho (midfielder) and Faouzi Ghoulam (left-back) joined the club during the January transfer window. Napoli spent 38 million on Higuaín alone. The flavor of the squad had certainly changed, and many wondered if such an overhaul would help Napoli, or hurt their chances due to having a squad filled with big personalities that were unable to gel in time to be successful, managed by a man carrying his own big ego and determined to reestablish his worth.

Napoli began the Serie A season with a seven-game unbeaten run, which came to a screeching halt when they visited Rome and lost to the *Giallorossi*, 2–0. Prior to the loss, Napoli had scored 18 goals, proving they were still the team to watch for those who wanted exciting football, and had been nipping at the heels of Roma, in 1st place. With ten wins and ten games, Roma had left both Napoli and Juventus five points behind. But, as Roma are wont to do, they began to slide, and by the end of the andata, Juventus topped the table, with Roma two points behind and Napoli, in 3rd, ten points adrift.

Meanwhile, the Champions League tournament was extremely painful for Napoli. In a group with Arsenal, Borussia Dortmund, and Olympique Marseille, it was always going to be tough to advance. Napoli beat Marseille twice (5–3 on aggregate), and emerged victorious over Arsenal (2–0) and Dortmund (2–1) at the San Paolo, while losing in England (2–0) and Germany (3–1). Dortmund, Arsenal, and Napoli all finished with four wins and two losses; in other words, twelve points each. To decide who would advance, the teams were ranked by their head-to-head goal difference, omitting Marseille from the equation. The group ended with Dortmund (+1 goal difference) and Arsenal (0 gd) moving on to the Round of 16, Napoli (-1 gd) dropping down to the Europa League Round of 32, and Marseille being knocked out of European competition altogether. This was especially frustrating for Napoli given that in six out of the seven remaining groups, the 2nd-placed team had scored fewer than twelve points.

In the second half of the season, Napoli lost just three times, and even beat Juventus 2–0 (one of just two defeats the Old Lady suffered) with goals from newcomers

Callejón and Mertens. They finished their season with a 5–2 win at Sampdoria and a 5–1 win over Hellas Verona at the San Paolo. But Juventus were untouchable that season, finishing with 102 points and a +57 goal difference. Roma, with 17 fewer points, finished 2nd, putting Napoli 3rd and into the Champions League playoff round.

Prior to the season's end, Napoli played in the Coppa Italia final. On their way, they had triumphed over Atalanta 3–1, with a brace by Callejón scored on either side of an Insigne goal, and Lazio 1–0, with a 82nd goal from Higuaín. Their next opponent was Roma, who won 3–2 at home, despite the fact that Morgan De Sanctis had very kindly gifted his former team with a goal (the second was scored by Mertens). At the San Paolo for the second leg, however, Napoli dominated, winning 3–0 with goals from Callejón, Higuaín, and Jorginho. From the Coppa Italia matches alone, it was clear that Napoli had made excellent moves in the transfer market that season.

The Coppa Italia final against Fiorentina was set for May 3. It would be a day of both heartbreak and jubilation for Napoli supporters.

The match, like most Coppa finals, took place at Rome's Stadio Olimpico. Roma ultras have a strong rivalry with the Napoli ultras, and various groups of Roma fans (or presumed as such) were present near the stadium and clashed with the Neapolitan fans who showed up to support their club.[28] They also pretended to be Napoli fans in order to attack Fiorentina fans. It didn't help that Napoli had beaten Roma 5–3 on aggregate in the semifinals.

A few hours before the game, a number of Roma fans attacked a group of Napoli supporters on the outskirts of the city, and three Napoli fans were shot and injured — one, Ciro Esposito, hit by Giallorossi ultra Daniele De Santis, severely. Esposito would later die from his injuries; De Santis was eventually sentenced to 16 years in prison.[29] Kickoff was meant to be at 9pm, but the news spread quickly throughout the stadium, alongside a wave of confusion over what exactly had happened. When players entered the pitch for their warm-up, about 20 minutes before the game was meant to start, the fans were going wild, each thinking the other team's supporters were attacking theirs. Firecrackers, smoke bombs, and whistles rained down as the clubs requested the start be delayed.

Napoli captain Marek Hamšík approached the Napoli curva to speak to capi-ultra Gennaro De Tommaso. Genny "'A *Carogna*" (Genny the swine), the son of a Camorra member, wore a t-shirt reading "*Speziale Libero*" (Free Speziale), indicating his solidarity with the 17-year-old Catania fan Antonino Speziale, who had killed policeman Filippo Raciti in 2007. 'A Carogna was no stranger to violence and believed that Esposito had been murdered. Marekiaro explained that the shots had come from a Roma ultra and that no one had died, so there was no need to continue to disrupt their chances to beat Fiorentina. The Napoli ultras, trusting their beloved *capitano*, backed off, allowing play to begin. Given that Esposito eventually passed away, the score seems almost irrelevant, but Napoli did win 3–1, with a brace from Insigne and a late goal from Mertens. Rafa had secured a trophy, just as he had promised. Despite the Napoli win, Fiorentina fans hung banners when the Partenopei next came to town, expressing their sorrow for the death of Esposito.

The next season saw a dip in Napoli's fortunes, thanks to a poor season in the transfer market (sporting director Riccardo Bigon had left that summer) and Benítez's decision to accept an offer from Real Madrid before the season had ended (although he did not officially announce his move until June). The more foolish acquisitions of the summer—Jonathan de Guzmán from Villarreal for 7.25 million, Michu from Swansea on loan—were made on the advice of Rafa. Neither did much of anything during their time at Napoli. However, Napoli got one thing right during the window—they snapped up Kalidou Koulibaly from Genk for 8 million euros; he would go on to become one of the world's best defenders over the next decade.

Napoli began their season by failing to make it to the Champions League group stage, holding Athletic Bilbao to a 1–1 draw in Naples but falling 3–1 in the second leg of the qualifying round. That sent them down to the Europa League group stage. Unlike other managers, Rafa took the Europa League seriously, and Napoli topped their group with four wins, one loss, and one draw, for 13 points. That was just enough to edge out Young Boys, with 12 points, for the number one spot. Sparta Prague earned ten points while Slovan Bratislava failed to notch a single one, and both were eliminated from the tournament.

With Napoli in 3rd place at the end of the andata, Benítez chose to remain focused on Europe. The club had beaten Juventus on penalties in the Supercoppa in December, and that, combined with their league position, convinced the coach they could make it all the way to the final of the Europa League. They could even make a strong run in the Coppa Italia, Benítez felt, after they scored all five of their penalties to knock out Udinese (ironically, the *Zebrette* player who failed to score the final penalty, Allan, would soon join Napoli), and then took out Inter 1–0 with a 93rd minute goal by Gonzalo Higuaín. Back in Europe, they defeated Turkish club Trabzonspor 5–0 on aggregate, and felled Dynamo Moscow 3–1 over two legs in the Round of 16. Prior to the meetings with Dynamo, Napoli had drawn with Lazio 1–1 in the Coppa. But while Napoli were focusing on those matches, they went without a win in the league for five rounds, dropping to 6th. Then everything else started to crumble. On April 8, Lazio beat Napoli 1–0 for a 2–1 win on aggregate, wrecking Napoli's hopes for yet another trip to the Coppa Italia final. They did manage to get past Wolfsburg in the Europa League thanks to a 4–1 win in Germany, followed by a 2–2 draw at the San Paolo. That put Napoli in the semifinals, the furthest they'd progressed in Europe since winning the UEFA Cup with Maradona.

Unfortunately for Napoli, Rafa's focus on the cups backfired. Ukrainian club Dnipro Dnipropetrovsk arrived in Naples for the first leg of the Europa League semifinal, and while the hosts opened the scoring in the 50th minute through a goal by David López, a late goal from Dnipro striker Yevhen Seleznyov not only ended the game at 1–1, but also gave the Ukrainians a crucial away goal. Then, in a frustrating match in Kyiv, Dnipro booked their place in the Europa League final with a single goal from Seleznyov.

Napoli not only lost out on two trophies, but ended the season in 5th, meaning another year of Europa League play. Lazio, in 3rd, qualified for the Champions League playoff round with just six more points than Napoli. Given their last match was a 4–2 loss to the Biancocelesti, with a penalty miss by Higuaín and two late Lazio goals from Ogenyi Onazi and Miroslav Klose, missing out on 3rd was a difficult blow. Were the Azzurri aiming in the right direction?

With that question on Aurelio De Laurentiis' mind, the owner made a complete 180, pivoting from an experienced European coach concentrated on playing in a manner that would earn his squad at least one trophy, to a banker who had never played professional football, and had just come off a season in which he led Empoli to promotion to Serie A. And while the new coach wanted results, Maurizio Sarri was also determined to play brilliant, entertaining football. (Like Walter Mazzarri, Sarri is also a brilliantly entertaining individual. And like Mazzarri, Sarri huffed and puffed his way through the stress of Napoli matches. RB Leipzig were brilliant Europa League hosts in 2018, creating a temporary smoking room for the Napoli coach in the away team's dressing room.)

Sarri has a distinctive style ("Sarriball") focused on a strong back line, positioning his players up the pitch to better spring an offside trap and pressing them high to regain possession. His approach requires high pressing and intelligence on the ball and off, the need for each player to know where his other teammates are. His full-backs are given full rein and often make overlapping runs. He uses a deep-lying playmaker to direct the action, and expects the team to retain possession. Like the full-backs, he uses dynamic wingers willing to overlap, attacking from the flank with short, low passes. His usual formation is a 4-4-3; one in which players are not shackled to their positions, but encouraged to move around the pitch and play possession-based football.

However, Sarri is also known to be tactically inflexible, and once teams figure out his tactics, they can find ways to fight back. In addition, he does not like to rotate his players, and when he left Napoli after three seasons, some of the players were almost broken from the number of games they'd played during those years. But that first year, 2015–2016, his fun-to-watch style of play brought 20% more fans to the San Paolo, exciting the Napoli fans immensely as they watched their team attack and attack.

It also produced results; though the team had to first overcome some bumps in the road, they did so rather quickly. Napoli lost to Sassuolo in their first match of the season—a 2–1 comeback by the *Neroverdi* after Marek Hamšík scored in the third minute. The Azzurri then went on to draw 2–2 to Sampdoria at home—another comeback from the opposition after Gonzalo Higuaín had scored a brace. They then drew again—this time 2–2 with Empoli—although at least Napoli had the final word thanks to a goal from newly purchased midfielder Allan. With those blips out of the way, Napoli conquered Club Brugge 5–0 in their first Europa League group stage match a few days later, with a brace apiece for José Callejón and Dries Mertens, and a goal for Marek Hamšík, who'd been shifted back into more of a deep-lying playmaker position. The next Serie A round was just as satisfying, thrilling the San Paolo fans with another 5–0 win, this time over Lazio, and, in the spirit of Sarriball, spreading the goals throughout the squad—two for Higuaín, and one each for Allan, Lorenzo Insigne, and Manolo Gabbiadini.

It turned out scoring five goals would become something of a trend for this Napoli squad; they did it twice more in the Europa League group stage, against Midtjylland and Legia Warsaw. Needless to say, Napoli conquered their group with 6 wins and 18 points, setting a record for goals scored with 22. Unfortunately, their luck changed in the next round where they were knocked out by Villarreal, 2–1 on aggregate.

That just gave Napoli more time to focus on the league. After a disappointing goalless draw with minnows Carpi, in their first season in the league, the Partenopei went on to beat Juventus 2–1 at the San Paolo, with goals from Insigne and Higuaín. With only one loss after the initial defeat to Sassuolo—a disappointing 3–2 failure at Bologna—Napoli were crowned winter champions, a feat they hadn't accomplished since 1989–90, the season they won the second scudetto. While the winter title is a journalistic award not accompanied by a handsome piece of silverware to store in the trophy cabinet, finishing at the top at the end of the andata means, more often than not, that a team will lift the scudetto at the end of the season.

The end of the season was tense for Napoli—despite losing just three games in the ritorno, they were nearly neck-and-neck with Roma. Going into the last round, they knew they needed a win to ensure they'd hold on to 2nd place and thus start in the Champions League group stage. The last game was at home against Frosione, who had already been relegated, and managed to pick up a red card 15 minutes in. Still, the visitors kept Napoli fans on the edge of their seats until captain Marek Hamšík scored just before the break, setting off a deafening roar in the San Paolo. The entire crowd knew the meaning of that goal. Gonzalo Higuaín followed up with a 20-minute hattrick in the second half, not only securing the win but, with 36 league goals, breaking Gunnar Nordahl's record for most goals scored in Serie A season (35 in 1949–1950) and of course earning the title of league Capocannoniere—yet another Argentine hero.

But Napoli's hero turned out to be a villain. The mercenary Higuaín, currently the most despised former Azzurri player, sought a move to Juventus for the 2016–2017 season. At least the sale brought in 90 million euros, the highest transfer fee ever by an Italian club, but Sarri did not want to spend it on another pricey forward. Instead, the club spent 32 million to buy Arkadiusz Milik from Ajax, who then picked up a serious knee injury on international duty in October and was unable to play again until mid-February. Piotr Zieliński also joined in the midfield, as did Amadou Diawara, both of whom helped to strengthen the center. The former, a skilled and versatile player, had been under Sarri's tutelage while on loan at Empoli, but he would truly blossom later into his Napoli career, becoming one of the club's most reliable players.

The club signed a few others, spending money on those who had good reputations in Italy, but they never quite fit with the Sarri system. The best example was spending 18 million to bring Leonardo Pavoletti from Genoa midseason, looking for a replacement for Milik. However, Sarri had moved Dries Mertens into the center, using him as a false 9. It worked extremely well—the little Belgian netted 34 goals in all competitions—and rendered Pavoletti's services surplus to requirements. He appeared just six times in a Napoli jersey, meaning they paid 3 million each time he stepped on the pitch.

Though as previously stated, teams who had cottoned on to Sarri's tactics could find ways to best his squad. Napoli began the Serie A season with a six-game unbeaten run, scoring more than two goals per match. The run ended days after Napoli beat Benfica 4–2 in the Champions League group stage; in their first match, they had come back to triumph 2–1 over Dynamo Kiev. Then, days after losing 3–1 to Roma in the league, they fell 3–2 to Beşiktaş at the San Paolo, after coming from behind twice to equalize. Although the Coppa Italia wouldn't begin until January, this roller coaster made some fans uneasy as to how many tournaments Sarri could handle, particularly without implementing much squad rotation.

Napoli wound up topping their Champions League group, with three wins, two draws, and a loss, for a total of 11 points. But just their luck, they were up against Real Madrid in the Round of 16, and they didn't even have the advantage of knowing what Rafa Benítez would do, as he'd lasted just half a season before Real replaced him with Zinedine Zidane. The Spanish side easily outpaced and out-maneuvered Napoli in both legs for an aggregate score of 6–2.

Now, it was time to focus on the league. And the Coppa. Serie A didn't actually require that much attention; Napoli had moved into 3rd place in Round 17, and would remain there until the end of the season. In their 1st round of the Coppa Italia, the Vesuviani easily brushed aside Spezia, 3–1, in a rare instance of Sarri rotating out some of his starters. The next round wasn't quite as simple, however. It took 71 minutes of Napoli's famous attacking and pressing for José Callejón to score, lifting Napoli to a 1–0 victory. Then, as was nearly inevitable for Napoli, they were pitted against Juventus in the semifinals. Napoli had lost to Juventus in the andata but drew 1–1 with them in the ritorno, thanks to a goal by Marek Hamšík. However, the first leg took place before their second league meeting, and Juventus dismantled Napoli 3–1, with two penalties converted by Paulo Dyabla and, of course, a goal by Higuaín. Napoli's lone goal came from Callejón once more. The second leg, however, came just days after the Partenopei had held Juventus to a draw, and their renewed confidence showed. Hamšík, Mertens, and Insigne all scored. Unfortunately, the viper in their nest, the ultimate traitor, scored a brace. With the two goals from Higuaín, Juventus won 5–4 on aggregate, moving on to the final.

We had the greatest expectations

Perhaps it wasn't the most successful season for Napoli, now that they had recent seasons that had lifted them to 2nd in the league, carried them to the semis of European competition, and brought additional silverware to the trophy cabinet, but their 3rd-place finish and qualification to the Champions League playoff round was perfectly adequate. Even if another win would have lifted them above Roma, who had just one more point, and again into the Champions League group stage, Napoli fans should've been pleased. In addition, Napoli set a record for the most points they'd collected in a single season with 86, and they were the only team in Europe with four players who scored 10 or more goals (Mertens, Hamšík, Insigne, and Callejón).

Yet it seemed as though Napoli had forgotten those lean, hungry years, the difficulties after Maradona left, the bankruptcy and demotion to Serie C1. The supporters now expected consistent Champions League play, trips to the Coppa Italia final, world-class coaches, and—most of all—a decent chance at the scudetto.

The next season, Sarri again lifted Napoli to 2nd place, but now that the rules had changed so the top four finishers from the top four European football associations would automatically be entered into the Champions League group stage, it didn't seem to matter so much. What mattered is that they finished four points off Juventus, when they could have clinched the scudetto if it had not been for a mental lapse at the end of April.

No significant changes had been made to the squad. Sarri remained, the top scorers stayed put . . . only the players Napoli had gambled decent money on and lost left. The club paid 12 million euros to Bordeaux for Adam Ounas who made just 25 appearances in 2 seasons, then started being sent out on loan. The most they spent was 21 million for Nikola Maksimović, who had been an incredible defender for Torino, but ended up being rather inconsistent with Napoli, interspersing moments of brilliance with what sometimes looked like pure incompetence. They also picked up Mário Rui from Roma for what would eventually be seen as a bargain, but at the time he was often defensively unstable on the left side, and prone to selfishness.

Napoli failed to make it out of the Champions League group stage, earning just 6 points, while Manchester City topped the group with 15, followed by Shakhtar Donetsk with 12. After dropping down to the Europa League knockout round, Napoli were quickly eliminated by RB Leipzig. Though the score was 3–3 on aggregate, Leipzig's three away goals to Napoli's two allowed them to progress to the next round.

But the real story of the 2017–2018 Napoli season occurred on April 29, 2018. Juventus had lost 1–0 to Napoli in Turin the previous week, courtesy of a 90th minute goal by Kalidou Koulibaly. In Tuscany, the team was watching Juve play Inter the night before their match against Fiorentina. Inter were up 2–1, but Juventus came back and won in the final minutes thanks to an own goal and an 89th minute winner from Higuaín. This is said to have greatly damaged the Neapolitan side's confidence, already dinged from having learned Sarri was leaving at the end of the season. The side turned up at the Stadio Artemio Franchi looking bedraggled, and Giovanni Simeone took advantage, scoring a hattrick unanswered by Napoli. Funnily enough, some Fiorentina fans were upset by the result, given they were unlikely to finish in the European places, and they wanted *someone* to end Juve's title-winning streak. Most of the calcio world suffered right along with Napoli supporters that night, knowing that Juventus were about to win a record-breaking seventh Serie A in a row.

The calcio world had already come together in suffering earlier that season. On March 4, 2018, Fiorentina captain Davide Astori passed away in his sleep in a hotel in Udine, where the Viola were staying prior to their match against Udinese. Only 31 years old, he had died from cardiac arrest. The games to be played that day in the top three calcio divisions were postponed, and supporters from all teams grieved along with the Viola fans.

Maurizio Sarri's success with Napoli caught Chelsea's attention, and the manager departed for London in the summer. De Laurentiis decided that it was time for another world-class manager with an abundance of silverware to his name to come to Naples. Carlo Ancelotti would take his place on the Napoli bench for 2018–2019.

The former Milan coach had guided the team to two Champions League victories. He had also coached Real Madrid in 2013–2014, when they had lifted the same trophy. His most recent stint was at Bayern Munich, where he led his side to the Bundesliga title in 2016–2017. Ancelotti had captured domestic titles coaching in each of the five big leagues—Italy, England, Germany, Spain, and France. If anyone could get this Napoli side to win the scudetto, it was *Carletto*.

The big transfer news of the summer felt inevitable weeks before it was officially announced. Jorginho would follow Sarri to Chelsea, who were willing to pay 60 million euros for the midfielder. The money was spent on goalkeeper Alex Meret, defender Kévin Malcuit, midfielder Fabián Ruiz, and forward Simone Verdi. Meret, snagged from Udinese, traded off with David Ospina between the sticks. Malcuit, for whom they paid around 12 million to Lille, made 30 appearances in two-and-a-half seasons with Napoli, before being loaned to Fiorentina in January 2021. Verdi, signed from Torino for 25 million, scored 3 goals in 22 appearances. In 2019, he was sent back to Torino on loan for 3 million with a 20 million obligation to buy. Only Fabián Ruiz, bought from Real Betis for 30 million, can be said to have lived up to the expectations of the club—and the fans.

Known for tactical versatility and the ability to get the best out of his players, Ancelotti was also a fan of the forward press. However, his style was often slower than Sarri's was, and while Napoli may have finished 2nd, they did not come close to challenging Juve for the title. Nor did they play such exciting, attractive football. But Ancelotti did get results. A 2–1 victory over Lazio to start the season, with goals by Milik and Insigne had the Napoli fans eager to get to the San Paolo to watch their new tactician face off against his former side Milan. By the 50th minute, Milan were leading 2–0, and more than a few Napoli fans were already set to write off the coach, and the season (Vesuviani supporters have a tendency toward passion, theatrics, and pessimism, not necessarily in that order). But in the 53rd minute, Piotr Zieliński put Napoli on the scoreboard, in the 67th, he equalized for his side. Then, with ten minutes left to go, Dries Mertens completed the comeback, lifting Napoli 3–2 over Milan and causing those same fans to wonder if they'd been a bit hasty in their judgment. Although they lost 3–0 to Sampdoria the next week, an Insigne goal made a difference at home to Fiorentina, where Napoli won 1–0. From that 4th round on, Napoli remained in 2nd place for the rest of the season. Juventus officially won the title in Round 34. Although Milik had returned from injury to score 17, the most Napoli goals in the league, only 3 of those goals were game-changers. Mertens and Insigne were more able to change the game than the striker. Another piece would need to be found before Napoli could win the scudetto.

It didn't help that in February, Napoli icon and captain Marek Hamšík left quietly for China. At age 31, having worn out his body during the Sarri years and finding himself under-used in Ancelotti's teams, he saw no future for himself with the Partenopei, and had little confidence in Ancelotti being able to guide the side to a trophy. When Marek said his quiet goodbyes in winter 2019, he held the record for most goals and most appearances for Napoli, beating Maradona in both categories. After nearly 12 seasons, in almost all of which he played a decisive role in the squad, he was also a huge fan favorite, and many were heartbroken. Fans in Naples were particularly upset that he had left so quickly and that they hadn't had time to properly thank him for his service to the club. There is talk of a testimonial match to be played in his honor, but at the time of writing he had just won his first title, conquering the 2021–2022 Süper Lig with Turkish club Trabzonspor, and fans and press alike immediately began to speculate that the beloved Marekiaro could now return. Hamšík did use the *Players Tribune* platform to write a love letter to Napoli fans two seasons before, one that would make all but the most hardened tear up and convince them that their capitano must have hit the breaking point if he was leaving Naples.[30] (Unfortunately, almost the same situation arose with Hamšík's successor; Lorenzo Insigne, having finally realized he would never be enough for many hometown fans, announced he would be passing off the captain's armband in summer 2022, bound for Toronto FC.)

The 2018–2019 season was successful in some ways for the Vesuviani, in that they placed 2nd and qualified directly for Champions League play. However, they finished 11 points off scudetto winners Juventus, a team they could have challenged much more had they not dropped points against teams they should have beaten, like Torino, Bologna, Sassuolo, Genoa (who only survived by virtue of beating Empoli on head-to-head points), and bogey team Chievo Verona (who finished dead last in Serie A).

But something no one could have envisioned was intent on disrupting football the next season. It would infect all clubs and alter the way football was both played and consumed. It also gave supporters time to consider the direction they wanted their teams to take.

[1] Peredo, S. T., & Kostopoulos, P. (2020, July 27). What became of the Intertoto Cup? 12 years without the mythical summer competition. *Marca*. https://www.marca.com/en/football/international-football/2020/07/27/5f1f11cc46163fe3568b4585.html

[2] Calaiò in comproprietà. (2008, July 1). *AC Siena*. https://web.archive.org/web/20081116064458/http://www.acsiena.it/comunicatidetail.php3?id=6359

[3] Donadoni replaces Reja at Napoli helm. (2009, March 10). *UEFA*. https://www.uefa.com/insideuefa/news/01d7-0f85b16b5f65-a0a3cdd5da9b-1000--donadoni-replaces-reja-at-napoli-helm/

[4] Azzi, M. (2009, April 4). L' ordine di De Laurentiis tutto il Napoli sott' esame. *La Repubblica*. https://ricerca.repubblica.it/repubblica/archivio/repubblica/2009/04/21/ordine-di-de-laurentiis-tutto-il.html

[5] Cappella, F. (2009, June 1). Chievo battuto, Napoli contestato I tifosi abbandonano lo stadio. *La Repubblica*. https://ricerca.repubblica.it/repubblica/archivio/repubblica/2009/06/01/chievo-battuto-napoli-contestato-tifosi-abbandonano-lo.html

[6] Fabio Quagliarella al Napoli: "Si avvera il mio sogno." (2009). *SSC Napoli*. https://web.archive.org/web/20090603185300/http://www.sscnapoli.it/client/render.aspx?root=707&fwd=2533&content=0

[7] #8 Luca Cigarini. (n.d.). *Transfermarkt*. Retrieved September 21, 2020. https://www.transfermarkt.com/luca-cigarini/transfers/spieler/33218/transfer_id/296505

[8] Donadoni sacked. (2009, October 6). *Eurosport*. https://www.eurosport.com/football/serie-a/2009-2010/donadoni-sacked_sto2084649/story.shtml

[9] Pierpaolo Marino: "Ecco perchè lasciai il Napoli." (2015, September 29). *Calcio Mercato*. https://www.calciomercato.com/news/pierpaolo-marino-ecco-perche-lasciai-il-napoli-684236

[10] Cimmino, F. (2009, September 28). Ufficiale l'addio di Marino: Comunicato di Aurelio De Laurentiis e Pierpaolo Marino. *TuttoNapoli*. https://www.tuttonapoli.net/in-primo-piano/ufficiale-l-addio-di-marino-comunicato-di-aurelio-de-laurentiis-e-pierpaolo-marino-38166

[11] President hints at Donadoni sack. (2009, October 5). *Sky Sports*. https://www.skysports.com/football/news/11854/5608803/president-hints-at-donadoni-sack

[12] Donadoni sacked.

[13] Sergio Buso: "Tra Donadoni e la società Napoli mancava intesa." (2009, November 6). *Tutto Mercato*. https://www.tuttomercatoweb.com/altre-notizie/sergio-buso-tra-donadoni-e-la-societa-napoli-mancava-intesa-178430

[14] Vitale, F. (2009, June 1). Mazzarri lascia C' è Del Neri. *Gazzetta dello Sport*. http://archiviostorico.gazzetta.it//2009/giugno/01/Mazzarri_lascia_Del_Neri_ga_10_090601041.shtml

[15] Bruno F. (n.d.). *Commento Carlo Alvino autorete aronica (parz. Napoli Lazio 2 a 2)* [Video]. YouTube. https://www.youtube.com/watch?v=vxPG6X8aD2M&ab_channel=BrunoF

[16] Landolina, S. (2010). Napoli striker Ezequiel Lavezzi makes special allowances for Edinson Cavani. *Goal*. https://www.goal.com/en/news/11/transfer-zone/2010/07/15/2026877/napoli-striker-ezequiel-lavezzi-makes-special-allowances-for

[17] Current points to coefficient rankings. (updated 2022, May 28). *Football Coefficient*. https://www.football-coefficient.eu/

[18] UEFA Champions League and Europa League changes next season. (2018, February 27). *UEFA*. https://www.uefa.com/uefachampionsleague/news/0242-0e16a85014cb-5ffe48223b95-1000--champions-league-and-europa-league-changes-next-season/

[19] Nee, C. (2020, March 23). Hamsik, Lavezzi and Cavani: The three tenors that made Napoli sing. *Sphinx Football*. https://www.sphinxfootball.com/football-posts/2020/3/23/hamsik-lavezzi-and-cavani-the-three-tenors-that-made-napoli-sing

[20] Azzi, M. (2012, April 11). Champions League con Mazzarri. *La Repubblica*. https://ricerca.repubblica.it/repubblica/archivio/repubblica/2011/04/12/champions-league-con-mazzarri.na_034champions.html

[21] PTAFOOTBALLBLOG. (2011, June 11). Napoli unveil signings the fun way—Gokhan Inler [Video]. YouTube. https://www.youtube.com/watch?v=kGDd1ugpK0Y&ab_channel=PTAFOOTBALLBLOG

[22] Christianson, M. (2012, February 20). Chelsea beware—Edinson Cavani has faith in Napoli's success. *Guardian*. https://www.theguardian.com/football/2012/feb/20/edinson-cavani-napoli-chelsea-champions-league

[23] Italian players' strike delays start of Serie A season. (2011, August 26). *BBC*. https://www.bbc.com/sport/football/14672599

[24] MacAree, G. (2012, May 20). Juventus vs. Napoli, 2012 Coppa Italia final: Napoli cruise to 2–0 win. *SB Nation*. https://www.sbnation.com/soccer/2012/5/20/3032795/napoli-vs-juventus-2012-coppa-italia-final-final-score

[25] Potts, R. (2017, April 26). Zdeněk Zeman and Pescara: The enduring legacy of the 2011/12 promotion season. *Gentleman Ultra*. https://gentlemanultra.com/2017/04/26/zdenek-zeman-and-pescara-the-enduring-legacy-of-the-2011-12-promotion-season/

[26] Gaetani, M. (2012, August 11). Supercoppa alla Juve; Ma il Napoli è furioso. *La Repubblica*. https://www.repubblica.it/sport/calcio/coppe/2012/08/11/news/supercoppa_alla_juve_ma_il_napoli_furioso-40787300/

[27] Come, P. (2013, July 17). Una lettera d'addio per Cavani al Psg. *La Repubblica*. https://ricerca.repubblica.it/repubblica/archivio/repubblica/2013/07/17/una-lettera-daddio-per-cavani-al-psg.html

[28] Gunfire injures three Naples fans ahead of Italy final. (2018, May 3). *BBC*. https://www.bbc.com/news/world-europe-27269959

[29] Roma, morte Ciro Esposito: definitiva la condanna a 16 anni per De Santis. (2018, September 28). *Corriere della Serra*. https://roma.corriere.it/notizie/cronaca/18_settembre_25/roma-morte-ciro-esposito-definitiva-condanna-16-anni-de-santis-6c4dc14e-c0ea-11e8-8c2f-234b69fe8a3d.shtml

[30] Hamšík, M. (2017, May 29). For Naples. *Players Tribune*. https://www.theplayerstribune.com/articles/marek-hamsik-napoli

Finale

L'ho provato sulla mia pelle

I have experienced that in my own skin.

I am a young fan in terms of my years following Napoli. I adopted the side in 2010, after the pleasure of the Maradona years, after the pain of the club falling to Serie C1, after the joy of watching them rise back up to Serie A.

But what I have been lucky enough to understand is perhaps the most important thing about Napoli, in all its forms throughout its history: the uniqueness of the club and the city in which it inhibits creates such a marvelous maelstrom that not only fans, but the players themselves do not feel able to extract all that is *Azzurri* from their hearts.

It is ironic that I write this final chapter as Lorenzo Insigne accepts a move to Toronto FC, a Major League Soccer (MLS) club based in Canada. Years ago, after starring for a subversive Parma side, Sebastian Giovinco transferred to Juventus, after which he accepted a move to Toronto, aged just 28. At the time, accepting an offer from an MLS side was tantamount to a player declaring his career to be over. And even seven later, Insigne faced criticism from many of the brightest voices in European football, questioning why a 30-year-old who clearly had a number of good years left in his legs would leave for an "inferior" league.

Regardless of whether soccer fans feel that MLS is a retirement league (an argument far too detailed and, yes, too boring to recount in a book about Napoli), Insigne made the correct and—I would argue—best decision he could when deciding to leave his boyhood club. What many do not understand is that Napoli fans, particularly those in and from Naples itself, are harshest on the players from the area. Although Napoli does not have one of the best academies in *calcio*, fans still expect those who come through the club to possess great talent. More than that, they must demonstrate, week in and week out, how they give their all for Napoli, both on and off the pitch. A *Napulitano* must speed, not trot, and heaven forbid not walk. He must execute every tackle, connect with every pass, hit the back of the net each time. He cannot let his smile drop, even when he fails to do one of these things and the fans jeer in response. He is required to trot out the correct lines at every press conference and media appearance—not just about the club, but about the city, too. And he can never, ever appear to be entertaining an offer from another team, especially another Italian team.

To his credit, Insigne never appeared to seriously consider transferring to another Serie A side, despite being linked with big names like Inter and Juventus.

Throughout the course of Napoli's history, staying with the club was not unusual; Antonio Juliano and Ciro Ferrara, both native Neapolitans, made more than 200 appearances in the Azzurri shirt. But the football world has changed, and not just in that it is much more difficult for a son of Naples to be showered with love and affection. A quick internet search will show that the vast majority of players with more than 200 games with one club signed on with that club before the 21st century began. At Napoli, however, the turnover rate since 2006, when the team returned to Serie A after its bankruptcy in 2004, has been remarkably low. Despite not having won the *scudetto* since 1990, despite playing under an often eccentric, often difficult owner, despite living in one of the poorer cities to boast a team in one of Europe's top five leagues, players at Napoli *want* to stay.

And that, my friends, has made all the difference.

Of the Napoli squad who returned to Serie A in 2007–2008, 11 of the 29 had played 100 or more league games with the club. In Walter Mazzarri's first full season in charge, 2010–2011, 13 Napoli players would make more than 100 appearances. The squad that Rafa Benítez took over and made his own in 2013–2014 had 8 players with 100 or more appearances, as well as 5 with 200 or more. Six players on Maurizio Sarri's team that finished 2nd in 2016 made 200 appearances in the Napoli shirt. The next season, it was seven players, with seven more that appeared in at least a hundred games for the *Partenopei*. But after Sarri left for English Premier League club Chelsea in the summer of 2018, after a legitimate title challenge and another 2nd-place finish, the gang began to break up. Defensive midfielder Jorginho, with 160 appearances, followed Sarri to Chelsea, as veteran trophy-collector Carlo Ancelotti came in to coach Napoli. Goalkeeper Pepe Reina left for Milan after donning the gloves for Napoli 139 times. Loyal right-back Christian Maggio went to Serie B side Benevento after a decade in Naples, and more than 300 appearances. To the heartbreak of many (particularly this author), captain Marek Hamšík suddenly left in February 2019, after setting the record for all-time appearances with Napoli at 520, and beating Maradona's number of goals scored (115) by netting 121 in all appearances.

The next to go, in July 2019, as Gennaro Gattuso built the squad to his satisfaction, was Spanish central defender Raúl Albiol, seen in 236 games, a figure that would've been higher had he not frequently succumbed to injury. In summer 2020, José Callejón, loyal and respected enough to take the pitch for Napoli 349 times, went to Fiorentina on a free. Premier League club Everton paid 25 million euros plus bonuses for Allan, a tremendous central midfielder who appeared more than 200 times for Napoli. Arkadiusz Milik, whose injuries had blighted a player whom Napoli had hoped would be the second coming of Edinson Cavani, still made over 100 appearances for the club, but was sent to Marseille on loan, with an obligation to buy. And then we arrive at 2021–2022, when at first it appeared new manager Luciano Spalletti would make few significant changes to the squad. But Aurellio De Laurentiis had other ideas, and by December Napoli fans knew Insigne would be off to Toronto

FC come summer 2022, while Dries Mertens, who gave his heart and soul for both Napoli and Naples, would be let go on a free transfer.

For Napoli fans, much of this is heartbreaking. The 2021–2022 season began on a high, the first time in years that it seemed the team could challenge for the scudetto right from the off. No crucial players had left; defender Elseid Hysaj had been an important contributor, but his talent had tapered off after Sarri moved on. The squad won six out of seven of their preseason friendlies. They won their first eight Serie A matches and went undefeated until Round 13, remaining top for two more rounds. Napoli finished 2nd in their UEFA Champions League group, moving on to the Round of 16. But as often happens when Napoli plays in Russia, things went downhill around the time of their game in Moscow—they lost 3–2 to Inter the week before, then went on to lose 2–1 to Spartak, and came home to pick up just four points from four games, bracketed around the must-win Champions League match against Leicester City, where they were successful.

Frissons began to develop, however. Seemingly half the squad contracted COVID-19, forcing Napoli to field half their bench for various matches. It became clear that 30-year-old Insigne was being forced out. Dries Mertens, who may have been 34 but was still hitting the back of the net, as well as being a force in the locker room, would be let go on a free. There was even talk of tremendous center-back Kalidou Koulibaly being sold. Midfielder Fabián Ruiz, with 150 Napoli appearances, was a hot commodity in the January transfer market; Manchester United paid particular interest.

Yet the majority of the fanbase harbors no ill will to those who have left the squad over the past decade. Gonzalo Higuaín was an exception. The forward is a mercenary who followed Rafa Benítez, then answered Juventus' siren song after scoring a hattrick in the final game of the 2015–2016 season, putting Napoli into the Champions League group stage and winning the league's *Capocannoniere* title by netting 36 goals. Any player who leaves for Juventus will earn the Napoli fans' ire, but having done it when it seemed that the team was in a position to challenge for the title the following season caused them to feel utter contempt for Higuaín. In fact, toilet paper bearing the disgraced star's face quickly became a hot commodity. Most other players, however, retained the supporters' respect, as most know their reasons for leaving were not solely about money or trophies.

What matters to the fans is that these players have not just become the shirt, but have become Napulitano. They have taken both Napoli the club and Napoli the city into their hearts. They have experienced them in their own skin. And when talking about having the chance to stay, or the possibility of leaving, their love for all things Partenopei glows like the lights on the ships in the harbor.

This is what the people of Naples wanted from Higuaín. He was another Argentine, a supposed second coming of Diego Armando Maradona, there to not only lift Napoli but become an adopted citizen, just as his compatriot had. It's been suggested that the way the city opens its arms to Argentines was one of the factors that convinced Higuaín to turn down other lucrative offers.[1] Surely he felt smug as hundreds of Neapolitans met him at the Rome airport, while at the same time, former Napoli star Edinson Cavani was greeted in Paris by only a few reporters covering Paris Saint-Germain. According to calcio journalist James Horncastle, Cavani, who had been with Napoli for three seasons and was the side's last Capocannoniere, was jeered from Naples as coverage of his arrival streamed, "Look what you're leaving behind, Edi! . . . You were a God in Naples. Now you're just a regular guy in Paris."[2]

Cavani was a true god of Naples, so much so that pizzerias named pies after him, bakeries constructed cakes of his visage, baristas even formed his face in the foam atop cappuccinos.[3] He never received that sort of attention in Paris, where he could actually walk the streets without the fans who "breathed football" chasing him down to smother him with adulation. Yet his leaving never caused Napoli fans to rage as they did when Higuaín jumped to Juventus three years later. Nearly a decade after he said goodbye, now aged 34, supporters still clamor for Cavani's return during every transfer window. He's still beloved because he didn't necessarily want to go. Aurelio De Laurentiis had inserted a 63 million euros release clause into his contract, believing it unlikely to be met—but he did not consider the billions Qatari-backed PSG had at their disposal. When Edi left, he took out a full page ad in *Corriere dello Sport*, thanking both Naples and the Napoli faithful.[4]

They may love him even more because he said, upon returning to Naples in 2021 to pay tribute to Maradona, "I will always be grateful to Napoli and their fans. That's why I rejected other Italian clubs like Juventus or Inter. I couldn't do that to the Neapolitans."[5] A clearer dig by a Uruguayan to an Argentine can rarely be found.

Even players who barely saw the pitch wanted to remain at Napoli. Duvan Zapata, bought by Napoli in 2013, remained constantly overlooked by Rafa Benítez. Despite four appearances in the 2014 Europa League group stage, in which the Colombian scored twice and provided one assist, he remained relegated to the bench. While Zapata eventually made his way to Atalanta, scoring 78 goals in 3-and-a-half seasons, while he was waiting for a chance to shine at Napoli, his agent made it clear that there was where he wanted to remain.[6] It also didn't require that a player remain in Naples for quite some time for him to become devoted to the city. Jorginho, brought in at the same time as Zapata and also overlooked by Rafa, kicked off the 2014–2015 season by playing football in front of a cathedral with a group of boys, who were astonished when he asked them for the ball, then posed for photos after dribbling it around.[7] Upon following Sarri to Chelsea in summer 2018, Jorginho said:

> I will miss Naples a lot, after four years of living here . . . I will remain bound to this city. I will give your hellos to Mister Sarri. Who will play in my spot? Eh, that's a good question (laughs).[8]

Jorginho could well have been modeling himself on Napoli *capitano* and hero Marek Hamšík who, in the summer after he helped Napoli to a 3rd-place finish, scoring 11 goals and 6 assists in that 2010–2011 season, played football with a group of kids and showed no mercy. The Slovakian scored against the children after about 20 seconds, then celebrated his goal by running through the small shop next to the "pitch."[9] Marek was welcomed by Napoli fans from the moment he arrived in Naples. As he wrote, "'Napoli fan' is almost meaningless in Naples. If you're from Naples, you are a Napoli fan."[10] Hamšík may have been overlooked elsewhere, his orchestral skills in midfield often taking a backseat to flashier players, but in Naples he was held in higher regard than anyone since the great Maradona, his instantly recognizable face adorning everything from ten-foot walls to clocks made from old CDs. His nickname quickly became "Marekiaro," after the Marechiaro neighborhood in Naples. Marek took that love and poured it right back into the city, as evidenced by his love letter in *The Players' Tribune*:

> Football is important to me, and to be able to play for Napoli for 10 years has been one of the greatest honors of my life. But the reason I have stayed so long is about more than football. In Naples, I am part of community — a family — that holds a very special place in my heart. I need to have more than just a paycheck and trophies. I need to feel something in my soul.
>
> Naples gave me that, and I am forever grateful.[11]

It might now be said that Dries Mertens became as beloved in Naples as Marek Hamšík—in fact, one of his first acts as a Napoli player was to come to the Halloween party dressed like the midfielder. Mertens, whose love for Naples is evident on his Instagram feed, where he shows his glorious life in the neighborhood of Posillipo, won over the Napoli fans slowly but surely. He arrived in 2013, but did not really begin to shine until 2016, when Higuaín left and Maurizio Sarri unleashed Dries' potential. As his play on the pitch flourished, so did his love for Naples; as he says, "People who aren't in love with Napoli have simply never seen it as I have. And the people who are in love know this love is eternal." In return, the supporters bestowed upon him a great honor, naming him "Ciro," a common nickname in the South, reflecting the lovely Neapolitan accent Mertens has picked up. And while fans now know that despite Ciro's own desire to stay at Napoli until he retires, his contract will not be renewed and the bond between club and player will be severed, they also know their own relationship with Napoli's all-time appearance holder will never be broken. After all, Ciro "chose Napoli with [his] heart."

Insigne, too, spoke about wanting to stay with Napoli for life. Raúl Albiol, a Napoli central defender for six seasons, commented on how different the support was in Naples, even compared to his former team, mega-club Real Madrid:

> It is always an incredible emotion to play in our stadium, fans support as in no other place and I think it is very tough for the opponents to play against us. Even Juve and Real Madrid suffered here. Our supporters are unique.[12]

In the same interview, fellow Spaniard José Callejón was asked, "What will you take with you forever from Naples?" He responded:

> My sons because they were born here, they will be Neapolitans forever. And the people. A magnificent population who lives the football as a religion. I will carry the Neapolitans in my heart forever.[13]

Despite the interview title, "Napoli is wonderful, we want to keep on staying here for a long time and we'll bring this city and its supporters in our heart[s] forever," Albiol moved to Villarreal in summer 2019; a year later, Fiorentina scooped up Callejón. At the start of the 2021–2022 season, the only one who remained was Kalidou Koulibaly, who promised, "Napoli have given me so much. After the initial skepticism, their fans appreciated me. I would like to give them a trophy back."[14] It is difficult for many football fans to fall in love with a solid central defender rather than an elegant forward. It is even more difficult for many Italians to unconditionally accept a black player into their hearts; racism, unfortunately, runs deep in the country, particularly in the North. But Napoli supporters adore Kouli, admiring his strength at the back as well as his vulnerability off the pitch, which came to light in the elegant essay he wrote for *The Players' Tribune* after experiencing racism in Serie A, particularly at Lazio. Toward the end of the piece, he writes:

> Napoli is a city that loves people. It reminds me of Africa because of all of the warmth. People do not just look past you. People want to reach out and touch you, they want to talk to you. The people don't tolerate you, they love you. My neighbors, they consider me like a son. Since I have been in Napoli, I am like another man. I am really at peace.
> I am Muslim. I am Senegalese. I am French. I am a Neapolitan.[15]

He is yet another adopted Neapolitan. And as much as he wants to win the scudetto for his city, he is almost certainly on his way out in the 2022 *mercato*. Still, the city wants to see that trophy lifted for him. For the others they have adopted. For Ciro. For Marekiaro. For José and for Raúl. Possibly even for Jorginho and Duvan. Not for Higuaín, but for Edi, especially if he comes back. Even for Lorenzo, even for the born-and-bred Neapolitan who is taking his leave. That time, then, must be soon.

The global pandemic certainly has not helped Napoli's game over the past couple years, and has likely contributed to the off-pitch decisions made by players and the management team. But Naples has survived centuries of squabbling over its rule, outbreaks of cholera, typhus, and bubonic plague, prejudicial decisions made by its own country's government, bombings that nearly leveled the city, years in which San Gennaro's blood did not transform, a major earthquake, the Camorra, waste that filled the streets, the judgment of those unwilling to visit—all while sitting beneath the still-active Mount Vesuvius. Naples will survive the pandemic. Napoli will survive the ones who leave. Supporters will adore those whose words stay. And Naples and Napoli will remain intertwined, always, a city at once magical and heartbreaking, a club both fantastical and soul-crushing, with fans that cut themselves open, bleeding Azzurri, screaming "See Naples and Die!"—for that is the only way to truly live.

[1] Horncastle, J. (2013). Higuain continues Napoli's Argentine tradition. *ESPN*. https://www.espn.com/soccer/blog/name/93/post/1840711/headline

[2] Id.

[3] Schlewitz, K. (2014, July 22). Edinson Cavani reveals that PSG just don't love him enough. *The Siren's Song*. https://www.thesirenssong.com/2014/7/22/5925755/edinson-cavani-paris-saint-germain-napoli-psg

[4] Cavani saluta i tifosi azzurri con una pagina sul Corriere dello Sport. (2013, August 9). *Corriera della Sera*. https://corrieredelmezzogiorno.corriere.it/napoli/notizie/sport/2013/9-agosto-2013/cavani-saluta-tifosi-azzurri-una-pagina-corriere-sport--2222556318824.shtml

[5] Cavani's knocks Higuain for his betrayal of Napoli. (2021). *Be Soccer*. https://www.besoccer.com/new/cavani-s-knocks-higuain-for-his-betrayal-of-napoli-927951

[6] Dowley, C. (2014, December 4). Duvan Zapata wants to stay with Napoli. *The Siren's Song*. https://www.thesirenssong.com/2014/12/4/7332721/duvan-zapata-wants-stay-napoli-transfer-rumors-playing-time-agent-quotes

[7] McKay, G. (2014, September 5). Video: Napoli midfielder surprises street kids. *Forza Italian Football*. https://forzaitalianfootball.com/2014/09/video-napoli-midfielder-surprises-street-kids/

[8] Vivard, L. (2018, July 12). Jorginho a CN24: "Napoli mi mancherà tantissimo, vi saluto mister Sarri. Chi giocherà al posto mio in azzurro? Bella domanda." *Calcio Napoli 24*. https://www.calcionapoli24.it/primo_piano/jorginho-napoli-mi-mancher-tantissimo-vi-saluto-mister-sarri-chi-gioca-al-n365972.html

[9] Imperatore1926. (2011, August 29). Marek Hamsik gioca con dei bambini ed esulta come un PAZZO [Video]. *YouTube*. https://www.youtube.com/watch?v=6kHq31z0Jg4

[10] Hamšík, M. (2017, May 29). For Naples. *The Players' Tribune*. https://www.theplayerstribune.com/articles/marek-hamsik-napoli

[11] Id.

[12] Albiol and Callejon, "Napoli is wonderful, we want to keep on staying here for a long time and we'll bring this city and its supporters in our heart forever." (2017, May 17). *SSC Napoli*. https://www.sscnapoli.it/static/news/Albiol-and-Callejon-Napoli-is-wonderful,-we-want-to-keep-on-staying-here-for-a-long-time-and-we-ll-bring-this-city-and-its-supporters-in-our-heart-forever-11840.aspx

[13] Id.

[14] Kalidou Koulibaly Quotes. (n.d.). BrainyQuote.com. Retrieved January 28, 2022, from BrainyQuote.com Web site: https://www.brainyquote.com/quotes/kalidou_koulibaly_1112792

[15] Koulibaly, K. (2019, June 27). We are all brothers. *The Players' Tribune*. https://www.theplayerstribune.com/articles/kalidou-koulibaly-napoli-we-are-all-brothers

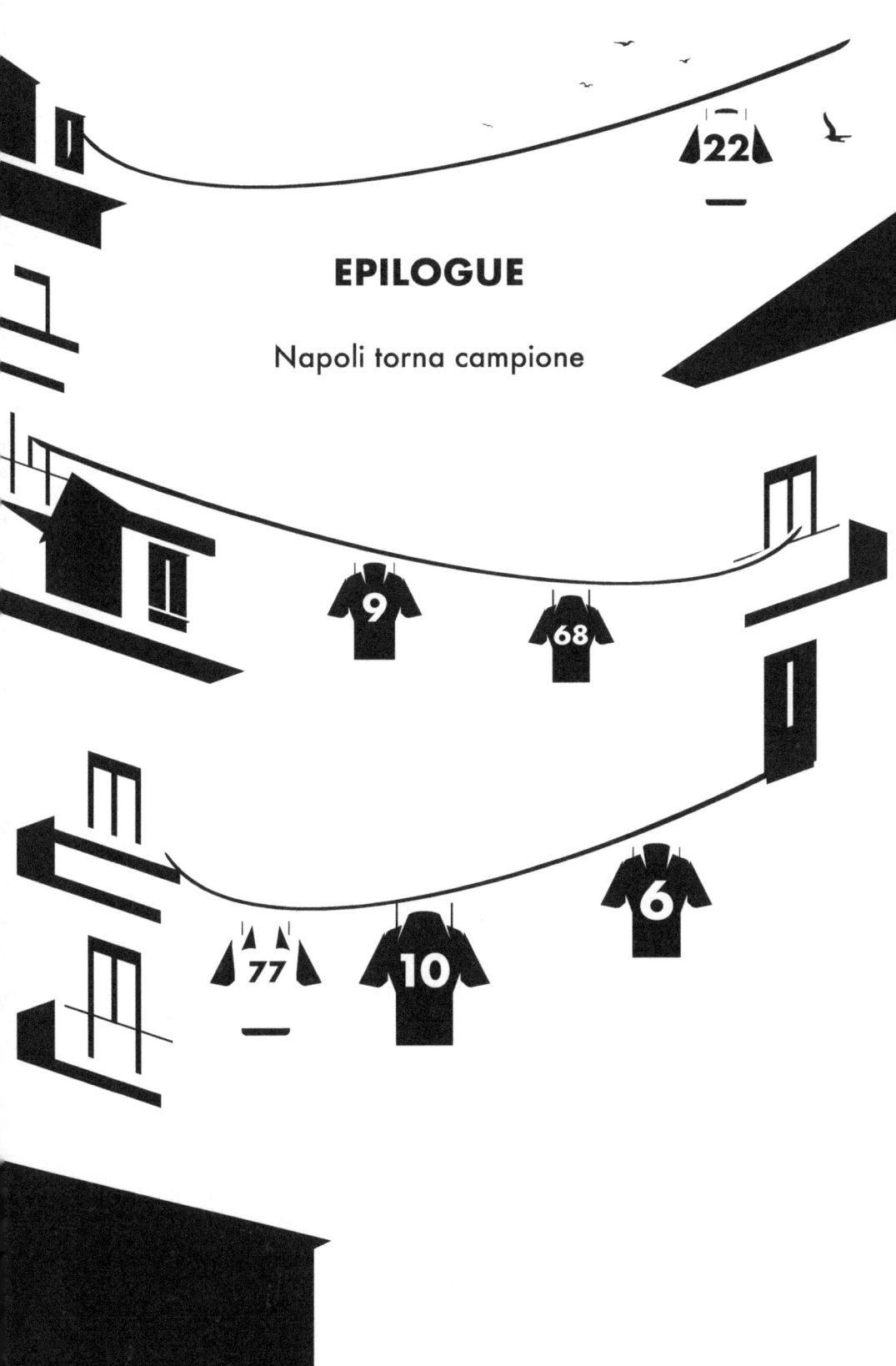

EPILOGUE

Napoli torna campione

Epilogue

When I started writing this book three years ago, I called it "More Than Maradona" because it seemed that all anyone knew of Napoli's story involved Diego. But now that the *Partenopei* have secured their third title, I understand that Napoli the club can never be more than Maradona. There can be before Maradona, there can be in addition to Maradona, there can be because of Maradona. What remains constant is that Diego Maradona is Napoli. This amazing, incredible, fantastic, astonishing 2023 side, a side that exceeds superlatives and defied expectations, would not exist if it hadn't been for Diego Armando Maradona.

While in Naples for the 2023 *scudetto* celebrations, I couldn't help wondering what non-Napoli fans who had wandered into the city thought. Although the title had been clinched on May 4 in a draw at Udinese, the emotive outpouring still overflowed on June 4, the final day of the season. Nearly every street was criss-crossed with sea-blue ribbons. Rather than laundry hanging from the buildings, there were banners thanking the squad, thanking the *Mister*, even thanking the owner. Seemingly every shop and cafe displayed a *tricolore* "3" plaque. Flags flew from balconies. And everywhere, everywhere, figures in teal t-shirts, "Buitoni" on the front, "Maradona" on the back, sipping spritzes, zipping past on Vespas, snapping photos, shopping for canned tomatoes and toilet paper. The unsuspecting tourist could be forgiven for thinking that Napoli's No.10 was the one who led the team to victory for the third time.

And in a way, he was. Before Diego, Napoli came close to conquering Italy, but had never quite managed it. Maradona was the one who gave both the club and the city hope that they could cling to when they seemed haunted by darkness and despair. Aurelio De Laurentiis may have provided the financial means to bring the club back from the brink of ruin, but only their patron saint could get the city to believe again. It was his spirit that guided the feet of this squad, drawing a line from the heady days of the late 1980s right through to the hands of those lifting the trophy on June 4, 2023.

The God of Naples may not have been physically present, but even the most casual of observers could see that his spirit hung above every joyful celebration of this third title. Many *calcio* fans easily grasped the idea that Diego was with this side long before the scudetto came within reach. But what wasn't clear except to those who love Napoli is just how much this squad dovetailed with others who had come tantalizingly close to being crowned champions, how much the presence of other venerated players could be felt alongside Maradona.

Even before the preseason kicked off in Dimaro, where Napoli retreat to every summer to wade in icy rivers and thrash low-level sides, many of the team's stalwarts had left. Neapolitan native Lorenzo Insigne had signed with MLS club Toronto FC in January, and officially left on July 1. Left-back Faouzi Ghoulam, who had joined the Partenopei in the winter of 2014, was let go in the summer. Experienced goalkeeper David Ospina left on a free transfer. Heartbreakingly, Dries Mertens, who had wanted to retire in Napoli blue, was not offered a new contract; many supporters cried when their Ciro said goodbye. Before August arrived, Chelsea finally got their hands on Kalidou Koulibaly, having wooed him practically since he joined Napoli in 2014. Paris Saint-Germain paid Napoli 25 million euros for excellent central midfielder Fabián Ruiz. The team lost four starters, their most potent game-changer in Mertens, their hope that Ghoulam would return to replace the ever-disappointing Mário Rui, and a solid goalscoring sub with the loan of Andrea Petagna to Monza. The blow was devastating to fans, who considered it the end of an era: only Mário Rui and Piotr Zieliński remained of the squad who had lit up the pitch under Maurizio Sarri, the team that had so very nearly led Napoli to their third title in 2018.

And who replaced these blue stalwarts? Alex Meret, the 25-year-old goalkeeper who'd started just seven times the previous season, donned the gloves trophy-winner David Ospina had shed. Kim Min-jae was an impressive pickup from Fenerbahçe, but he had to step in for Koulibaly, widely regarded as one of the best central defenders in the world. A loanee from Tottenham Hotspur, Tanguy Ndombele, was supposed to help fill the Fabián-shaped hole in the center of the pitch. And with Insigne, Mertens, and even Petagna gone, along with Fabián's seven goals, that was 32 Serie A goals snatched away. Napoli brought in Giovanni Simeone on loan from Hellas Verona and Giacomo Raspadori on loan from Sassuolo—neither exactly Serie A powerhouses—and 21-year-old Khvicha Kvaratskhelia from Georgian club FC Dinamo Batumi. When no one else arrived in the final days of the August transfer window, many supporters were resigned to the fact that it would be a season spent rebuilding. Even qualifying for next year's UEFA Champions League spots looked difficult, never mind challenging for the title. Fans had already been upset enough with Spalletti: they'd stolen his Fiat Panda back in October 2021, then hung a banner outside the Maradona during the Coppa final in Rome (which defending champions Napoli had crashed out of during the semis), promising to return it if he left Naples.[1] Third place—never mind that they'd finished with four straight wins, conceding just one goal in the process—wasn't good enough, and the fans' whistles upon the presentation of the 2022–2023 squad in July made it clear they doubted this team could do better.

Yet, in what was one of the most improbable events in football history, this ragtag bunch of misfits swooped in and snatched up the title.

It could've gone either way, really. Just as in Diego's first year, Napoli's first match of 2022–2023 was away at Hellas Verona. The last time the team had visited the Bentegodi, back in March, they were greeted by a banner suggesting Russian and Ukrainian forces bomb Naples.[2] Throughout the match, many of the home supporters hurled racist chants at Koulibaly and Victor Osimhen, reminding the world that a particularly vocal section of the Verona faithful had changed little since Maradona stared up at the "Welcome to Italy" banner on his first visit. The Partenopei had emerged victorious in the spring, but this time around, it seemed as though this new-look Napoli would flail and fall. Just as in 1984, Verona opened the scoring just before the half-hour mark. This time, however, "Kvaradona" was unfazed, answering Kevin Lasagna's goal with a headed equalizer less than ten minutes later. Osimhen closed out the half by putting his side in front, after which he mocked the home fans by pretending to cry.

Verona came back after the break, again exposing a wobbly Napoli defense. But the visitors then made their intentions known, not just for the match but for the season. Kvara teed up Zieliński to put them back on top. Mário Rui fed Stanislav Lobotka for Napoli's fourth goal. The fifth and final goal was a beautifully worked team effort—Matteo Politano racing up the right flank, passing to Giovanni Di Lorenzo, who sent it through to Osimhen, who, with one touch, gave it back to Politano to drive in the goal with his left foot. It was a glorious, beautiful, united show of strength, one that would come to symbolize so much about this Napoli side.

Unlike in the days of Maradona, the squad that won the 2023 scudetto was defined as a unit, not as an individual. In 38 Serie A rounds, the team scored 77 goals via 17 different goalscorers while conceding just 28. More than anything, this was Luciano Spalletti's squad, not the squad of a charismatic superstar or a fresh-faced wunderkind. Spalletti inherited a side that had coalesced under Sarri, but had been handled with kid gloves by Carlo Ancelotti. Don Carlo may be one of the best managers in the world, but many in the locker room rejected his methods, and the resulting rift was evident on the pitch. After seven straight games without a win in fall 2019, Aurelio De Laurentiis sacked Ancelotti and replaced him with Gennaro Gattuso. The infamously tough defensive midfielder reined in the unruly players and got them playing well enough that they lifted the 2020 Coppa Italia. By the next season, Gattuso's inexperience had begun to show, and the side's lack of consistency led to them missing out on the Champions League spots. ADL brought in Spalletti, who, before taking two years off, had brought Inter back to two consecutive 4th-place finishes during Napoli's heady Sarri days.

Spalletti had clearly studied up on his opponent, having been one of the few coaches to consistently shut down Maurizio's sides. He moved the team back toward more of a Sarri style, with a bit of Mazzarri thrown in for good measure. His side showed an almost preternatural awareness of where each player was on the pitch, the way the Holy Trinity of Edinson Cavani, Ezequiel Lavezzi, and Marek Hamšík had looked in the early 2010s. Spalletti cultivated Lobotka in midfield the way Sarri did Jorginho as a deep-lying playmaker. The manager may not have the gentle touch of Carlito, but his willingness to stand up for the side's new players gave Kim Min-jae the confidence to take up Koulibaly's mantle as Serie A's best defender. Having worked with Mário Rui back in his Roma days, Spalletti knew the fullback wasn't living up to his full potential. The new coach reminded him that while he excels at reading the game as a forward, he needed to remember his defensive skills too, with well-timed interceptions and clearances. As Mário Rui finally settled into his left-sided role, he began to evoke shades of Christian Maggio, the longtime Napoli right-back who loved to get forward but knew how to stop an attack.

Mazzarri and Sarri had enjoyed the greatest successes of any managers since 1990, but certain traits—Walter's emotional responses and defensive lapses, Maurizio's refusal to rotate his side—left their talented squads vulnerable. Spalletti, with his unruffled demeanor, stocked bench, and insistence on both physical and technical fitness—had the potential to rectify their mistakes. But after the flood of departures in the summer of 2022, most fans and pundits insisted that Napoli needed to bring in a massive star with plenty of experience. Spalletti stood firm: he would trust his young players to integrate into his system quickly, and rely on his few veterans—namely Di Lorenzo, Zieliński, and Mário Rui—for leadership.

It was a big risk. Insigne had captained the squad, Koulibaly filled in, and Mertens was third in the captain pecking order. Losing Dries was perhaps the biggest blow to the team's leadership and the squad's outlook. The honorary Neapolitan loved his city almost as much as he loved his club. He was a positive, outspoken personality in the dressing room and on the bench, giving everything he had even after losing out on his starting role. To fill this gaping hole, Spalletti handed the captain's armband to Giovanni Di Lorenzo—not a local boy, not a legendary player, and certainly not an extroverted personality. The manager was breaking with tradition in a big way. Yet the 29-year-old represented the best of what Spalletti wanted in his squad: humility, hard work, dedication, willingness to adapt, the ability to communicate. Prior to Napoli, Di Lorenzo had bounced around the Italian football system, from Serie B to C1 and down to D before being signed by Empoli. He helped the Tuscan club get promoted and earned his first Serie A start at age 25. It was this journey that helped him develop his professionalism and gain mental strength. And, like Dries (and so many other Napoli players), he has fallen deeply in love with Naples, stating that he now understands why so many stay for so long, and that he can't imagine himself leaving unless something changes drastically.[3]

With the benefit of hindsight, it's clear Spalletti made the right decision. Ciro Ferrara, another full-back who stepped into the captain's role after a charismatic leader—*the* charismatic leader—left, would not only call Di Lorenzo's leadership skills "key" to the title win, but label him the most representative player in the squad.[4] At the time, though, choosing Di Lorenzo to lead was a risky decision. Fortunately for this special Mister, Napoli won their first two matches and drew the next two, taking them to August 31st and deadline day. Had they not been sitting in 2nd, Napoli may well have brought in Cristiano Ronaldo, a global superstar who almost certainly would've wreaked havoc on the dressing room dynamics and overridden the team's egalitarian approach on the field.

The unbeaten five-game run was surprising, but it was the incredible performance in the season's first Champions League match that shocked even Spalletti's most vocal detractors shocked into silence. The city of Naples was already fired up in anticipation of last year's runners-up coming to town—a few days before Liverpool FC had released a video urging visiting fans to stay off the streets of Naples because the city was so dangerous. The Reds were the clear favorites, but the fans wanted to show the world that they, at least, would not be cowed. To their surprise and delight, the boys on the pitch were determined to do the same. Victor Osimhen hit the post after just 42 seconds, and Liverpool quickly disintegrated. Inside five minutes, James Milner had conceded a penalty off a handball, which Zieliński coolly converted. Ten minutes later, alarmed after an André-Frank Zambo Anguissa pass put Osimhen free on goal, Virgil van Dijk brought down the Nigerian. After Alisson pushed away Osimhen's penalty kick, Di Lorenzo sent his shot over the bar, and Liverpool breathed a sigh of relief.

They couldn't get in enough air to truly challenge this Napoli side, however. In the 27th minute, Spalletti's high press got Napoli close, Osimhen snatching the ball from Joe Gomez and sliding to Kvaratskhelia, only for van Dijk to clear. Then, in the 31st, Napoli doubled their lead. This time it was Kvara who stole from Gomez and sent to Anguissa, who performed a quick one-two with Zieliński before netting his first Champions League goal. The home side were up 2–0, but they could've easily had two or three more. Even after Osimhen came off injured, Napoli didn't stop. His replacement, Giovanni Simeone, scored the third in the 44th, minutes after coming on, bursting into tears and kissing the Champions League tattoo he'd had inked on at the age of 13. Even up 3–0, Napoli kept coming after the break. Gomez had been replaced at halftime, but the Partenopei still tore up the Liverpool defense with their perfectly timed runs up the right flank, Simeone getting on the end of a long ball from Anguissa to slot over to Zieliński. Alisson blocked his first attempt, but he had no trouble knocking in the rebound. 4–0. Luis Díaz quickly responded with a powerful shot Meret was unable to save, but Napoli collected themselves shortly after, and came away with the 4–1 win.

It was that day, September 7, 2022, when Napoli fans dared to dream. Kvaratskhelia had dribbled like a deity, earning him the nickname "Kvaradona." Zieliński rose to the challenge and stormed away triumphantly. Kim looked up to the job of filling Koulibaly's boots in central defense. And Osimhen, well . . . they were worried about Victor's injury, but he'd disrupted Liverpool even while hurt. It gave them a chance to see how important the depth Spalletti had accrued over the summer was, with young Giacomo Raspadori stepping in up top just fine. Napoli won every game without Osimhen, including a 2–1 victory at Milan and a 6–1 drubbing at Ajax, in which Jack Raspadori scored twice.

That said, Osimhen deserves a huge amount of credit for what he brought to this Napoli side. With 26 goals, he won the 2023 *Capocannoniere*, the first African ever to do so; he'd also become the highest-scoring African player in Serie A history, overtaking George Weah when he scored his 47th goal. Yet the Nigerian is more—far more—than just a set of statistics. His attributes stretch beyond those automatically attributed to African players by lazy analysts: strength, physicality, sheer athleticism. Victor boasts these traits, sure—no one wears a Batman mask throughout the season if they haven't already injured themselves smashing their face into opponents a time or two—but what truly sets him apart is his ability to remain alert and read the game. He is brilliant with the ball, but he is a genius off of it. Osimhen will not be caught loitering; instead, he is constantly seeking space, catching defenders off-guard with quick bursts of speed that find him alone in scoring positions, often at the far post. Spalletti's commitment to quick-but-direct passing and spatial awareness means the rest of the team know exactly where their striker is heading and can direct the ball into his path. From there Osi can pounce, twisting his body to find the best angle to knock the ball into the net.

Yet Osimhen, while revered, was not celebrated like Maradona during the campaign for the third title or the celebrations after it was awarded. Perhaps the reason he is not being elevated as a deity is due to his exceedingly humble positioning of his own self, always giving credit first to his own God in an unabashed display of faithfulness, and then lifting up his teammates. One might consider racism a factor in Victor not reaching the legendary heights of Maradona; indeed, it is difficult to argue that it played no role in Serie A naming Kvaratskhelia its Player of the Year rather than Osimhen. As a white woman who admittedly adores Kvara, I feel unqualified to speak to the degree to which racism plays a role in Victor not being made the symbol of the 2023 side. But Osimhen himself, subject to repeated, offensive taunts elsewhere in Italy, brushes aside the idea that racism factors into how he is treated in Naples, saying, "Since I came to Napoli, I have not experienced such in the city because everyone—black or white—takes each other like brothers and sisters."[5] Indeed, apart from Maradona, it was his name that stretched across most backs as Naples celebrated, his name emblazoned on almost every child's shirt, regardless of skin color.

Kvaratskhelia, too, appeared throughout the city, on kits and walls and stretched across streets. One reason Kvara is so adored is because he has the flair Napoli fans love so much. His ability to dribble circles around opponents and score dramatic goals is what earned him the "Kvaradona" nickname, and his memorable celebrations, like blowing a kiss to the crowd or feigning sleep, combined with his roguish appearance, easily endear him to football fans. Prior to Kvaratskhelia, Osimhen was a strong player with the potential to be fantastic; the Georgian helped him truly shine. With Kvara looking to move inside, Osimhen has a predictable partner who allows him to either peel off to the back post or run straight toward the keeper to net a goal. Kvaratskhelia's tendencies have also helped Mário Rui to finally become the dominant left-back Napoli fans have been hoping to find for years, giving him space to run forward and pump crosses into the area. On the opposite flank, rotating right wingers Hirving Lozano and Matteo Politano, as well as right-back Giovanni Di Lorenzo, also benefited from the

Kvaratskhelia–Osimhen partnership, as they had plenty of opportunities to get into the penalty box and help create or make goals. It is this type of play that takes some of the attention off Victor Osimhen as the single star of this Napoli squad.

The glory of this Napoli squad—with Osimhen's goals a driving force, Kvaratskhelia's shine sparkling in most games, the wide players consistently supporting the two primary goalscorers, and the midfield conducting it all—was on full display when Napoli kicked serious Juventus butt. The *Azzurri* went into this January 13 matchup a bit hesitant: after going unbeaten in the 15 games prior to the World Cup break, and having won their previous 11, they fell 1–0 to Inter in the first game of the calendar year. Midweek, they won 2–0 in Genoa, but considering Sampdoria had been in the relegation zone all season and ended up dead last with 47 goals conceded, the victory felt rather cosmetic. Juventus, meanwhile, had won their last eight, and kept a clean sheet throughout. Felling the *Bianconeri* incites immense joy in Napoli fans no matter what the season; this time, it felt like edging out the Italian giants would signal that Napoli were really doing the thing.

Prior to the match, pundits pitted the coaches against one another, hailing the "style" of Spalletti's side but wondering whether it would measure up against the "substance" of Massimiliano Allegri's approach, which may not have been pretty but was certainly getting results. By the end of 90 minutes, Napoli had made it abundantly clear that their manager could back up his style with substantial results.

Just 15 minutes in, Napoli's high-pressing, free-passing, full-on exuberant joy had Allegri looking like he wanted to spit nails. A Politano dink set up Kvaratskhelia. Wojciech Szczęsny managed to block the Georgian's attempted scissor kick, but Osimhen was right there to head in the rebound. Spalletti looked on impassively as his team celebrated, trying to convey, telepathically, that he expected nothing less than a thorough dismantling. Unable to handle the pressure, Amir Rrahmani got caught out in possession, allowing World Cup winner Ángel Di María to go it alone. The Argentine hit the crossbar, and somehow Meret withstood multiple Juventus chances before Napoli charged forth for their second. Osimhen could've taken a shot, but like a true Ted Lasso disciple, he made the extra pass for Kvara, who was totally open and scored easily.

Juventus quickly proved why 2–0 is the most dangerous score in football, pressuring the Napoli defense until the ball went pinging around the area and fell at Di María's feet. With the visitors back in the game, Rrahmani buckled again, very nearly equalizing for the opposition. Fortunately, Meret made an outstanding save on his own teammate, and the halftime whistle bailed him out further. The Kosovar then had something to prove, and redeemed himself in the 55th minute, getting on the end of a corner kick from Kvaratskhelia and slicing the ball low through the Juventus defense, into the opposite corner of Szczęsny's net for his first goal of the season.

The 3–1 lead wasn't enough for Napoli; they wanted Juve's *soul*. Osimhen grabbed his second ten minutes later, off a cross from Kvara. Then others got in on the action, Di Lorenzo sending the ball up the right flank to Eljif Elmas, who turned Federico Chiesa upside down and inside out before netting Napoli's fifth. The two celebrated by pretending to play cards, relaxed and at ease while the Maradona lost its mind. That 5–1 victory over Juventus, not even halfway through the season, was when even the most superstitious of Napoli fans started to whisper, "This just might be our year."

After destroying Juventus, Napoli won their next six matches, an eight-game winning streak in which Osimhen scored at least once each round. Their most significant victory came against Roma, with Simeone coming off the bench to score the winner in the 86th minute—shades of Mazzarri time peeking through. The forwards were incredible, the depth from the bench was what the Azzurri had been lacking for over a decade . . . but what really made this team into a powerhouse was the strength of the midfield.

Having lost Marek Hamšík in early 2019, the team was in desperate need of a central player with vision, range, and attacking prowess. What Spalletti realized is what Napoli fans already knew: the incomparable Hamšík could not be replaced by one mortal stepping into his scrunched-down socks. They brought in a new Slovak in January 2020, one who had begun his career as a classic No. 10, tasked with getting forward and creating connections between the defense and the attack. However, Spalletti knew that just because Stanislav Lobotka was from the same country as Marek Hamšík, his strengths on the field didn't align with those of his compatriot. The manager began teaching him not to take risks by physically going forward, but rather to hang back and move the attack forward. He attracts pressure from the opposition, but his strength, his low center of gravity, and his incredible ball-control skills enable him to hold them off, while his perfect passes propel his side forward.

Establishing Lobotka in this role is what truly moved this Napoli side forward. Now Zieliński was free to do what he does best: pressure the defense, find space to run into, show off his fantastic technique with the ball at his feet, and set up goals—eight assists this season, a stat bettered only by Kvaratskhelia. The Polish attacking midfielder had always shown sparks of true greatness, but after seven seasons with Napoli, he'd finally settled into a comfortable, consistent groove. As Zieliński went forward and Lobotka held back the opposition press, Anguissa was there to lure the other side into a false sense of complacency. Built like a typical defensive player, attackers always seemed surprised when the Cameroonian displayed impressive footwork and beautiful agility. And, almost compulsory for this Spalletti team, he also loves to get forward, lingering around the penalty box to set up his teammates and even score a few of his own goals. With these three all accepting a role in conducting their side's play, Napoli became nearly indestructible.

By the 25th round, kicking off the month of March, it was clear Napoli felt they had the title well in hand. Despite playing host to 3rd-place Lazio, the Partenopei gave them free rein from the start; unable to successfully implement their usual high press against their former manager's side, Lazio walked away with the 1–0 victory. Then April hit and, like clockwork, Napoli collapsed again. Milan—who they were about to face in the Champions League quarterfinal, the furthest Napoli had progressed in the competition—completely dismantled the home side at the Maradona, a 4–0 route that could've easily been more. Unsurprisingly, Napoli also lost the first Champions League leg in Milan, following it up with a 1–1 draw at the Maradona that tipped them out of the tournament. In Serie A, they beat Lecce, drew with Verona, and did the double over Juventus, setting the stage for their title win.

Brimming with confidence, the Azzurri were ready to make history, becoming the first team to claim the scudetto with six games to spare. All that was needed was a win over little Salernitana. Napoli blazed with confidence. The Maradona was full, the stage literally set for the post-game celebration. Osimhen very nearly headed in a goal in the second minute. Then the Seahorses, aided by veteran goalkeeper Guillermo Ochoa, closed ranks. Osimhen's shot was blocked. Zieliński's ball skimmed the crossbar. Ochoa saved an attempt from Anguissa. On and on it went until the 63rd minute, when Mathías Olivera, possibly the most unlikely of scorers in this Napoli side, headed in a corner from Raspadori. The Maradona lost its collective mind as blue smoke engulfed the outside of the stadium, where fans had gathered in hopes of partying all night. What the locals in Fuorigrotta had temporarily forgotten was that Salernitana were their rivals, hailing from less than 60 kilometers away. Not on their watch. Boulaye Dia easily dodged Juan Jesus, Napoli's weak link at the back, to put away the equalizer and put the kibosh on the party. Another year, another April for the boys in blue.

After all that, the actual title win almost came as a letdown. The team was in Udinese, where they'd need just a draw to clinch the title. Thirteen minutes in, Sandi Lovrić rolled the red carpet back up when his blazing shot put the hosts in front. When the hour mark had passed with no Napoli goals, the Partenopei fans began chewing their nails. Yes, they'd still have four games after this to ensure that scudetto was theirs, but old habits die hard.

And then it came, in possibly the most appropriate way for this Napoli side. From an Elmas corner, Olivera headed the ball on to Anguissa's feet. The Cameroonian neatly threaded it through the mass of players in front of goal to where Kvaratskhelia was standing open on the edge of the area. Goalkeeper Marco Silvestri blocked his shot but it fell directly to Osimhen, who nearly blew out the side of the net with his goal. His resounding roar echoed that of millions of Napoli fans across the globe, fans who had been waiting 33 years for this exact moment. This time, their savior had come in the form of a Black man from Nigeria, supported by more underdogs from Georgia, Cameroon, North Macedonia, Slovakia, Kosovo, Poland. When the whistle sounded, the players and staff embraced and broke into tears. The fans invaded the pitch, turning the Friuli into a cloud of blue. Back home, fireworks lit the sky, supporters flooded the streets, chants and whistles rang out from every balcony. The North's domination had finally come to an end. Naples, chaotic, glorious, joyful Naples, was set to party all night.

In fact, they partied all month. They could still be partying today—unfortunately, I had to leave on June 5, joining others sleeping on the floor of Napoli Centrale before the first train pulled away that Monday morning. But being there when the third scudetto was presented to the club will forever remain one of my life's highlights. If you weren't lucky enough to be there, go find photos, videos, podcasts, anything. I'll wait.

Sorry for sending you away. The trouble is that I can't describe what it was like, living through that moment, a moment that fused together history and culture and place, and hope and ecstasy and future and pure satisfaction. A mass of humanity, snuggled together, all painted blue, all lifted in one voice. It was a moment when the faithful could see their gods, and the disbelieving understood how one could put their trust in unseen forces. As this Napoli squad lifted their third scudetto, anyone present knew that Maradona had been in Naples since its first club was founded in 1904, that he'd come down to the pitch to bless the city in 1984, that he'd been watching and nudging those Mazzarri and Sarri squads, that he'd been resurrected to guide this 2023 side . . . and that he'd forever be Napoli.

More than Maradona? Impossible.

[1] Friggi, Francesco. (2022, June 12). Serie A—Napoli, Striscione shock contro Luciano Spalletti: "Ti restituiamo la Panda ma vattene." Eurosport Italia. https://www.eurosport.it/calcio/serie-a/2021-2022/serie-a-napoli-striscione-shock-contro-luciano-spalletti-ti-restituiamo-la-panda-ma-vattene_sto8921371/story.shtml

[2] Verona handed partial stadium ban for racist chants, bombing banner. (2022, March 15). ESPN. https://www.espn.com/soccer/story/_/id/37626412/verona-handed-partial-stadium-ban-racist-chants-bombing-banner

[3] Fanso, A. (2023, April 11). Giovanni Di Lorenzo e un incantesimo napoletano. Rivista Undici. https://www.rivistaundici.com/2023/04/11/intervista-giovanni-di-lorenzo-napoli/

[4] Noto, Antonio. (2023, May 7). Dazn, Ferrara: "Ok Osi e Kvara, ma è Di Lorenzo il più determinante per il Napoli." Tutto Napoli. https://www.tuttonapoli.net/le-interviste/i-dazn-i-ferrara-ok-osi-e-kvara-ma-e-di-lorenzo-il-piu-determinante-per-il-napoli-539716

[5] Shehu, Idris. (2023, March 7). It can make one commit suicide, says Osimhen on racism in Italy. The Cable. https://www.thecable.ng/it-can-make-one-commit-suicide-says-osimhen-on-racism-in-italy

Glossary

ADL: Nickname for Aurelio De Laurentiis, Napoli owner since 2004.

Andata: The word can be translated as "going" or, more specifically, "outbound journey," but in the football context it refers to the first half of the season, in which each team plays each of its opponents once—at present, the first 19 games of Serie A.

Autogol: Italian for "own goal."

Azzurri: The "sky-blues refers to both the Italian national team and Napoli, who don a pale blue as their home stripe.

Biancocelesti: Not to be confused with the sky-blues, Lazio are the white-and-sky-blues.

Bianconeri: Typically this refers to Juventus, though other "black and whites" also use the nickname.

Calcio: Italian for football/soccer/fútbol/fußball/etc.

Calciopoli: Italian football scandal of 2004–2005 and 2005–2006, in which multiple clubs (most famously Juventus) were punished for negotiating the selection of favorable referees.

Camorra: The organized criminal gang in Campania.

Capocannoniere: "Top scorer," the award presented by the FIGC to the player who scores the most goals in each season of Serie A.

Carabinieri: The Italian military police who, perhaps surprisingly to some readers, carry out the majority of the country's police functions, including crowd control at stadiums.

El Castor: Fans referred to Daniel Fonseca as "the beaver."

Catenaccio: "Door-bolt," the system of play in which the *libero*, granted freedom in the backline, mops up behind a three-man defense. Helenio Herrera's *Grande Inter* is one of the best examples.

Ciro Mertens: Dries Mertens, the diminutive attacker who spent nine seasons with Napoli, fell profoundly in love with the city. Supporters adored the Belgian both for his play and for the fact that he'd adopted Naples as his own, and so they nicknamed him "Ciro," a classic Neapolitan nickname.

Ciucciarelli: "Little donkeys," one of Napoli's nicknames.

Coppa Italia (Coppa): Italy's domestic cup competition, involving the top four divisions.

Curva: Named after the narrower curves on either end of an oval-shaped stadium, the curva is where the most passionate supporters watch from.

La Dea: "The Goddess" is one of Atalanta's nicknames, and one of the most distinctive in calcio.

Derby della Campania: Match between two teams from the Campania region, a meeting that rarely occurs for Napoli, since most teams in the South play in the lower divisions.

Il Duce: "The Leader," Benito Mussolini's nickname and self-proclaimed title.

Fédération Internationale de Football Association (FIFA): The "international governing body" of football, FIFA organizes tournaments such as the men's and the women's World Cups.

Federazione Italiana Giuoco Calcio (FIGC): The Italian Football Federation governs calcio.

Forza: Literally "force" or "power," it's better understood as a reminder that someone or something has the strength to finish out a task—an English equivalent might be "Come on" or, even better, "You can do it!" Combined with a club's name, it's used by fans prior or during a match; when "sempre" is added, the connotation is something like, "Win or lose, I will support this club forever."

Friulani: Nickname for Udinese, derived from the club's location in the Friuli region in northeastern Italy. Also referred to as the *Zebrette*.

Gemellaggio: "Twinning" describes a friendship between two Italian clubs.

Giallorossi: The "yellow-and-reds" most commonly refers to AS Roma.

Granata: Most commonly translated as "grenade," it also refers to the color garnet, which is the reason it is a nickname for Torino FC.

Gufare: "To root against," meaning that while a fan might not support, say, Empoli, they will cheer them on against the likes of Juventus (sometimes called "*tifare contro*").

Lazzaroni: Historical term for the lowest class in Naples, typically beggars.

Libero: The "free" one, who plays the sweeper position, behind the defense in *catenaccio*.

MaGiCa: Refers to Diego Maradona, Bruno Giordano, and Careca, the front line of the Napoli squad that won the first Scudetto in 1987.

Marekiaro: Nickname for Marek Hamšík, a play on the Naples' neighborhood of Marechiaro, a seaside village with gorgeous views of the city and Mount Vesuvius.

El Matador: While Edinson Cavani's nickname was bestowed on him due to his love of hunting, animal rights enthusiasts can be forgiven for believing it's used because, like a bullfighter shaking a red cape, Edi draws the eye of the fan and the roar of the crowd, while attracting yet enraging defenders.

Mezzogiorno: Italian for "midday," this is the Southern half of the Italian state, the area once under control of the Kingdoms of Naples and Sicily, as well as Sardinia.

Il Mister: Name for manager/head coach (these terms are often used interchangeably when talking about *calcio*).

Napulitano: The Neapolitan language (*not* dialect) and term for a Neapolitan citizen.

Nerazzurri: The black-and-blues is a nickname for Internazionale Milano, shortened to "Inter" in English (preferably minus the "Milan" that so many outlets include).

The Old Lady: Another nickname for Juventus, this pun arose alongside their habit of buying older players in the 1930s, despite their name stemming from the Latin word for "youth."

Oriundi: Used for players of Italian heritage, living and playing in Italy but born in South America. The term fell out of fashion after 1966, when *calcio* stopped admitting foreign-born players for 14 years after Italy's disastrous World Cup.

Partenopei: One of Napoli's nicknames, derived from the name for the original Greek settlement of *Parthenope*, which in turn was named for one of the sirens of Greek mythology, said to lure sailors to their deaths using her songs.

Petisso: Nickname for Bruno Pesaola, a longtime Napoli winger and three-time manager of the side.

El Pocho: Ezequiel Lavezzi, one of the "Three Tenors" under Walter Mazzarri, was called "The Chubby One," a nickname that belied his speed and agility.

Quattro giornate di Napoli: Four days in September 1943 in which the people of Naples rose up against Nazi German occupation forces.

Regista: More than a mere midfielder, the regista is the creative player in a *calcio* squad, tasked with dictating the game, providing the link between the defense and attack, and distributing the ball around the pitch. Andrea Pirlo may be the most famous, but one of the best in the current game, Jorginho, developed at Napoli under Maurizio Sarri.

Risorgimento: Meaning "rising again," the 19th-century movement, stemming from reforms introduced to Italy by the House of Bourbon, resulted in unification of the Italian states and the modern country of Italy.

Ritiro: This "retreat" is a tradition for Italian clubs who have run up against a brick wall, the idea being that sequestering themselves for a short period of intense training and team-building will improve their play. Napoli go to their ritiro at Dimaro Folgarida, in the north of the country; photos of the players rolling up their shorts to cool off in the river are plentiful in the summer.

Ritorno: The "comeback," or the second half of the *calcio* season.

Rosanero: With just one "s," the nickname is the pink-and-blacks rather than the red-and-blacks, referring to Palermo FC. Sadly, few other clubs in Italy wear pink.

Rossoblù: The red-blues often refers to Bologna FC or Genoa FC, but it's also a nickname of other clubs such as Cagliari and Crotone.

Rossoneri: The red-and-blacks is one of the nicknames for AC Milan, who are also called *Il Diavolo*, or "The Devil."

San Gennaro: The patron saint of Naples. His blood, stored in the Naples Cathedral, is said to liquify three times per year. The liquefaction is said to be a reminder of Gennaro's continued protection of Naples; when it does not liquify, the city may be in danger.

Sarriball: Named for coach Maurizio Sarri, Sarriball is a flexible and fluid method of play in which teams build the attack from the back, employing a high press where two players often exchange quick passes while a third slips, unnoticed, away from the opposition defense.

Scudetto: The "little shield" worn on the kits of the team who won the previous season's tournament, but more often means the Serie A trophy (plural: *scudetti*).

"Southern Question": Phrase to describe the dualism that exists between the North and the South of Italy, inquiring as to why the latter features far less industrial development and far more poverty.

Stadio Arturo Collana: Built in 1925 as the Stadio del Vomero, and renamed for one of Napoli's founders, a sports journalist. The stadium was used by the club from 1933–1959 and is now a multipurpose sports facility located in the Vomero neighborhood.

Stadio Diego Armando Maradona: The current Napoli stadium, named after Naples' most beloved football god, began life as the Stadio del Sole in December 1959, became the San Paolo in 1963, and was renamed in December 2020 after Maradona's death.

Stadio Giorgio Ascarelli: Constructed in 1929, the stadium bore the name of the first Napoli president. The name was changed to the Stadio Partenopeo, as Italy were hosting the 1934 World Cup, and the fascist regime would not permit one of their grounds to be named after a Jewish man. Bombarded in 1942, the now-demolished stadium lives on in name only, the "Rione Ascarelli" neighborhood having grown up around the site.

Supercoppa Italia: Played between the winners of the *scudetto* and the Coppa Italia, unless they are one and the same, in which case the Coppa Italia runners-up face off against the champions. Now being played outside Italy to generate more support for the country's clubs.

Terroni: A classist term used to refer to the group of people who work the land, implying poverty, filth, and the general belief that those using the word believe themselves to be better than the "peasants" they are insulting.

Tifosi: Literally "infected with typhoid" or "feverish," the word is used to describe a group of supporters of a sports team, primarily football clubs (masculine singular = *tifoso*, feminine singular = *tifosa*).

Totonero: The 1980 match-fixing scandal involving 12 clubs and more than 20 players;

Napoli's Giuseppe Damiani was suspended for three months, the shortest sentence imposed on any player, while the club itself was acquitted.

Totonno: Antonio Juliano, the extremely popular Neapolitan playmaker who went on to secure Maradona's signature with Napoli.

Ultras: Fanatics. The diehard fans. The ones who fill the *curva*, chanting, waving banners, setting off flares. Although the term can be used to signify a supporter who engages in violence and risky behaviors, the vast majority of ultras simply want to cheer on their team while creating an intimidating atmosphere for visiting sides.

Union of European Football Associations (UEFA): The governing body of European football, which hosts tournaments such as the Champions League, the Europa League, and the Euros.

Il Veltro: "The Greyhound," Attila Sallustro, played for Napoli in 1926–1937, scoring 107 goals in 258 appearances.

Vesuviani: Yet another nickname for Napoli and their fans, referring to *Vesuvio* the active volcano beneath which Naples rests.

Viola: Nickname for Fiorentina, whose home strip is a fetching shade of purple.

Zona Cesarini: Named for Juventus midfielder Renato Cesarini, who often scored at the end of a game, this term is used to indicate that a team is particularly successful in scoring goals during the final minutes of a match. For Napoli fans, the expression "Mazzarri Time," named after 2010s manager Walter Mazzarri, is equivalent.

Author

Kirsten Schlewitz fell in love with soccer when she was only four years old, running around the dirt fields of Central California. After a decade spent working in sports media, she cofounded Unusual Efforts, a community dedicated to elevating the voices of women—cis and trans—and nonbinary persons in soccer. Writing is her passion project; her pieces have appeared at ESPN and the Guardian, among many other outlets, and she often consults on soccer-related stories for international media. Kirsten is part of the Far From Vesuvius network, which provides content to Napoli fans, as well as The Gentleman Ultra, a site for all things interesting about Serie A. Currently, she is tying together her BA in International Political Economy and her JD in International Law as she pursues a Master's in Justice and Reconciliation, applying her writing and editing skills to peacebuilding efforts in the Balkans. Kirsten firmly believes that sport can be used to establish and repair relationships between individuals, communities, and countries. She splits her time between Belgrade, Serbia, and the Pacific Northwest, but part of her heart always lingers in Naples.